THE SELF-POTENTIAL METHOD
Theory and Applications in Environmental Geosciences

The self-potential method is a simple yet innovative process, enabling non-intrusive assessment and imaging of disturbances in electrical currents of conductive subsurface materials by measuring the electrical response at the ground's surface or in boreholes. It has an increasing number of applications, from mapping fluid flow in the subsurface of the Earth to understanding the plumbing systems of geothermal fields, and detecting preferential flow paths in earth dams and embankments.

This book provides the first full overview of the fundamental concepts of this method and its applications in the field. It begins with a historical perspective, provides a full explanation of the fundamental theory, laboratory investigations undertaken, and the inverse problem, and concludes with chapters on seismoelectric coupling and the application of the self-potential method to geohazards, water resources, and hydrothermal systems.

End-of-chapter exercises aid practitioners and students in developing and testing their understanding of the theory and its applications. Additionally, data sets and analytical software are made available online for the reader to put the theory into practice and analyze their own data. This book is a key reference for academic researchers and professionals working in the areas of geophysics, environmental science, hydrogeology, hydrology, environmental engineering, and geotechnical engineering. It will also be valuable reading for related graduate courses.

ANDRÉ REVIL is Associate Professor at the Colorado School of Mines and Directeur de Recherche at the National Centre for Scientific Research (CNRS) in France. His research focuses on the development of new methods in petrophysics, and the development of electrical geophysical methods applied to geothermal systems, water resources, and oil and gas reservoirs. Dr. Revil is presently Editor of the *Journal of Geophysical Research – Solid Earth*, Associate Editor of the *Vadose Zone Journal*, and Editorial member of the *International Journal of Mathematical Modelling & Numerical Optimization*, and has published 168 peer-reviewed papers. He received the Bronze Medal from the CNRS for outstanding research in 2003 and an award of excellence from the American Rock Mechanics Association (ARMA) in 2004.

ABDERRAHIM JARDANI is Associate Professor at the University of Rouen, where he also obtained his PhD in Geophysics in 2007. His research interests center on environmental geophysics, mathematical modeling of hydrologic systems and inverse problems. Dr. Jardani has authored 27 peer-reviewed publications in the field of hydrogeophysics, especially in the theory of self-potential signals and its application to environmental problems. He is a member of the American Geophysical Union.

THE SELF-POTENTIAL METHOD

Theory and Applications in Environmental Geosciences

ANDRÉ REVIL

Colorado School of Mines

ABDERRAHIM JARDANI

University of Rouen

CAMBRIDGE
UNIVERSITY PRESS

CAMBRIDGE
UNIVERSITY PRESS

University Printing House, Cambridge CB2 8BS, United Kingdom

One Liberty Plaza, 20th Floor, New York, NY 10006, USA

477 Williamstown Road, Port Melbourne, VIC 3207, Australia

4843/24, 2nd Floor, Ansari Road, Daryaganj, Delhi - 110002, India

79 Anson Road, #06-04/06, Singapore 079906

Cambridge University Press is part of the University of Cambridge.

It furthers the University's mission by disseminating knowledge in the pursuit of
education, learning and research at the highest international levels of excellence.

www.cambridge.org
Information on this title: www.cambridge.org/9781108445788

© André Revil and Abderrahim Jardani 2013

First published 2013
First paperback edition 2017

A catalogue record for this publication is available from the British Library

ISBN 978-1-107-01927-0 Hardback
ISBN 978-1-108-44578-8 Paperback

This book is dedicated to the memory of the Spanish-born physicist Albert Tarantola. Albert's influence on geophysics was profound and immense, especially regarding his work on the inverse problem based on probabilistic foundations and his criticism of the Tikhonov approach (reviewed in this book), still popular nowadays. Albert was a generous man with a strong character and an amazing creativity. We miss him a lot. We invited him in 2008 to give a lecture at the Department of Geophysics of the Colorado School of Mines and this was the lecture that opened our eyes on the use of stochastic approaches, especially Bayesian methods, to self-potential problems as discussed in several chapters of this book.

Albert Tarantola, on his last day at Stanford in 2008. Photo taken by André Journel.

Contents

Foreword by Susan S. Hubbard   *page* ix
Preface   xi

1   Fundamentals of the self-potential method   1
    1.1   Measurements   1
    1.2   The electrical double layer   11
    1.3   Brief history   14
    1.4   The Poisson equation   16
    1.5   Sources of noise   17
    1.6   Conclusions   19
          Exercises   19

2   Development of a fundamental theory   23
    2.1   Non-equilibrium thermodynamic   23
    2.2   Upscaling: from local to macroscopic equations   44
    2.3   The geobattery and biogeobattery concepts   68
    2.4   Conclusions   75
          Exercises   77

3   Laboratory investigations   82
    3.1   Analyzing low-frequency electrical properties   82
    3.2   Investigating the geobattery concept in the laboratory   99
    3.3   Conclusions   104
          Exercises   105

4   Forward and inverse modeling   110
    4.1   Position of the problem   110
    4.2   Gradient-based approaches and their limitations   114
    4.3   Fully coupled inversion   131

	4.4	Conclusions	148
		Exercises	149
5	Applications to geohazards		154
	5.1	Landslides and flank stability	154
	5.2	Sinkhole detection	160
	5.3	Detection of cavities	167
	5.4	Leakages in dams and embankments	171
	5.5	Conclusion	191
6	Application to water resources		192
	6.1	Pumping tests	192
	6.2	Flow in the vadose zone	209
	6.3	Catchments hydrogeology	219
	6.4	Contaminant plumes	232
	6.5	Conclusions	243
		Exercises	243
7	Application to hydrothermal systems		245
	7.1	Stochastic inversion of temperature and self-potential data	245
	7.2	The Cerro Prieto case study	261
	7.3	Gradient-based approach applied to hydrothermal fields	268
	7.4	Conclusions	282
		Exercises	282
8	Seismoelectric coupling		284
	8.1	Position of the problem	284
	8.2	Seismoelectric theory in saturated media	286
	8.3	Numerical modeling	291
	8.4	Application in saturated conditions	293
	8.5	Seismoelectric theory in unsaturated media	298
	8.6	Application in two-phase flow conditions	319
	8.7	Localization of hydromechanical events	326
	8.8	Seismic beamforming and the formation of electrical bursts	335
	8.9	Conclusions	338
		Exercises	339
	Appendix A *A simple model of the Stern layer*		342
	Appendix B *The* **u**–*p formulation of poroelasticity*		345
	References		348
	Index		367

The color plate section can be found between pages 178 and 179.

Foreword

As an environmental scientist who uses geophysical methods to quantify subsurface properties and processes, I recognize the challenges of extracting quantitative information from self-potential data. Although self-potential data are easy to acquire and often provide good qualitative information about subsurface flows and other processes, a quantitative interpretation is often complicated by the myriad of mechanisms that contribute to the signal. Like many others in the community, more than once I have chosen to not collect self-potential data or have abandoned a quantitative interpretation of acquired datasets due to the mechanistic complexity of self-potential signals.

André Revil and Abderrahim Jardani offer a panacea for this common problem through an unparalleled and cutting-edge treatise on the self-potential method, a book that lays a solid foundation for quantitative use of the method to characterize and monitor the Earth's subsurface. The foundation is established through describing the history, physics, and several inversion approaches clearly and comprehensively, and through walking the reader through examples of the use of self-potential data to quantify geotechnical hazards, vadose zone and groundwater processes, and flow in geothermal fields. The book also describes the use or extension of the method to interrogate phenomena not typically explored with self-potential methods, including redox zonation and hydromechanical disturbances associated with fracking. The inclusion of forward and inversion modeling software and a data reduction tutorial will render the book useful for teaching a graduate level course on self-potential. True to André Revil's renaissance style, the book describes many unifying theories and concepts that connect different geophysical signatures and petrophysical properties and that portray a tantalizing vision of future applications using the self-potential method.

The authors have integrated their formidable experience and insights into a valuable book, which marks and propels a quantum leap forward for the self-potential method. It will undoubtedly lead to more acceptance of the self-potential as a quantitative approach and importantly, to new applications that will increase our understanding of subsurface processes and management of precious subsurface resources. This book deserves a central spot on the bookshelves of students, research geophysicists, hydrogeologists, engineers, and professionals.

Susan S. Hubbard
Senior Scientist, Lawrence Berkeley National Laboratory

Preface

Ten years ago, a good fraction of geophysicists considered passive geophysical methods as having a limited pedigree in comparison to active methods. In the past decade, a breakthrough has occurred in the use of passive seismic methods to image for instance oil and gas reservoir and to monitor dynamic processes (e.g., oil recovery duing water flooding). The same type of revolution has started with the electromagnetic methods, but has lagged behind, limited by the resources of a smaller community. The self-potential method, sometimes nicknamed the ugly duckling of environmental geophysics (Nyquist and Corry, 2002), has been typically used for a broad range of applications, but mostly qualitatively. The self-potential method is, however, one of the oldest of all geophysical methods and is characterized by more complicated (and rich) physics than those used to describe active methods like the d.c. resistivity or seismic methods. It is therefore much more challenging to develop a complete understanding of this method than other classical geophysical methods. The poor understanding of the complex causative sources of self-potential signals has slowed down the quantitative use of this method. The self-potential method is usually not described in geophysicsal textbooks and, when it is described, it is usually with mistakes in the description of the physics or the basic equations. Therefore, we aim to provide here a fundamental description of its electrochemical roots and the quantitative applications of the self-potential method in geophysics and hydrogeophysics, which has been missing to date.

We believe, for the reason mentioned above, that the self-potential method has been underused and its full potential (pun intended) has not been reached. Clearly, a fundamental description of the physics of self-potential signals and the principles of measurements in the field could be used to unleash this method and turn this ugly duckling into a beautiful swan. A few years ago, we also recognized how the self-potential method is similar to another method used in medical imaging: the electroencephalographic (EEG) method. The EEG method has been a breakthrough method in the past decade for mapping brain functions and understanding the dynamics of its electrical activity. The same type of evolution is expected for the self-potential method for a broad range of applications, from exploration geophysics to the production of geothermal systems; from hydrogeophysics to civil and environmental engineering problems. We expect this book will help the community to move forward in this direction. We also invite feedback from readers to help us improve this book in future editions.

We thank also many students and researchers who have directly or indirectly contributed to the present book including Stéphanie Barde-Cabusson, Jean Paul Dupont, Anthony Finizola, Nicolas Florsch, Allan Haas, Becky Hollingshaus, Joyce Hoopes, Scott Ikard, Damien Jougnot, Marios Karaoulis, Vincenzo Lapenna, Niklas Linde, Harry Mahardika, Angela Perrone, Paul Sava, Justin Rittgers, Myriam Schmutz, Magnus Skold, Abdellahi Souied, Bill Woodruff, and Junwei Zhang.

This book is composed of eight chapters that are now briefly discussed.

Chapter 1 introduces the basic concepts of the method. The self-potential method is a passive geophysical method involving the measurement of the electric potential at a set of self-potential stations. The sampled electrical potential can be used to obtain important remote information pertaining to ground water flow, and hydromechanical and geochemical disturbances in the conducting ground. In this chapter, we discuss the principle of the measurements, strategies to map or monitor the self-potential field and the origin of spurious signals and noise. We also discuss the electrical double layer coating the surface of the minerals, which is responsible for an excess of electrical charges (sometimes a deficiency) in the pore water. Finally, we provide a short history of the self-potential method.

The fundamental theory of self-potential signals in porous media is covered in Chapter 2. Our goal is to provide an in-depth understanding of the causes of self-potential signals in deformable porous rocks. We start with an introduction to non-equilibrium linear thermodynamics, which provides the form of the macroscopic constitutive and continuity equations. To gain some knowledge about the material properties, we need to upscale local equations (valid in each phase of a porous composite) to the scale of a representative elementary volume of porous material. These two approaches can be combined to give explicitly the contributions entering into the source current density responsible for self-potential anomalies. The contribution to the self-potential signals associated with the transfer of electrons is investigated separately, as this contribution can be non-linear. This last contribution provides a theoretical basis for the geobattery and biogeobattery models used to localize ore bodies and, more recently, to explain some strong self-potential anomalies observed, under some specific circumstances, over contaminant plumes.

Chapter 3 investigates two types of measurements that can be made in the laboratory to get a better insight on the processes responsible for self-potential anomalies. The first type of measurements concerns core sample measurements. We present electrical conductivity, permeability, and streaming potential measurements and we show how these measurements can be considered into a unified framework of petrophysical properties. Such a unified framework is of paramount importance in considering the natural complementarity of d.c. resistivity, induced polarization and self-potential in solving hydrogeophysical problems. In this chapter, we also discuss a sandbox experiment investigating the geobattery concept and its predictions for ore bodies. We are especially interested in the occurrence of a dipolar anomaly and the role of the redox potential distribution in this behavior.

In Chapter 4, we introduce a finite element formulation to solve the forward self-potential problem associated with ground water flow (primary flow problem). As geophysicists, we are also interested in the inverse problem. We introduce two very distinct ways to invert self-potential data. One approach is to invert the source current density distribution and then to interpret this source current density in terms of relevant parameters (hydraulic head, redox potential, salinity). The second approach is to fully couple the self-potential inverse problem with the physics of the primary flow problem (solving the non-reactive or reactive transport equations and performing the inversion with either deterministic gradient-based or stochastic Bayesian-type approaches).

In Chapter 5, we describe four applications of the self-potential method to geohazard problems including (i) the use of self-potential signals to understand the ground water flow pattern associated with landslides and flank stability, (ii) the detection of sinkholes and cryptosinkholes in karstic environments, (iii) the detection of caves, and (iv) the study of leakages in dams and embankments using salt tracer tests and self-potential monitoring. In each case, we develop a specific approach to interpret self-potential data, and we provide insights regarding the physical mechanisms at play.

In Chapter 6, four additional applications of the self-potential method to hydrogeological problems are described. These applications include (i) pumping tests, (ii) the flow of water in karstic aquifers, (iii) vadose zone hydrogeology, and (iv) the delineation of contaminant plumes associated with a landfills. Each case is illustrated with a different approach, in terms of interpreting the self-potential data. This not only shows the versatility of the applications of the self-potential method in hydrogeology, but also its limitations. The self-potential method should come as a complementary method to in-situ measurements and other (active) geophysical methods.

The self-potential method can also be applied to understand the flow pattern as well as the flow magnitude (to some extent) in hydrothermal and geothermal fields. In Chapter 7, we first propose a stochastic inversion of borehole temperature and surface self-potential data to determine the flow pattern at a depth of several kilometers in a geothermal field. This shows that the self-potential method can be used for deep applications (down to 3 or 4 km). This approach is used to understand the Cerro Prieto geothermal field in Baja California (Mexico), one of the biggest geothermal fields in the world in terms of the production of electrical power. We also develop a gradient-based approach to invert the flow rate of thermal water along faults and we show how this approach can be combined with d.c. resistivity data on two geothermal fields in Colorado and Oregon.

In Chapter 8, we first describe the seismoelectric theory in fully water-saturated conditions. The seismoelectric theory is an extension of the streaming potential theory to the frequency domain, including inertial terms in the momentum conservation equation of the skeleton and pore fluid. It explains how electromagnetic disturbances can be created by the propagation of seismic waves. Then, we provide an extension of this theory to the frequency domain accounting for partial saturation of the water phase. We provide a numerical example related to the water flooding of an oil reservoir and the electrical disturbances associated with the passage of the seismic waves at the oil/water encroachment front. This approach may offer a completely new way to monitor changes in saturation, in both near-surface and deep applications. The passive record of electrical signals can be used to track hydromechanical disturbances in a cement block during the rupture of a seal associated with a fracking experiment. In the last section of this chapter, we briefly show a new approach using beamforming of seismic waves to extract the electrokinetic properties of the point where the seismic field is focused. We believe that this approach could be used to monitor change of saturation in the vadose zone, contaminated aquifers as well to monitor oil and gas reservoirs during their production.

1

Fundamentals of the self-potential method

The self-potential method is a passive geophysical method, like the gravity and magnetic methods. It involves the measurement of the electric potential at a set of measurement points called self-potential stations. The sampled electrical potential (or electrical field) can be used (inverted) to determine the causative source of current in the ground and obtain important information regarding ground water flow, hydromechanical and geochemical disturbances. In this chapter, we discuss the principle of the measurements, strategies to map or monitor the self-potential field and the origins of spurious signals. We also provide a short overview of the electrical double layer coating the surface of the minerals. Indeed, a good understanding of the electrical double layer is of paramount importance in the study of self-potential signals. We also provide a short overview of the history of the self-potential method.

1.1 Measurements

1.1.1 Equipment

Self-potential measurements are performed using non-polarizing electrodes connected to a voltmeter. For example, Figure 1.1a shows a multichannel voltmeter used to record the voltage of 80 non-polarizing electrodes at a frequency of one sample per minute. A non-polarizing electrode is formed by a metal in contact with its own salt (e.g., silver in a silver chloride solution, or copper in a copper sulfate solution). An example of non-polarizing electrode, the Petiau Pb/PbCl$_2$ electrode, is shown in Figure 1.1b. The difference of the electric potential between two electrodes is measured by using a voltmeter with a high sensitivity (at least 0.1 mV), and high input impedance (typically ~10–100 MOhm, for soils to 1000 GOhm on permafrost). Figure 1.1c shows the record of the electrical potential difference between one electrode, located close to an injection well, and a reference electrode,

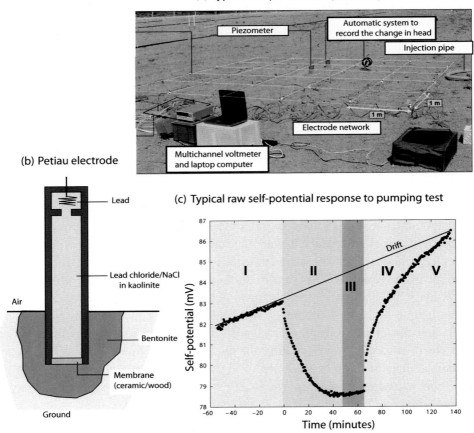

Figure 1.1. Typical recording system to monitor the self-potential response associated with a pumping test. (a) Picture showing the network of electrodes and the recording multielectrode voltmeter. (b) Petiau non-polarizing electrodes. (c) Typical raw data for the self-potential (modified from Jardani *et al.*, 2009a). Phase I corresponds to the data obtained prior to the start of the pumping test; II is the transient phase during pumping; III is the steady-state phase; IV is the rapidly changing portion of the recovery phase; and V corresponds to the slowly changing to steady-state portion(s) of the recovery phase. The line shows the drift of the potential of the scanning with respect to the reference electrodes (both electrodes experiencing different temperatures over time). Before being analyzed, these data need to be detrended, filtered, and shifted in such a way that the potential prior to the start of pumping ($t = 0$) is equal to zero for each electrode (static removal).

located several tens of meters way (further than the radius of influence of the well). The data are the raw (unfiltered) data. We can see the self-potential response associated with the pumping test and the recovery following the shutdown of the pump. In other words, the self-potentials are remotely sensitive to the ground water flow triggered through the pumping test, a point that makes the self-potential method a non-intrusive flow sensor.

To perform accurate self-potential measurements, the impedance of the voltmeter has to be at least ten times higher than the impedance of the ground between the two electrodes, in order to avoid leakage of current in the voltmeter. A voltmeter with an internal impedance of 100 MOhm would be high enough for most applications. However, working over very resistive materials ($>100\,000$ Ohm m, e.g., ice, permafrost, crystalline rocks) requires the use of a voltmeter with much higher internal impedance (some voltmeters can be made with an internal impedance of 10^{12} Ohm). The voltmeter, like all instrumentation in geophysics, has to be calibrated regularly against known resistances to check its accuracy over a broad range of resistance values.

1.1.2 Self-potential mapping

The oldest approach in self-potential is to establish a map of the distribution of the electrical potential at the ground surface. In this case, a reference electrode is used as a fixed base, and the second electrode is used to scan the electric potential at the ground surface. The fixed (reference) electrode is kept in a small hole filled with bentonite mud. Because the presence of the mud modifies the electric potential at the contact between the electrode and the ground, the potential of this station is arbitrary, and should not be used as a point (with zero potential) in performing the self-potential map. Adding salty water to improve the coupling between the electrode and the ground should be also avoided, especially for monitoring purposes, because evaporation of the water changes the salinity of the pore water, generating spurious potential changes over time in the vicinity of the surface of the electrodes. These spurious potentials are due to diffusion potentials that will be explored further in Chapter 2. The roving (scanning) electrode is used to measure the electric potential at a set of stations, referenced in space with a GPS (2 m accuracy in x and y is usually good enough for most applications). Both prior to and after the measurements, the difference of voltage between the reference electrode and the scanning electrode has to be checked by putting the electrodes one against the other, contacted through their porous membranes (for instance in Figure 1.1b, the porous membrane is made of wood), and measuring the difference in electrical potential. The drift of the voltage between the electrodes should be kept as small as possible over time (e.g., < 2 mV per day is an acceptable drift for self-potential mapping). The potential map is, therefore, a map relative to the (unknown) potential at the base station. Actually, like all scalar potentials in physics, the self-potential is defined to an additive constant. Only the electric field (which is the gradient of the electric potential in the low-frequency limit of the Maxwell

Loop network approach

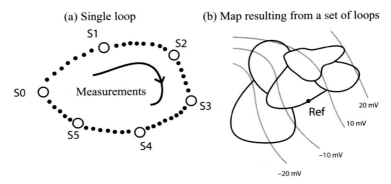

Figure 1.2. Large-scale self-potential mapping over the ground surface using a close loop approach combined with the leap-frog approach. (a) Sketch showing a single loop. S0 denotes the first base station. Measurements are performed along the self-potential stations characterized by the small black circles. At some point, a new base station is established S1 and so on. The potential distribution can be reconstructed along the loop respecting the fact that the self-potential loop should be closed. The value at S0 should be close to zero. If this is not the case, a closure error should be applied to the data to "close the loop." (b) A self-potential map is built by combining the information on several loops and using one of the base stations as reference for the entire map. The black plain lines denote the self-potential loops while the gray lines denote the electrical equipotentials obtained for instance by kriging the self-potential data.

equations) is well-defined. At the interface between the ground and the atmosphere, the electric field is tangential to the ground surface, because air can be considered as an electric insulator. A tutorial, made by S. Barde Cabusson and Anthony Finizola on data reduction for self-potential mapping is provided on the website associated with this book (file SP_Processing_tutorial, provided with permission of the authors).

For large-scale mapping of the self-potential, several strategies are possible. The first has been used by numerous researchers: one base station is chosen as the reference and measurements are performed with scanning electrodes at different secondary stations. To extend the measurement array, the initial base station is removed and the reference electrode is transplanted, or "dropped," at the position of the last measurement station (a leap-frog approach can be used instead). This operation is repeated to close a loop. This approach is called the loop network. The circulation of the electrical field should be zero along a closed loop performed at the ground surface; in other words, the sum of the drop potentials along a closed loop is zero. This strategy is described in Figure 1.2. It is very important to close the loop when making self-potential measurements, in order

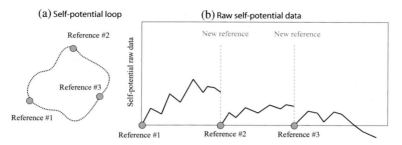

Figure 1.3. Typical self-potential profile along a closed loop with a total of three base stations used as references. (a) Closed loop. (b) Raw self-potential profile (modified from Barde-Cabusson and Finizola, 2012, unpublished material, see file SP Processing_tutorial, provided with permission of the authors).

to check the closure error (due to the propagation of errors in the changes of the reference electrode) and to correct the self-potential measurements from this closure error. If this is not done, there is a risk of accumulating errors toward the end of the profiles (some published self-potential maps in volcanic areas clearly show huge accumulated errors at the end of profiles that were misinterpreted as ground water flow pattern). An example of application of the loop approach to map the large-scale self-potential anomalies downstream a landfill can be found in Naudet *et al.* (2003, 2004). An excellent description of the procedure of self-potential mapping and reduction of closure errors can be found in Minsley *et al.* (2008).

The corrections along a close loop with several changes of self-potential base stations are explained in Figures 1.3 and 1.4. In Figure 1.3, we show that the self-potential measurements are by nature discontinuous each time a base station is used (the potential at the base station is setup to zero). The first correction is naturally to reestablish the continuity of the electrical potential using the first base station of the profile as a global base station for the entire loop. The final step is to correct for the closure error along the loop as mentionned above.

A completely different strategy for self-potential mapping is called the "star network." In this approach, we first determine the difference of potential between a set of base stations separated from each other by several hundred meters (up to a kilometer, see Figure 1.5). As the wires used to measure a difference of potential between two points of the Earth are usually not shielded, they can be subject to induction effects; some fluctuations in the readings are typically observed when the cable used for the measurements is too long. In subsequent steps, each base station is used as the local reference of profiles that are more or less radially distributed about this station. This approach was, for instance, followed by Fournier (1989) to get a large-scale self-potential map over tens of square kilometers.

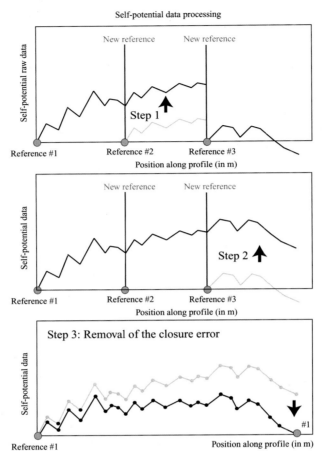

Figure 1.4. The corrections for the raw self-potential profile along a closed loop involve a series of steps. In the first set of steps, the potential is set continuous at each change of base station. The final step corresponds to the closure error (modified from Barde-Cabusson and Finizola, 2012, unpublished material, see file SP Processing_tutorial, provided with permission of the authors).

A third strategy, which has been used rarely in the literature, is to directly measure the electrical field (not the electrical potential) at a set of stations (Figure 1.6). However, measuring the first spatial derivative of a noisy field can be difficult, as the gradient measurement may amplify the noise. The electrical field can be measured at a station P by measuring the potential difference in two normal directions. The electrical field due to a buried current source is tangential to the ground surface. In order to get a reliable estimate of the electrical field at this point, it is necessary to average the estimated electrical field magnitude over a set of several points. This method does not require long wires and is easy to carry out

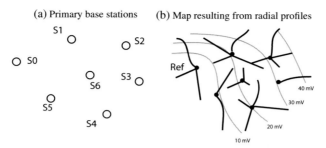

Figure 1.5. Large scale self-potential mapping using a star approach. (a) A set of base stations is chosen and prepared with bentonite plug setup in the ground. The difference of potential between these stations is measured. (b) Each of these stations is used as a secondary reference and radial profiles are performed from this station. The self-potential map is built by using one of the base stations as a reference for the entire survey. The black plain lines denote the self-potential profiles from the base stations while the gray lines denote the electrical equipotentials obtained for instance by kriging the self-potential data.

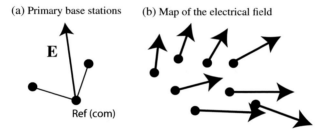

Figure 1.6. Large-scale mapping of the electrical field. (a) Each station is formed by three measurement points (L shape) to determine the two components of the electrical field, which is locally tangential to the ground surface. (b) Such a pair of measurements is repeated at various stations at the ground surface of the Earth and a map of the electrical field (a vector map) is built. Note that this approach is reference free but the electrical field can be very noisy.

in the field. However, a careful check of the reliability of the measurements is necessary to avoid interpreting spurious effects, especially associated with the very heterogeneous nature, in terms of resistivity distribution, of the shallow subsurface.

Any fieldwork requires the use of a good notebook (if possible waterproof), in which all information pertaining to the survey is recorded. This includes weather conditions (both during and prior to the survey), the position of the self-potential stations (using GPS, for instance), the value of the self-potential data, and any notes pertaining to understanding the survey, including a description of the equipment and any technical difficulties met during the survey. The notebook should also include some basic drawings of the profiles and, if possible, some printed pictures taken during the survey. After the completion of a day working in the field, it is recommended to take pictures of the relevant pages of the notebook, store the information in a memory stick and store the memory stick in a safe place.

1.1.3 Self-potential monitoring

It is possible to use the self-potential method as a monitoring method to track the changes of variables of interest like the Darcy velocity, the moisture content, and salinity. We will show numerous examples of self-potential monitoring in this book, for a variety of problems, especially in Chapters 5–8. In monitoring applications, a multi-electrode array is connected to a multichannel or multiplexed voltmeter, as shown in Figure 1.1a, and distributed along the Earth's surface over a target of interest. This approach is completely analogous to what is done in electroencephalography, for medical applications. In electroencephalography, a network of electrodes is used to monitor the change in the distribution of the electric potential on the scalp in order to localize the active part of the brain where electrical currents are manifested along the synapses between neurons; see Grech *et al.* (2008). Even if the electrical potentials are measured at several hundred hertz in electroencephalography, the same fundamental (quasi-static) elliptic equation applies as discussed below in Section 1.3. Because the main sources of currents in the ground are closely connected to the existence of the electrical double layer at the pore water–mineral interface, we discuss this electrical double later in the next section (Section 1.2). Some important, relevant publications about this methodology applied to the long-term monitoring of a site include Perrier *et al.* (1998) and Trique *et al.* (2002).

1.1.4 Electrode drift

The potential of the electrodes is always temperature dependent. Even the Petiau $Pb/PbCl_2$ electrodes have non-negligible temperature dependence on the order of 0.2 mV $°C^{-1}$ despite the fact that they were designed to minimize such effect. Indeed, other non-polarizing electrodes have, easily, a temperature dependence comprised between 1 and 2 mV $°C^{-1}$. This has many implications for both mapping and monitoring. In the case of self-potential mapping, having a reference electrode in the cold ground and scanning the potential at the ground surface with a warmer electrode (e.g., one held in a hand) can easily generate a difference of potential higher than 10 mV. In these cases, the voltage of the reference electrode should not be used in building a self-potential map as already mentioned above. Measures should be taken to avoid such a temperature differential between the electrodes; for instance, the scanning electrode can be attached to a stick avoiding direct contact between the scanning electrodes and the hands. The temperature of the inner part of the electrode can be measured and a temperature correction can be applied.

In monitoring, the shallow subsurface is always characterized by diurnal temperature variations in the ground down to 30–50 cm. Hence, the monitoring electrodes should be installed at a depth of 30–50 cm, or temperature sensors need to be utilized in the vicinity of the electrodes to apply a post-correction of the effect of the temperature. Burying the electrodes has several advantages, as it limits the influence of external electromagnetic noise (sensitivity decreases rapidly with depth), reduces the risk of desiccation, and avoids the spurious electrical signals associated with the roots of vegetation. The drift shown in Figure 1.1c was due to a variation of temperature between one of the monitoring electrodes and the reference electrode during a transient hydraulic pumping test in Boise (Idaho, USA) in June 2007.

For both mapping and monitoring, it is strongly recommended to avoid the use of salty water. The presence of salty water between the electrodes and the ground creates a highly variable, localized diffusion potential that is hardly the same from one place to the other. To make matters worse, this potential is expected to change with time, because of the drying of the saline solution and the concomitant increase in salinity. Our experience indicates that the use of salty water yields unreliable self-potential mapping, and that unpredictable changes and drifts in the recorded potentials are observed during monitoring. For mapping, it is much better to dig a small hole until the presence of moisture is recognizable by visual inspection, and firmly apply the end-face of the electrode against the ground. The contact resistance can be measured with the voltmeter from time to time, to check that the resistance of the ground between the reference electrode and the scanning electrodes remains low, with respect to the internal impedance of the voltmeter. Bentonite (silt and smectite powder) can be used at each station, usually the stations need to be prepared at least 10 minutes before the measurements in order to stabilize. The advantage of bentonite is that it will keep the potential drop between the electrodes and the ground constant in space and time (in this case the potential drop between the bentonite and the electrode is called a membrane potential and it is always constant). Figure 1.7 shows two profiles. The first profile (Figure 1.7a) indicates the level of repeatability expected in the field under normal conditions. The second profile (Figure 1.7b) shows a discrepancy between the prediction of a model, based on the computation of the ground water flow, and the measured self-potential profile. This discrepancy was due to a problem with the scanning electrode that was not in thermal equilibrium with the ground (and the reference electrode) at the beginning of the survey.

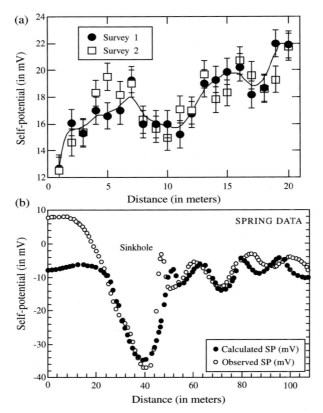

Figure 1.7. Example of high resolution self-potential profile. (a) Repeat of a high resolution self-potential survey. The repeatability is better than 2 mV. The electrodes were just put in contact with the ground after a small hole was dug. (b) High resolution self-potential (SP) survey associated with a sinkhole. Note the discrepancy between the model-predicted and measured self-potential values at the beginning of the profile. Thermal equilibrium between the reference and scanning electrodes was not reached at the beginning of the profile. Modified from Jardani *et al.* (2006a).

In some cases, it may be necessary to reoccupy exactly the same stations in order to accurately repeat a profile or a map over weeks, months or years. In this case, the best practice is to avoid leaving the electrodes in place, because of their tendency to drift over time. This drift may be difficult to estimate. A way to overcome this issue is to prepare self-potential stations with a bentonite plug put in the ground at a depth of about 30 cm. The bentonite plug (obtained by mixing water, salt and bentonite) needs to be capped. A plastic cap would protect the bentonite pot from be washed out by rainwater infiltration (Figure 1.8). This approach can be extremely precise (accuracy on the order of 1 mV or better) for repeating surveys over time.

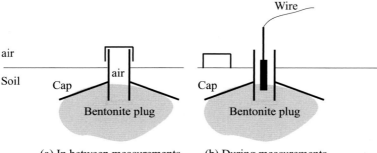

(a) In between measurements   (b) During measurements

Figure 1.8. Example of self-potential station that can be reoccupied for long term or even short term monitoring. This approach avoids dealing with the drifting of the electrodes and we are sure that the same points are measured again and again. (a) Between surveys, the porous plug of moist bentonite remains in the ground at a depth of at least 30 cm (deeper in dry areas). The plastic cap avoids the washing of the bentonite by the infiltration of the rainwater. (b) During measurements, the cap of the tube is removed and the non-polarizing electrode is introduced through the plastic tube down to the bentonite plug. The difference of potential between this station and a reference station can be recorded.

A correction must be applied to the self-potential measurements, due to the slow drift and electrode polarizations. If the electrodes are not stored together in a salt solution (of the same salt as that inside the electrode) and effectively short-circuited, they will develop a distinct inner potential when aging. This drift can reach several tens of millivolts after several years, which may be higher than the desired signal. Therefore, the operator should measure the potential of all the electrodes with respect to the reference electrode prior and after a survey; these differences of potentials need to be removed from the measurements. It is also very important to properly label all the electrodes used in the field.

1.2 The electrical double layer

Before proceeding with the description of the self-potential signals, it is worth describing the concept of the electrical double layer at the pore water–mineral interface, because of its importance in all that follows. Figure 1.9 sketches a silica grain coated by an electrical double layer. When a mineral like silica is in contact with water, its surface becomes charged due to chemical reactions between the surface sites and the pore water. For instance, the silanol groups, $>SiOH$ of the surface of silica (where $>$ refers to the mineral framework), behave as weak acid–base (amphoteric sites). This means that they can lose a proton when in contact with water to generate negative surface sites ($>SiO^-$). They can also gain a proton to become positive sites ($>SiOH_2^+$). The resulting mineral surface charge is therefore

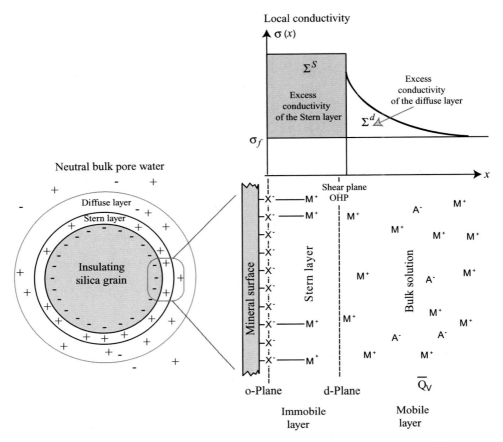

Figure 1.9. Sketch of the electrical double layer at the pore water–mineral interface coating a spherical grain. Modified from Revil and Florsch (2010). The local conductivity $\sigma(\chi)$ depends on the local distance χ from the charged surface of the mineral. The pore water is characterized by a volumetric charge density $\bar{Q}_V$ corresponding to the (total) charge of the diffuse layer per unit pore volume (in coulombs (C) m^{-3}). The Stern layer is responsible for the excess surface conductivity Σ^S (in siemens, S) with respect to the conductivity of the pore water σ_f while the diffuse layer is responsible for the excess surface conductivity Σ^d. These surface conductivities are sometimes called specific surface conductance because of their dimension but they are true surface conductivities. The Stern layer is comprised between the o-plane (mineral surface) and the d-plane, which is the inner plane of the electrical diffuse layer (OHP stands for Outer Helmholtz plane). The diffuse layer extends from the d-plane into the pores. The element M$^+$ stands for the metal cations (e.g., sodium Na$^+$) while A$^-$ stands for the anions (e.g., chloride Cl$^-$). In the present case (negatively charged mineral surface), M$^+$ denotes the counterions while A$^-$ denotes the coions. The fraction of charge contained in the Stern layer is called the partition coefficient f.

pH dependent, and typically negative at near neutral pH values from 5 to 8. For silica, a complete description of the electrical double layer can be found in Wang and Revil (2010).

At near-neutral pH values, the negative surface charge of the mineral attracts the ions of positive sign (counterions) and repels the ions of the same sign (coions). It results the formation of a diffuse layer characterized by an excess of counterions and a depletion of coions with respect to the free pore water located in the central part of the pores (Figure 1.9). In addition, some ions can be sorbed directly on the mineral surface forming the Stern layer; see Stern (1924). The Stern layer is located between the o-plane (mineral surface) and the d-plane, which is the inner plane of the electrical diffuse layer (Figure 1.9). The term "electrical double layer" is a generic name describing this electrochemical system coating the surface of the minerals. The term electrical Triple Layer Model (TLM) is often used in electrochemistry when different types of sorption phenomena are considered at the level of the Stern layer. Electrochemists make the distinction between inner-sphere complexes for ions strongly bound to the mineral surface (e.g., Cu^{2+}) and losing their mobility along the surface; the term outer-sphere complex is used to characterize ions that are weakly bound to the mineral surface (e.g., K^+, Na^+), and generally keep their hydration layer and a certain mobility along the mineral surface as discussed below in Chapter 3. Note that in smectitic clays, among other type of clay minerals, the solid particles themselves are characterized by a negative charge, directly related to isomorphic substitutions in the crystalline network. This charge acts in superposition to an amphoteric (pH-dependent) surface charge density associated with amphoteric reactions at the edge of the crystalline planes.

There are three fundamental consequences associated with the existence of the electrical double layer: (1) the pore water is never neutral, (2) there is an excess of electrical conductivity in the vicinity of the pore water–mineral interface responsible for the so-called surface conductivity, and (3) the double layer is responsible for the (non-dielectric) polarization of the porous material. The first consequence is fundamental to understanding the nature of electrical currents associated with the flow of the pore water (termed streaming currents); therefore, the occurrence of self-potential signals associated with ground water flow. The second consequence is crucial to the understanding of electrical conductivity in porous materials, and to the interpretation of d.c. resistivity data (curiously, this contribution is nearly always neglected in the EM community in the interpretation of large-scale resistivity data, and abusively neglected by a good fraction of hydrogeophysicists). The third consequence is important to the understanding of induced polarization, which translates into the frequency dependence of the electrical conductivity. As discussed in the next chapters, d.c. resistivity and induced polarization are complementary geophysical methods to self-potential surveys in hydrogeophysics and recent advances in their modeling should lead soon enough to a unified model for these properties.

1.3 Brief history

The long history of the self-potential method begins with the first measurements made by Fox (1830) in Cornwall (UK) over sulfide vein mineralizations. It is, however, unclear if Fox measured truly some self-potential signals or spurious signals associated with his equipment. Nevertheless, the self-potential method strated to be used as a qualitative prospection method to locate ore bodies; the reason being that in this application, strong self-potential anomalies (on the order of hundreds of millivolts) can be observed at the ground surface. The development of non-polarizing electrodes for geophysical applications, by M. C. Matteucci who was working at the Greenwich Observatory (1865), drastically improved field measurements. Thanks to non-polarizing "pot" electrodes, Matteucci observed, in 1847, that earth currents in telegraph wires were correlated with the aurora borealis; see Spies (1996). Non-polarizing electrodes were later used by C. Barus for the exploration of ore bodies, especially the Comstock gold lode in Nevada; see Barus (1882) and Rust (1938). In 1906, Norway and Muenster made the first discovery of an ore body of commercial interest with the self-potential method; see Rust (1938). Sato and Mooney (1960) synthesized all the available information related to the occurrence of self-potential signals associated with ore bodies and proposed the famous "geobattery model." In this model, the corrosion of an ore body and the distribution of the redox potential with depth (associated with the change of oxygen activity or concentration with depth) explain the occurrence of self-potential anomalies at the ground surface. This model was refined later by many researchers including Stoll *et al.* (1995) and Bigalke and Grabner (1997), using a non-linear model known to electrochemists as the Butler–Volmer model; see Bockris and Reddy (1970) and Mendonça (2008). Recently, a similar battery model was proposed by Arora *et al.* (2007), Linde and Revil (2007), and Revil *et al.* (2010) to explain the occurrence of self-potential anomalies over organic-rich contaminant plumes. In this case, a sequence of redox reactions occurs, and biotic electronic pathways termed nanowires exist between bacteria in a biofilm. This process may be responsible for a long-range transport of electrons on the order of millimeters and even centimeters; see Risgaard-Petersen *et al.* (2012). This process can lead therefore to half-redox reactions that are decoupled over space. The difference between the abiotic and biotic models of geobatteries is that, in the biotic model, there are no potential losses between the electron donors and electron acceptors, and the electronic conductor. The bacteria play the role of catalysts in the transfer of electrons between electron donors and acceptors, lowering energy potential barriers; see Revil *et al.* (2010). These concepts of geobatteries were also tested and confirmed in the laboratory by Timm and Möller (2001), Naudet and Revil (2005), and Castermant *et al.* (2008), and will be extensively discussed below in Chapters 2 and 3.

A second contribution to the development of the self-potential method is related to concentration gradients of ionic species in the pore water, the so-called diffusion current or potential. It was first commercially used in geophysics by Conrad Schlumberger (1920), especially as a downhole measurement tool for diagraphy; see also (Schlumberger *et al.*, 1932, 1933). In this case, the contribution to the measured electrical potential comes from the difference of ionic concentrations between the mud filling the well and the pore water of the formations. This contribution is known as the diffusion potential in a number of texbooks and an extreme case of diffusion potential is called the (perfect) membrane potential in presence of clays. An additional streaming potential contribution can also arise in boreholes when water flows between the well and the formations.

A third contribution to the method accounts for the occurrence of self-potential signals is precisely related to ground water flow as briefly discussed above (streaming potential). As shown in Figure 1.7, most natural porous media have an excess of charge in their pore water to balance the fixed charge occurring on their mineral surface (Figure 1.9). The streaming current associated with the drag of the excess of charge by the flow of the pore water corresponds to an advective flow of electrical charges and therefore a net source current density. Indeed, the advection of a net electrical charge in a fixed framework attached to the grains is, by definition, a current density (flow of electrical charges per unit surface area per unit time). The physics of the streaming potential takes its roots in the experimental work done initially by Quincke (1859). Helmholtz (1879) obtained a theoretical expression, 20 years later, for glass capillaries; see also von Smoluchowski (1903). In the meantime, Bachmetjew, in 1894, was probably the first to observe a self-potential field associated with the motion of the ground water through sands in Germany. Nourbehecht (1963) developed some linear constitutive equations for the coupled generalized Darcy and Ohm laws in porous media based on non-equilibrium thermodynamics; see Prigogine (1947). Later, Pride (1994) gave a comprehensive theory of these "electrokinetic" effects by upscaling the Nernst–Planck and Navier–Stokes equations using a volume-averaging method. Bogoslovsky and Ogilvy (1973) gave a semi-quantitative treatment of the self-potential signature of preferential drainage near dams; see also Gex (1993). Rizzo *et al.* (2004) and Maineult *et al.* (2008) applied this approach to pumping tests following the ideas developed first by Semenov (1980) in Russia. The self-potential method has also been applied successfully in geothermal exploration. See a review of earlier works by Zablocki (1976), Corwin and Hoover (1979) and Jackson and Kauahikaua

(1987), and the more recent works by Ishido (1989, 2004), Aubert and Atangana (1996), Ishido and Pritchett (1999) and Richards *et al.* (2010). Streaming potential can be used in the field to characterize preferential ground water flow paths; see Wilt and Corwin (1989), Wilt and Butler (1990) and Panthulu *et al.* (2001).

This brief history of the self-potential method is probably incomplete but provides a quick view of the method and its contributions. In Chapter 8, we will discuss an extension of the self-potential method called the seismoelectric method, which has its own rich history.

1.4 The Poisson equation

In this section, we describe the fundamental equation used to interpret self-potential signals in the quasi-static regime of the Maxwell equations. The total macroscopic electric current density $\mathbf{J}$ (expressed in A m^{-2}) represents the flux of electrical charges; therefore, the amount of electrical charge passing per cross-sectional surface area of the porous material per unit time (expressed in C m^{-2} s^{-1} or equivalently in A m^{-2}). It is given by the sum of two terms, one conduction current density (described by the classical Ohm's law) and a source current density $\mathbf{J}_S$. The total current density is, therefore, given by

$$\mathbf{J} = \sigma_0 \mathbf{E} + \mathbf{J}_S, \tag{1.1}$$

where $\mathbf{E}$ is the electric field (in V m^{-1}) (in the quasi-static limit of the Maxwell equations written as $\mathbf{E} = -\nabla\psi$, where ψ is the electric potential expressed in V), σ_0 is the d.c. electrical conductivity of the porous material (in S m^{-1}), and $\mathbf{J}_S$ is a source current density (in A m^{-2}) associated with any disturbance that can affect the movement of charge carriers. Chapter 2 will be dedicated to the various mechanisms generating such a source current density. Equation (1.1) stands for a generalized Ohm's law, which is valid for all brine conductivities: the first term on the right-hand side of Eq. (1.1) corresponds to the classical conduction term (classical Ohm's law defining the conduction current density).

In addition to the constitutive equation, Eq. (1.1), we need a continuity equation for the current density in order to determine a field equation for the electrostatic potential ψ. In the magnetoquasi-static limit of the Maxwell equations, for which the displacement current is neglected, the continuity equation for the total current density is

$$\nabla \cdot \mathbf{J} = 0. \tag{1.2}$$

Equation (1.2) means that the total current density is conservative (all the current entering a control volume must also exit in the absence of sources and sinks; there is no storage of electrical charges inside the control volume).

Combining Eq. (1.1) with the continuity equation for the charge, Eq. (1.2), the self-potential field ψ is the solution of the following elliptic (Poisson-type) equation

$$\nabla \cdot (\sigma_0 \nabla \psi) = \nabla \cdot \mathbf{J}_S, \tag{1.3}$$

where the source current density $\Im = \nabla \cdot \mathbf{J}_S$ (in A m^{-3}) denotes a volumetric current density. Equation (1.3) is the fundamental field equation in the interpretation of (quasi-static) self-potential signals. It states that an electrical potential distribution is created by a source term, corresponding to the divergence of a source current density. The electrical potential distribution is also controlled by the distribution of the electrical conductivity σ_0. Note that the existence of a source current density $\mathbf{J}_S$ requires that the system is in thermodynamic disequilibrium (see Chapter 2). It is especially important to discriminate between "steady-state" and "equilibrium" situations. Since self-potential signals are generally observed to be persistent and stable over years in a number of geological environments (especially in the case of ore bodies), these environments are usually working in a steady state of thermodynamic disequilibrium.

1.5 Sources of noise

To end this brief review of the self-potential method, and before entering into the various fundamental aspects of the theory, we now discuss the origin of electrical noise in the measurements of self-potential signals. In self-potential mapping, noise can originate from different transient sources. These include induction of telluric currents occurring inside the conductive Earth, due to transient current flow in the ionosphere, lightning, the presence of large cumulus clouds, and induction in the ground associated with power lines, as well as direct current injection associated with cathodic mise-à-la-masse (grounding) used to prevent corrosion of metallic pipes; see, for instance, Corwin and Hoover (1979). Some examples of spurious self-potential signals of cultural origin are shown in Figure 1.10. They can be related to corrosion of the metallic casing of a piezometer (Figure 1.8a) or cathodic protection (Figure 1.10b), for instance. Transient signals can be filtered if a fixed dipole is used to record the electrical signals during mapping (or, in the case of telluric noise, a magnetometer to record correlated magnetic signals, see any texbook on the magnetotelluric method).

Spatial noise can be associated with strong heterogeneity of the resistivity distribution in the shallow subsurface (comprising the first several centimeters to a few meters of the ground). This strong heterogeneity in resistivity can be due to dry conditions and/or the presence of resistive bodies (like stones) embedded in a more conductive (clayey) matrix, like a fine-grained soil with higher moisture

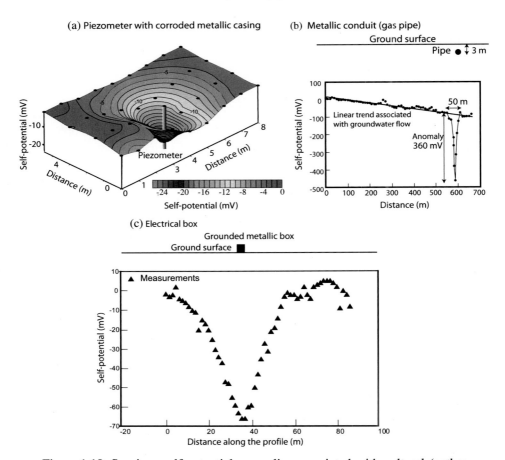

Figure 1.10. Spurious self-potential anomalies associated with cultural (anthropogenic) source mechanisms. (a) Example of self-potential anomaly associated with the corrosion of the metallic casing of a piezometer. (b) Self-potential anomaly associated with the cathodic protection of a gas pipe. (c) Self-potential anomaly associated with grounded monitoring instruments (grounding is used to avoid their damage by lightning) at the Bemidji USGS site in Minnesota (USA).

content. This spatial noise can be filtered out if measurements are made with a high spatial density (e.g., five measurements over a square meter at a given station). Generally, Fourier or wavelet-based filters can be used to remove such a noise in a self-potential map.

Other artifacts can be associated with the non-polarizing electrodes themselves, which have an inner potential that is temperature dependent, as explained above in Section 1.1. Therefore, having two similar electrodes at two different temperatures yields a difference in their electric potentials. This effect can be corrected if the temperature is measured at the place where the self-potential measurements are made, with the electrodes in thermal equilibrium with their environment, and if the temperature coefficient of the electrode type is known (e.g., 0.2 mV °C^{-1} for Petiau electrodes; see Petiau (2000)). Usually, Petiau electrodes (Figure 1.1b) are preferred to Cu/CuSO$_4$ electrodes, which have substantially higher temperature dependence (0.7–0.9 mV °C^{-1}; see Antelman (1989)).

1.6 Conclusions

The self-potential method is one of the oldest geophysical techniques to investigate the subsurface of the Earth. Self-potential anomalies observed at the ground surface can be associated with a causative distribution of current in the ground. However, getting good self-potential data is not easy, and great care needs to be taken to prepare a survey. A good understanding of the various sources of noise is required as well as strategies to mitigate or cancel out spurious signals. Finally, the mechanisms governing the occurrence of self-potential signals are fundamentally electrochemical in nature. This means that geophysicists have to make an effort to have better understanding of basic interfacial electrochemistry, as well as non-equilibrium thermodynamics and upscaling techniques, if they want to understand and use self-potential signals to their full extents.

Exercises

Exercise 1.1. Physics of the electrical diffuse layer. The electrical double layer is important for various geophysical problems (self-potential, d.c. resistivity, and induced poalrization). We consider below the case of a binary 1:1 electrolyte like NaCl (so $i = (+)$ for the cations and $(-)$ for the anions). We first define the electrochemical potential of cations $(+)$ and anions $(-)$ $\mu_{(\pm)} = \mu_{(\pm)}^0 + k_b T \ln C_{(\pm)} + q_{(\pm)} \psi$, where $\mu_{(\pm)}^0$ is the chemical potential of the ions in a reference state (this is a constant), k_b is the Boltzmann constant, T is temperature (in K), $C_{(\pm)}$ is the concentration of cations $(+)$ and anions $(-)$, $q_{(\pm)}$ is the charge of cations $(+)$ and anions $(-)$ (in C), and ψ is the electrostatic potential (in volts). We consider a charged mineral surface. Because we are working at small scales, we can neglect the radius of curvature of the mineral surface and the mineral surface will appear as a plane of mean constant charge density Q_0 (expressed in C m^{-2}) (we will assume that Q_0 is negative). The electrochemical potential of a ionic species is a potential like those seen in class (electrostatic potential or fluid pressure for instance). It means that the force acting on the ions corresponds to the gradient of this potential. We will not consider any kinetic effect here, only the equilibrium situation.

(1) Using the assumption of equilibrium (implying that the electrochemical potential of the ions is the same everywhere in the pore water), you can demonstrate that the concentration of cations and anions at a distance x from the mineral surface is related to the salinity C_f (concentration of salt in the free pore water) by:

$$C_{(\pm)}(x) = C_f \exp\left[-\frac{(\pm e)\psi(x)}{k_b T}\right], \tag{1.4}$$

where e is the elementary charge. Far from the wall, the concentration are $C_{(\pm)}(\infty) = C_f$. This type of distribution is known as a Boltzmann distribution. Another totally equivalent way to show this result is to say that at equilibrium the flux of ions due to the electrostatic field (due to the charged surface) counterbalances the diffusive flux of ions. These two fluxes can be respectively as

$$\mathbf{j}_{(\pm)} = (\pm)C_{(\pm)}\beta_{(\pm)}\mathbf{E}, \tag{1.5}$$

$$\mathbf{j}_{(\pm)} = -D_{(\pm)}\nabla C_{(\pm)}, \tag{1.6}$$

where $D_{(\pm)} = k_b T \beta_{(\pm)}/e$ is the diffusion coefficient of the ionic species (note the relationship between the diffusion coefficients and the mobilities of the ionic species, both are defined by two different phenomena, diffusion and electromigration, and both are interrelated because they express the movement of the ions in the water phase). Perform a dimensional analysis of Eqs. (1.5) and (1.6).

(2) Starting with Gauss's law $\varepsilon_f \nabla \cdot \mathbf{E} = \rho$ (where ρ is the charge density) for a medium (here the pore water) of dielectric constant ε_f and where $\mathbf{E}$ is the electrical field, demonstrate that the electrostatic potential in the vicinity of the mineral surface obeys the following partial differential equation called the Poisson–Boltzmann equation:

$$\frac{d^2\psi(x)}{dx^2} = \frac{2C_f e}{\varepsilon_f}\sinh\left[\frac{e\psi(x)}{k_b T}\right]. \tag{1.7}$$

To obtain this result, you need to express local charge density ρ as a function of the concentrations and charges of the ions in the diffuse layer.

(3) Under the assumption that the electrostatic potential is relatively small (smaller than 25 mV in absolute value), you should demonstrate that the electrostatic potential is approximately given by:

$$\psi(x) = \psi_0 \exp(-x/x_d). \tag{1.8}$$

Note: use a first-order Taylor expansion of $\sinh(X)$ and note that $k_b T/e$ has the dimension of a voltage (compute the value of this ratio in mV at 25 °C).

Equation (1.8) is called the Debye approximation. Can you give an expression for the characteristic distance x_d (which is called the Debye length, you can easily find an expression of the Debye length on the internet!). For a salinity of 10^{-4} mol l^{-1} (note you should compute the salinity in m^{-3} in the international system of units using the Avogadro number $N = 6.02 \times 10^{23}$ mol^{-1}), a temperature of 25 °C (to be converted into kelvin, K) (you can find the values of e and the Boltzmann constant on the internet), you can compute the value of x_d, which corresponds to a rough approximation of the thickness of the electrical diffuse layer. For a sand, the typical size of the pore is $R = 20$ μm (of course it depends on the size of the grains and the grain size distribution). Compare the size of the pores with the size of the diffuse layer.

Exercise 1.2. Self-potential mapping. Three self-potential profiles have been performed in the field. Profile A uses the reference at position A0, and is composed of five small profiles with all the same reference point A0. Profile B uses a reference at position B0 for all the stations. There is no change of base along this profile. Profile C uses the reference at point C0 for all the stations along this profile. The goal of this exercise is to create a self-potential map with C0 as the common reference. (1) Compute the values of the electrical potential at all the stations using C0 as a common reference. The base stations A10 and B12 are at the same location. Stations C6 and A25 are at the same locations. (2) Contour by hand the self-potential map to draw the equipotentials at 0, 10, 20, 30 and 40 mV.

Self-potential data position (x in meters, y in meters): measured self-potential value: Profile A [A0(ref, 90,160): 0 mV; A1(75,145): −13 mV; A2(60,130): −23 mV; A3(48,120): −31 mV; A4(30,105): −38 mV; A5(18,90): −45 mV; A6(0, 70): −45 mV; A7(100,145): −7 mV; A8(110,125): −13 mV; A9(120,108): −20 mV; A10(130,90): −23 mV; A11(140,70): −30 mV; A12(150,50): −35 mV; A13(160,25): −37 mV; A14(173,0): −40 mV; A15(112,160): −9 mV; A16(130, 160): −20 mV; A17(153,155): −30 mV; A18(175,152): −37 mV; A19(200, 150): −45 mV; A20(100,185): −13 mV; A21(108,203): −22 mV; A22(115,122): −27 mV; A23(120,240): −32 mV; A24(128,265): −38 mV; A25(130,285): −45 mV; A26(85,185): −15 mV; A27(80,200): −20 mV; A28(70,220): −23 mV; A29(60,250): −27 mV; A30(50,270): −30 mV; A31(40,290): −34 mV]. Profile B [B0(ref, 15,95): 0 mv; B1(20,128): 5 mV; B2(25,155): 7 mV; B3(35;185): 13 mV; B4(60,210): 22 mV; B5(85,222): 21 mV; B6(105,230): 17 mV; B7(130,220): 14 mV; B8(150,200): 12 mV; B9(160,170): 12 mV; B10(160,135): 13 mV; B11(95,105): 16 mV; B12(130,90): 22 mV; B13(105,75): 20 mV; B14(80,70): 15 mV; B15(50,80): 10 mV; B16(30,90): 5 mV]. Profile C [C0(ref, 5,240): 0 mV; C1(20,245): 10 mV; C2(35,250):12 mV; C3(60,260): 15 mV; C4(90,270): 12 mV; C5(115,280): 7 mV; C6(130,285): 0 mV].

Exercise 1.3. Stern layer complexation and CEC. This exercise deals with a simple, non-electrostatic, model of the Stern layer. We first start considering the sorption of sodium in the Stern layer of clays and the dissociation of protons according to $> SH^0 \Leftrightarrow > S^- + H^+$ and $> S^- + Na^+ \Leftrightarrow > S^-Na^+$, where $>S$ refers the surface sites attached to the crystalline framework, H^0 are protons (which are assumed to be immobile) while the weakly sorbed sodium counterions Na^+ are considered to be mobile in the Stern layer. The equilibrium constant for the two chemical speciation reactions given above are:

$$K_{Na} = \frac{\Gamma^0_{SNa}}{\Gamma^0_{S^-}[Na^+]^0},$$ (1.9)

$$K_H = \frac{\Gamma^0_{S^-}[H^+]^0}{\Gamma^0_{SH}},$$ (1.10)

where Γ^0_i denotes the surface concentration (in mol m^{-2}) of surface species i. The two previous equations are constitutive equations. To complete the description of the mineral surface, we need a conservation equation for the surface species $\Gamma^0_S = \Gamma^0_{SNa} + \Gamma^0_{SH} + \Gamma^0_{S^-} + \Gamma^0_X$, where Γ^0_S denotes the (known) total surface site density (including the charge associated with isomorphic substitutions in the crystalline framework Γ^0_X), Γ^0_{SNa}, $\Gamma^0_{S^-}$, and Γ^0_{SH} represents the surface charge density of the sites $> S^-Na^+$, $> S^-$, and $> SH^0$ respectively, and Γ^0_X represents the number of equivalent sites corresponding to isomorphic substitutions (all expressed in sites m^{-2}). To simplify the notation, we write $pH = -\log_{10}[H^+]$ and $[Na^+] = C_f$ denotes the salinity. We assume that the negative charge corresponding to the isomorphic substitutions are compensated in the basal plane by counterions located in the diffuse layer.

(1) Find the concentrations of the following species (Γ^0_{SH}, Γ^0_{SNa}, $\Gamma^0_{S^-}$) as a function of the salinity, the pH, the total site density Γ^0_S, and the density of the isomorphic substitutions Γ^0_X.

(2) Find an expression of the fraction of the counterions in the Stern layer defined by

$$f = \frac{\Gamma^0_{SNa}}{\Gamma^0_{SNa} + \Gamma^0_{S^-} + \Gamma^0_X}.$$ (1.11)

Simplify the obtained expression when the pH is high.

(3) Find an expression for the amount of counterions sorbed in the Stern layer, that is the cation exchange capacity as a function of the salinity (CEC $= e\Gamma^0_{SNa}S_{Sp}$, where S_{Sp} denotes the specific surface area of the material in m^2 kg^{-1} and e is the elementary charge).

2

Development of a fundamental theory

The readers who are interested only in the applications of the self-potential method can skip this chapter. The fundamental relationships needed for the environmental applications will be summarized in the following chapters. The goal of this chapter is to give an in-depth understanding of the causes of occurrence of self-potential signals. Section 2.1 starts with the macroscopic physics, which corresponds to non-equilibrium thermodynamics. This approach provides the form of the macroscopic constitutive and continuity equations; however, non-equilibrium thermodynamic tells us nothing about the relationships between the material properties entering into the macroscopic constitutive equations, the properties of the porous materials (texture and interfacial properties), and the properties of the constituents (solid and fluid phases). To gain knowledge of the material properties, we must upscale the local equations, valid in each phase of a porous composite. Section 2.2 provides a description of a simple upscaling approach that can be used to describe the relationships between the material properties on one hand and the texture of the material, the properties of the solid and fluid phases, and the properties of the solid fluid interface, on the other. This approach can be used to derive an explicit equation for the contributions entering into the source current density $\mathbf{J}_S$ mentioned in Chapter 1. The contribution to the self-potential signals associated with electron transfer is investigated in Section 2.3. This last contribution provides a theoretical basis for the geobattery and biogeobattery models.

2.1 Non-equilibrium thermodynamic

In this section, we present a generalization of the non-equilibrium thermodynamic approach of de Groot and Mazur (1984), which was restricted to non-porous visco-elastic materials. Our approach is also based on the non-equilibrium thermodynamic framework of Progogine (1947) developed in the vicinity of thermodynamic equilibrium (this concept of vicinity of thermodynamic equilibrium is discussed in more details below). De Groot and Mazur (1984) developed a thermodynamic model of a multi-component visco-elastic fluid. In our case, we are interested in describing the thermodynamics of a multi-component deformable porous continuum, with a non-hydrostatic stress tensor. The porous material is decomposed into a number of elementary volumes. These elementary volumes are large enough to define thermodynamic potentials with statistical significance and small enough so that in each elementary volume, the gradients of the thermodynamic potentials of interest can be considered negligible (quasi-thermodynamic equilibrium). The porous body is not in equilibrium, as thermodynamic potentials vary from elementary volume to elementary volumes. However, the local equilibrium assumption implies that the specific entropy is the same function of extensive variables, per unit mass, as it is in equilibrium; see Prigogine (1947) and Truesdell (1969). Macroscopic representative elementary volumes can be defined for the macroscopic quantities using a volume averaging approach as discussed later in Section 2.2.

2.1.1 Continuity of mass

We call component i below any element of the porous material (solid grains, water molecules, and ions). In an Eulerian framework (i.e., from the standpoint of a stationary observer), the local conservation of mass for a component i is:

$$\frac{\partial \rho_i}{\partial t} = -\nabla \cdot (\rho_i \mathbf{v}_i), \tag{2.1}$$

where ρ_i (in kg m^{-3}) represents the bulk density of partial component i ($i \in \{1, \ldots, Q\}$), $\mathbf{v}_i$ is its velocity (in m s^{-1}), and t is time (in s). The law of mass conservation is obtained by summing the previous equations over all the constituents of the system,

$$\frac{\partial \rho}{\partial t} = -\nabla \cdot (\rho \mathbf{v}), \tag{2.2}$$

$$\rho \equiv \sum_{i=1}^{Q} \rho_i, \tag{2.3}$$

$$\rho \mathbf{v} \equiv \sum_{i=1}^{Q} \rho_i \mathbf{v}_i, \tag{2.4}$$

where ρ is the total density of the porous system, the sum of all the partial mass densities ρ_i of its components, and $\mathbf{v}$ is the velocity of the barycentric center of mass. We introduce the substantial time derivative in a Lagrangian framework attached to the barycentric center of mass motion by

$$\frac{D.}{Dt} = \frac{\partial .}{\partial t} + (\mathbf{v} \cdot \nabla). \tag{2.5}$$

In Section 2.1.6 below, we will define another Lagrangian framework that will be relative to the mineral skeleton, and in which the observable (measurable) fluxes are usually defined and measured. The diffusion fluxes of the various species (ions, water, and solid grains) in the Lagrangian framework attached to the barycentric center of mass (superscript m) are defined by

$$\mathbf{J}_i^m = \rho_i(\mathbf{v}_i - \mathbf{v}). \tag{2.6}$$

The mass fraction (dimensionless) of species i is defined by $C_i = \rho_i/\rho$. Equations (2.2) to (2.6) yield expressions for the substantial time derivative of the mass density and concentrations:

$$\frac{D\rho}{Dt} = -\rho\nabla \cdot \mathbf{v}, \tag{2.7}$$

$$\rho\frac{DC_i}{Dt} = -\nabla \cdot \mathbf{J}_i^m. \tag{2.8}$$

Introducing the specific volume $\upsilon = 1/\rho$, Eq. (2.7) yields a mass conservation equation for an observer moving with the barycentric center of mass motion. This mass conservation equation is given by:

$$\rho\frac{D\upsilon}{Dt} = \nabla \cdot \mathbf{v}. \tag{2.9}$$

In addition, it follows from Eqs. (2.4) to (2.6), in the Lagrangian framework previously defined, the sum of all the fluxes is equal to zero,

$$\sum_{i=1}^{Q} \mathbf{J}_i^m = 0, \tag{2.10}$$

where Q denotes the total number of species. Another identity, used later, follows directly from the previous equations,

$$\frac{\partial a\rho}{\partial t} = \rho\frac{Da}{Dt} - \nabla \cdot (a\rho\mathbf{v}), \tag{2.11}$$

where a can be a scalar, a vector, or a tensor.

2.1.2 Continuity of charge

We denote q_i the charge per unit mass of component i ($q_i = 0$ if the species is neutral, e.g., water or neutral molecules). The total current density (in A m^{-2}) is defined as the net amount of charge flowing per unit surface area and per unit time:

$$\mathbf{J}^m = \sum_{i=1}^{Q} \rho_i q_i \mathbf{v}_i. \tag{2.12}$$

The total charge per unit mass of the system q (C kg^{-1}) is defined by

$$q = \frac{1}{\rho} \sum_{i=1}^{Q} \rho_i q_i = \sum_{i=1}^{Q} C_i q_i. \tag{2.13}$$

Using Eq. (2.6), the total current density is given by:

$$\mathbf{J}^m = \mathbf{J}_c^m + \rho q \mathbf{v}, \tag{2.14}$$

$$\mathbf{J}_c^m \equiv \sum_{i=1}^{Q} q_i \mathbf{J}_i^m, \tag{2.15}$$

where $\mathbf{J}_c^m$ represents the conduction current (in A m^{-2}). Therefore the total current density is equal to the conduction current density plus an advective term $\rho q \mathbf{v}$. The porous medium is also characterized by a global electroneutrality condition: the net charge of the grains is exactly counterbalanced by the net charge of the pore water as discussed in Chapter 1. Consequently $q = 0$ and therefore the convective term is equal to zero in the Lagrangian framework attached to the barycentric center of mass.

The continuity equation for the electrical charge is obtained by summing the continuity equations for the mass of the various charged components, Eqs. (2.8) and (2.15). This yields:

$$\rho \frac{Dq}{Dt} = -\nabla \cdot \mathbf{J}_c^m, \tag{2.16}$$

and therefore $\nabla \cdot \mathbf{J}_c^m = 0$ if there is a global electroneutrality in the system ($q = 0$). Neglecting the influence of the Lorenz force, neglecting electric and magnetic polarizations, and using the macroscopic electroneutrality condition ($q = 0$), the Maxwell stress tensor $\overline{\overline{T}}_M$ and the Maxwell body force $\mathbf{F}_M$ per unit mass are given by,

$$\overline{\overline{T}}_M = \varepsilon \left(\mathbf{E} \otimes \mathbf{E} - \frac{1}{2} \mathbf{E}^2 \overline{\overline{I}} \right), \tag{2.17}$$

$$\rho \mathbf{F}_M = \rho q \mathbf{E} = 0, \tag{2.18}$$

where ε is the dielectric constant of the porous body, $\mathbf{E} \otimes \mathbf{E}$ denotes a dyadic product and therefore a second order tensor ($\otimes$ denotes the tensorial product), and $\overline{\overline{I}}$ is the unit dyadic with elements δ_{ij} (Kronecker delta, $\delta_{ij} = 1$ if $i = j$ and $\delta_{ij} = 0$ if $i \neq j$). More general equations including electrical and magnetic polarizations can be found in de Groot (1961). A force balance equation (Newton's law with the inertial terms omitted) is given by

$$\nabla \cdot \overline{\overline{T}}_M = \rho \mathbf{F}_M (=0). \tag{2.19}$$

In the present book, we will neglect the electromagnetic force terms (other than Coulomb force) and the Maxwell stress tensor but an even more general theory for cross-coupled effects in charged porous media than the one presented below can be foreseen.

2.1.3 Force and momentum balance equations

We first specify the assumptions made in developing the model below; see also Revil (2007a). (1) The influence of the external magnetic field sources is ignored and the porous medium is assumed to be free of any magnetic polarizable particles. (2) Thermal free convection of the pore fluid is ignored. Only the case of scalar potential motion is considered in the present model. (3) We consider viscous-laminar flow of the pore fluid relative to the skeleton. (4) We keep the theory linear. (5) Non-linear redox porcesses and electron transfer are not accounted for and more generally chemical reactions are disregarded. The equation of motion of the system (Newton's law) is

$$\rho \frac{D\mathbf{v}}{Dt} = \nabla \cdot \overline{\overline{\sigma}} + \rho \mathbf{F}, \tag{2.20}$$

$$\rho \mathbf{F} \equiv \sum_{i=1}^{Q} \rho_i \mathbf{F}_i, \tag{2.21}$$

$$\overline{\overline{\sigma}} = \sum_{i=1}^{Q} \overline{\overline{\sigma}}_i, \tag{2.22}$$

where $\overline{\overline{\sigma}}$ (expressed in N m^{-2}) is the total (symmetric) Cauchy stress tensor (tension is considered positive). This stress tensor is equal to the sum of the peculiar (partial) stresses (positive for tension) of the individual components of the mixture $\overline{\overline{\sigma}}_i$. The vector $\mathbf{F}$ is the total body force per unit mass exerted on the system and $D\mathbf{v}/Dt$ represents the acceleration of the center of mass. The forces $\mathbf{F}_i$ entering Eq. (2.20) represent the gravity and the electrostatic body forces per unit mass exerted on component i. As these forces arise from scalar potentials ψ_i, $\mathbf{F}_i = -\nabla \psi_i$ (potential flow) and $\partial \psi_i / \partial t = 0$, the total potential ψ of the system is defined by

$$\rho\psi = \sum_{i=1}^{Q} \rho_i \psi_i. \tag{2.23}$$

Using Eqs. (2.2), (2.7), and (2.20), we write the balance equation for the momentum density $\rho\mathbf{v}$ as

$$\frac{\partial(\rho\mathbf{v})}{\partial t} = -\nabla \cdot (\rho\mathbf{v} \otimes \mathbf{v} - \bar{\bar{\sigma}}) + \rho\mathbf{F}. \tag{2.24}$$

The momentum flux associated with the momentum density $\rho\mathbf{v}$ is $(\rho\mathbf{v} \otimes \mathbf{v} - \bar{\bar{\sigma}})$ and the term $\rho\mathbf{F}$ therefore represents a source term of momentum.

2.1.4 Energy balance equation (first principle)

From here, we generalize the approach of de Groot and Mazur (1984) to thermo-poroelastic media rather than to a visco-elastic fluid. From Eq. (2.20) we have a conservation equation for the kinetic energy of the barycentric center of motion,

$$\rho\frac{D}{Dt}\left(\frac{1}{2}\mathbf{v}^2\right) = \nabla \cdot (\bar{\bar{\sigma}} \cdot \mathbf{v}) - \bar{\bar{\sigma}} : \nabla\mathbf{v} + \rho\mathbf{F} \cdot \mathbf{v}, \tag{2.25}$$

where the colon indicates a tensor dot product ($\mathbf{a}{:}\mathbf{b} = a_{ij}b_{ij}$ with the Einstein convention; note that symbols "." and ":" between tensors of various orders denote their inner product with simple or double contraction, respectively). This yields a conservation equation for the kinetic energy

$$\frac{\partial}{\partial t}\left(\frac{1}{2}\rho\mathbf{v}^2\right) = -\nabla \cdot \mathbf{W}_k - \bar{\bar{\sigma}} : \nabla\mathbf{v} + \rho\mathbf{F} \cdot \mathbf{v}, \tag{2.26}$$

$$\mathbf{W}_k = \frac{1}{2}\rho\mathbf{v}^2\mathbf{v} - \bar{\bar{\sigma}} \cdot \mathbf{v}, \tag{2.27}$$

where $\mathbf{W}_k$ denotes the kinetic energy flux in an Eulerian framework of reference (in J m^{-2}). We note u the internal energy of the system. The internal energy flux vector $\mathbf{W}_u$ is the sum of the heat flux plus energy change due to diffusion and local stresses $\bar{\bar{\sigma}}_i$,

$$\mathbf{W}_u = \mathbf{H}^m + \sum_{i=1}^{Q} (u_i\bar{\bar{I}} - \bar{\bar{\sigma}}_i/\rho_i)\mathbf{J}_i^m, \tag{2.28}$$

where u_i denotes the internal energy of species i, and $\mathbf{H}^m$ is the conductive heat flux including heat transported by mass diffusion (Nitao and Bear, 1996). A balance equation for the energy (kinetic plus internal energies) is therefore given by

$$\frac{D}{Dt}\left(\frac{1}{2}\rho\mathbf{v}^2 + \rho u\right) = -\nabla \cdot (\mathbf{W}_u - \bar{\bar{\sigma}} \cdot \mathbf{v}) + \rho T\Pi + \rho\mathbf{v} \cdot \mathbf{F} + \sum_{i=1}^{Q} \mathbf{J}_i^m \cdot \mathbf{F}_i, \tag{2.29}$$

where Π (positive or negative) denotes the power per unit mass and temperature. The term $\rho T \Pi$ corresponds to the energy sources provided to the system by external sources and is not accounted for by de Groot and Mazur, 1984). This yields the following local conservation equation for the internal energy;

$$\frac{\partial}{\partial t}(\rho u) = -\nabla \cdot \mathbf{W}_u + \bar{\bar{\sigma}} : \nabla \mathbf{v} + \rho T \Pi + \sum_{i=1}^{Q} \mathbf{J}_i^m \cdot \mathbf{F}_i, \qquad (2.30)$$

$$\mathbf{W}_u = \rho u \mathbf{v} + \mathbf{H}^m + \sum_{i=1}^{Q} (u_i \bar{\bar{I}} - \bar{\bar{\sigma}}_i/\rho_i)\mathbf{J}_i^m. \qquad (2.31)$$

Equation (2.30) generalizes Eq. (2.34) of De Groot and Mazur (1984) to thermoporoelastic bodies.

A local conservation equation for the potential energy is given by

$$\frac{\partial}{\partial t}(\rho \psi) = -\nabla \cdot \mathbf{W}_\psi - \rho \mathbf{v} \cdot \mathbf{F} - \sum_{i=1}^{Q} \mathbf{J}_i^m \cdot \mathbf{F}_i, \qquad (2.32)$$

$$\mathbf{W}_\psi = \rho \psi \mathbf{v} + \sum_{i=1}^{N} \psi_i J_i^m, \qquad (2.33)$$

where $\mathbf{W}_\psi$ is the potential energy flux vector in an Eulerian framework of reference.

The total energy e of the system per unit mass is the sum of the kinetic energy, potential energy and internal energy per unit mass. It is given by $e = \mathbf{v}^2/2 + \psi + u$, which defines the internal energy u per unit mass of the system. The internal energy corresponds to the energy associated with the short-range interactions and motions of the individual components of the porous mixtures. The internal energy is dependent on the temperature and density (pressure) of the system as discussed later. The balance equation for the total energy is

$$\frac{\partial}{\partial t}(\rho e) = -\nabla \cdot \mathbf{W}_e + \rho T \Pi, \qquad (2.34)$$

$$\mathbf{W}_e = \mathbf{W}_k + \mathbf{W}_u + \mathbf{W}_\psi, \qquad (2.35)$$

$$\mathbf{W}_e = \rho e \mathbf{v} - \bar{\bar{\sigma}} \cdot \mathbf{v} + \mathbf{H}^m + \sum_{i=1}^{Q} \left[(u_i + \psi_i)\bar{\bar{I}} - \frac{\bar{\bar{\sigma}}_i}{\rho_i} \right] \cdot \mathbf{J}_i^m, \qquad (2.36)$$

where $\mathbf{W}_e$ represents the total energy flux (energy per unit mass per unit surface area and per unit time). This energy flux includes three contributions: (1) a convective term $\rho e \mathbf{v} - \bar{\bar{\sigma}} \cdot \mathbf{v}$, (2) a conductive heat flux $\mathbf{H}^m$, and (3) a diffusive energy flux.

Subtracting the total energy given by Eq. (2.34) to the sum of the kinetic plus potential energy given by Eq. (2.29) respectively, and using Eq. (2.36) we obtain a balance equation for the internal energy u,

$$\rho \frac{Du}{Dt} = -\nabla \cdot \mathbf{W}_u + \bar{\bar{\sigma}} : \nabla \mathbf{v} + \rho T \Pi_T, \tag{2.37}$$

$$\Pi_T \equiv \Pi + \left(\sum_{i=1}^{Q} \mathbf{J}_i^m \cdot \mathbf{F}_i \right) / (\rho T), \tag{2.38}$$

where Π_T is the total power supplied to the system including the influence of the body forces. The stress tensor is decomposed into a reversible elastic component (superscript e) and a dissipative component (superscript d) corresponding to irreversible mechanisms of deformation, $\bar{\bar{\sigma}} = \bar{\bar{\sigma}}^e + \bar{\bar{\sigma}}^d$, which is consistent with the approach of Truesdell (1969). Using Eqs. (2.31), (2.37), and (2.38), we obtain a balance equation for the thermal energy,

$$\frac{Du}{Dt} = \frac{DQ}{Dt} + \upsilon\bar{\bar{\sigma}}^e : \nabla \mathbf{v} + \upsilon\bar{\bar{\sigma}}^d : \nabla \mathbf{v} + \upsilon \sum_{i=1}^{Q} \mathbf{F}_i \cdot \mathbf{J}_i^m + T\Pi, \tag{2.39}$$

$$\rho \frac{DQ}{Dt} = -\nabla \mathbf{H}^m. \tag{2.40}$$

The parameter Q is here the heat per unit of mass in the system and $\upsilon \equiv 1/\rho$ the specific volume. The terms of the left-hand side of Eq. (2.39) represent the rate of gain of internal energy per unit mass. These terms are: (1) the rate of internal energy input by conduction per unit mass DQ/Dt, (2) the reversible rate of internal energy increase per unit mass by compaction $\upsilon\bar{\bar{\sigma}}^e : \nabla \mathbf{v}$, (3) the irreversible rate of internal energy increase per unit mass due to dissipation associated with irreversible mechanisms of deformation $\upsilon\bar{\bar{\sigma}}^d : \nabla \mathbf{v}$, (4) the irreversible rate of internal energy increase per unit mass by the internal motion of the various components forming the system, and finally (5) the power per unit mass and temperature supplied to the system. The term $\upsilon\bar{\bar{\sigma}}^e : \nabla \mathbf{v}$ can be either positive or negative depending upon whether the porous system is expanding or contracting. This term is a reversible mode of exchange of energy between mechanical and internal energy. The term $\upsilon\bar{\bar{\sigma}}^d : \nabla \mathbf{v}$ is always positive and represents an irreversible degradation of mechanical to thermal energy.

For practical applications, it is convenient to recast the thermal energy equation in terms of temperature and heat capacity rather than in terms of internal energy. We first specify an equation of state for the internal energy $u(C_i, v, T)$ of the porous system as a function of the pressure and the temperature,

$$\frac{Du}{Dt} = \left(\frac{\partial u}{\partial v}\right)_{T,C_i} \frac{Dv}{Dt} = \left(\frac{\partial u}{\partial T}\right)_{v,C_i} \frac{DT}{Dt} + \sum_{i=1}^{N} \left(\frac{\partial u}{\partial C_i}\right)_{v,T} \frac{DC_i}{Dt}, \quad (2.41)$$

$$\frac{Du}{Dt} = \left[-P + T \left(\frac{\partial P}{\partial}\right)_T\right] \frac{Dv}{Dt} + C_v \frac{DT}{Dt} - v \sum_{i=1}^{Q} u_i \nabla \cdot \mathbf{J}_i^m, \quad (2.42)$$

where we have use an additional equation of state for the average pressure $P(v, T)$ defined below by Eq. (2.45) and the conservation equations for the various components of the system. The term $u_i = (\partial u/\partial C_i)_{T,v}$ is the partial internal energy per unit mass of component i, $C_v \equiv (\partial u/\partial T)_v$ (in J kg^{-1} °C^{-1}) represents the heat capacity (specific heat) of the system measured at constant volume per unit mass. The stress tensors $\bar{\bar{\sigma}}$ and $\bar{\bar{\sigma}}_i$ are decomposed into their isotropic (P and p_i) and deviatoric ($\bar{\bar{S}}$ and $\bar{\bar{s}}_i$) components,

$$\bar{\bar{\sigma}} = -P\bar{\bar{I}} + \bar{\bar{S}}, \quad (2.43)$$

$$\bar{\bar{\sigma}}_i = -p_i\bar{\bar{I}} + \bar{\bar{s}}_i, \quad (2.44)$$

$$P = \sum_{i=1}^{Q} p_i, \quad (2.45)$$

$$\bar{\bar{S}} = \sum_{i=1}^{Q} \bar{\bar{s}}_i. \quad (2.46)$$

For the pore water, the partial deviatoric stress is null. Combining Eqs. (2.39), (2.40), and (2.42) yields,

$$\rho C_v \frac{DT}{Dt} = -\nabla \cdot \left(\mathbf{H}^m - \sum_{i=1}^{N} u_i \mathbf{J}_i^m\right) - T \left(\frac{\partial P}{\partial T}\right)_v \nabla \cdot \mathbf{v} + D^S$$

$$+ \sum_{i=1}^{Q} \mathbf{F}_i \cdot \mathbf{J}_i^m + \rho T \Pi, \quad (2.47)$$

where $Dv/Dt = v\nabla \cdot \mathbf{v}$. The term $D^S = \bar{\bar{\sigma}}^d : \nabla \mathbf{v} \geq 0$ (in J m^{-3} s^{-1}) is the component of the total dissipation function associated with the dissipative deformation processes. Equation (2.47) states that the temperature of the system changes because of heat conduction, deformation effects, heating associated with irreversible deformation (e.g., viscous, plastic, or visco-plastic deformations), friction, and power supplied to the system. Equation (2.47) is written in a more familiar form as

$$\frac{D}{Dt}(\rho C_v T) = -\nabla \cdot \mathbf{H} + Q_S, \tag{2.48}$$

$$\mathbf{H} \equiv \mathbf{H}^m - \sum_{i=1}^{N} u_i \mathbf{J}_i^m, \tag{2.49}$$

$$Q_S \equiv -T\left(\frac{\partial P}{\partial T}\right)_v \nabla \cdot \mathbf{v} + D^S + \sum_{i=1}^{Q} u_i \nabla \cdot \mathbf{J}_i^m + \sum_{i=1}^{Q} \mathbf{F}_i \cdot \mathbf{J}_i^m + \rho T \Pi, \tag{2.50}$$

where $\mathbf{H}$ is the heat flux (in J m^{-2} s^{-1}) and Q_S is the heat source term (in J m^{-3} s^{-1}).

2.1.5 The entropy balance equation (second principle)

Prigogine (1947) showed that the classical definition of entropy in thermostatics applies to a system in the vicinity of equilibrium. We denote by s the entropy per unit mass of the system. The entropy is related to the internal energy u, to the specific volume v, and to the mass fraction C_i according to Euler relationship of thermostatics. In linear non-equilibrium thermodynamics, the Gibbs–Duhem relationship is assumed to hold,

$$\rho T \frac{Ds}{Dt} \rho \frac{Du}{Dt} - \bar{\bar{\sigma}}^e : \nabla \mathbf{v} - \sum_{i=1}^{Q} \bar{\bar{u}}_i : \nabla \mathbf{J}_i^m, \tag{2.51}$$

where the terms $P(Dv/Dt)$ and $\mu_i(DC_i/Dt)$ appearing in the classical formulation of the Gibbs–Duhem equation of Prigogine (1947) and de Groot and Mazur (1984) have been substituted by $-v(\bar{\bar{\sigma}}^e : \nabla \mathbf{v})$ and $\bar{\mu}_i \nabla \mathbf{J}_i^m$, respectively, to account for the tensorial nature of the elastic stress and chemical potential tensors in porous media.

The chemical potential tensor of species i is defined by Truesdell (1969) as:

$$\bar{\mu}_i = \Psi_i \bar{\bar{I}} - \bar{\bar{\sigma}}_i/\rho_i = (u_i - Ts_i)\bar{\bar{I}} - \bar{\bar{\sigma}}_i/\rho_i, \tag{2.52}$$

where $\Psi_i = u_i - Ts_i$ is the specific Helmholtz free energy of component i and s_i is the specific entropy of component i. Replacing the time derivative of the internal energy in the Gibbs–Duhem equation by its expression given by Eq. (2.37) yields

$$\rho T \frac{Ds}{Dt} = -\nabla \cdot \left[\mathbf{H}^m + \sum_{i=1}^{Q} \left(u_i \bar{\bar{I}} - \bar{\bar{\sigma}}_i/\rho_i \right) \mathbf{J}_i^m \right] + \bar{\bar{\sigma}} : \nabla \mathbf{v}$$

$$+ \sum_{i=1}^{Q} \mathbf{J}_i^m \cdot \mathbf{F}_i - \bar{\bar{\sigma}}^e : \nabla \mathbf{v} - \sum_{i=1}^{Q} [(u_i - Ts_i)\bar{\bar{I}} - \bar{\bar{\sigma}}_i/\rho_i]\nabla \mathbf{J}_i^m + \rho T \Pi.$$

$$\tag{2.53}$$

After long, but straightforward, algebraic manipulations, Eq. (2.53) can be recasted as a local continuity equation for entropy,

$$\rho \frac{Ds}{Dt} = -\nabla \cdot \mathbf{S}^m + \Theta + \rho \Pi, \tag{2.54}$$

$$\mathbf{S}^m = \frac{\mathbf{H}^m}{T} + \sum_{i=1}^{Q} s_i \mathbf{J}_i^m, \tag{2.55}$$

$$T\Theta = -\mathbf{H}^m \nabla \ln T - \sum_{i=1}^{Q} s_i \mathbf{J}_i^m \cdot \nabla T - \sum_{i=1}^{Q} (\nabla \cdot \bar{\bar{u}}_i - \mathbf{F}_i)\mathbf{J}_i^m + \bar{\bar{\sigma}}^d : \nabla \mathbf{v}, \tag{2.56}$$

where Θ is the entropy source (inner entropy production per unit volume and unit time).

2.1.6 Dissipation function

The equations described in the previous section were written in a Lagrangian framework attached to the barycentric center of mass of the system, which comprises Q components. These components are: (1) N components located into the pore water (anions, cations and neutral species), (2) the pore water itself (i.e., the neutral solvent), and (3) the grains (so $Q = N + 2$). The balance equation for the internal energy u for an open porous medium (first principle) is given by Eqs. (2.37). The local balance law for entropy (second principle) for an open system is given by Eqs. (2.54) to (2.56). The stress tensor is divided into a reversible elastic component (superscript e) and a dissipative component (superscript d) corresponding to irreversible mechanisms of deformation, i.e., $\bar{\bar{\sigma}} = \bar{\bar{\sigma}}^e + \bar{\bar{\sigma}}^d$. For an open porous material, the entropy production can be divided into two contributions $Ds/Dt = (Ds/Dt)_e + (DS/Dt)_i$. This expression is just an alternative form of the entropy balance equation; δs_e is the entropy exchanged between the system and its surrounding and δs_i is the entropy produced in the system itself. The term $(Ds/Dt)_e$ can be either positive or negative while $\Theta = \rho(Ds/Dt)_i \geq 0$, the inner entropy production of the system, is necessarily positive.

In linear non-equilibrium thermodynamics, the total dissipation of the system is equal to the product of the temperature with the entropy production term, $D = T\Theta$ and $D \geq 0$ because $\Theta \geq 0$ and $T > 0$. This yields

$$D = -\mathbf{H}^m \nabla \ln T - \sum_{i=1}^{Q} s_i \mathbf{J}_i^m \cdot \nabla T - \sum_{i=1}^{Q} (\nabla \cdot \bar{\bar{\mu}}_i - \mathbf{F}_i)\mathbf{J}_i^m + \bar{\bar{\sigma}}^d : \nabla \mathbf{v}, \tag{2.57}$$

$$D = -\mathbf{S}^m \nabla T - \sum_{i=1}^{Q} (\nabla \cdot \bar{\bar{\mu}}_i - \mathbf{F}_i)\mathbf{J}_i^m + D^S, \tag{2.58}$$

where we have replaced the heat flow (divided by the temperature) by the entropy flux and where $D^S = \bar{\bar{\sigma}}^d : \nabla \mathbf{v}$ represents the dissipation component associated with irreversible deformation processes (e.g., with visco-plastic deformation).

We have therefore derived above a general formulation for the two principles of thermodynamics in a Lagrangian framework attached to the barycentric center of mass of the porous material. Therefore all the components were treated in a symmetrical fashion and the equations have been stated in the framework associated with this barycentric center of mass. However, a practical Lagrangian framework to interpret laboratory measurements is the one associated with deformation of the skeleton; see, for example, Biot (1941). This yields a new Lagrangian derivative associated with the velocity of the solid phase (subscript s) according to:

$$\frac{d.}{dt} = \frac{\partial.}{\partial t} + \mathbf{v}_s \cdot \nabla. \tag{2.59}$$

The pore fluid (solvent plus ions and dissolved neutral species) are written with the subscript f. The fluid is assumed to be ideal, which implies that the peculiar stresses of the components of the pore electrolyte are assumed to satisfy the Dalton's law of ideal mixtures, i.e., $\bar{\bar{\sigma}}_i = -p_i \bar{\bar{I}}$, where p_i is the partial pressure of component i. We write p the pore fluid pressure of a reservoir in contact with the charged porous medium, and $\bar{p}$ the fluid pressure of the pore fluid in the charged porous medium (the charge is due to the electrical double layer described in Chapter 1). The pore fluid pressure p is defined thermodynamically as the pressure of an external fluid in local contact with the system and in thermodynamic equilibrium with it (see Section 2.2 below). A direct consequence of the previous assumptions is that the chemical potential of the components of the pore water are considered as scalars as discussed further below. The terms $\phi \rho_i$, $\phi \rho_w$ and $(1 - \phi)\rho_s$ denote the partial mass densities of component i, solvent and solid phase, respectively (φ is the connected porosity).

We assume that the forces acting on both the pore fluid and the mineral matrix are always in balance and thus the acceleration of the barycentric center of mass and inertia are neglected. Obviously the time-scale to reach mechanical equilibrium is always much shorter than the time-scale to reach the (thermostatic) equilibrium state. This mechanical balance assumption yields $\rho \mathbf{F} + \nabla \cdot \bar{\bar{\sigma}} = 0$. Prigogine's theorem states that in the mechanical equilibrium state, the barycentric velocity $\mathbf{v}$ entering into the definition of the diffusion flow in the dissipation function can be replaced by another arbitrary velocity. This theorem is extremely important for determining the dissipation function in the newly defined Lagrangian framework. It is demonstrated in Exercise 2.1. In this mineral matrix framework, the measurable fluxes are defined by

$$\mathbf{J}_i \equiv \rho_{(\pm)}\phi(\mathbf{v}_i - \mathbf{v}_g) = \mathbf{J}_i^m - \rho_i\phi(\mathbf{v}_g - \mathbf{v}), \tag{2.60}$$

$$\mathbf{J}_w \equiv \rho_w\phi(\mathbf{v}_w - \mathbf{v}_g) = \mathbf{J}_w^m - \rho_w\phi(\mathbf{v}_g - \mathbf{v}), \tag{2.61}$$

$$0 = \mathbf{J}_s^m - \rho_s(1 - \phi)(\mathbf{v}_s - \mathbf{v}), \tag{2.62}$$

$$\mathbf{S} = \mathbf{S}^m - \rho s(\mathbf{v}_s - \mathbf{v}), \tag{2.63}$$

$$\bar{\bar{\sigma}}^d : \nabla\mathbf{v}_s = \bar{\bar{\sigma}}^d : \nabla\mathbf{v} - (\mathbf{v}_s - \mathbf{v})\nabla \cdot \bar{\bar{\sigma}}^d, \tag{2.64}$$

where the subscripts i correspond to the component i, the subscript w corresponds to water, and the subscript g corresponds to the mineral grains. Prigogine's theorem can be generalized by combining the Gibbs–Duhem relationship, the assumption of mechanical equilibrium, and the dissipation function written in the new Lagrangian framework (see solution of Exercise 2.1).

The peculiar body forces per unit mass entering the dissipation function, see Eq. (2.58), include the electrical force plus the gravity force. For the components i dissolved in water, $i \in (1, \ldots, N)$,

$$\mathbf{F}_i = -q_i\nabla\psi/\rho_i + g \approx -q_i\nabla\psi/\rho_i, \tag{2.65}$$

while for water $\mathbf{F}_w = \mathbf{g}$. In these equations, q_i is the charge of species i, $q_i = (\pm e)Z_i$, e is the elementary charge (1.6×10^{-19} C), and Z_i is the valence of species i. Note that both forces are derived from a scalar potential as required by the theory (the gravitational acceleration is related to the gravitational potential $\varphi = g\, z$ by $g = -\nabla\varphi$, z positive upward). The effective potentials per unit mass of the ions and water are scalars defined by

$$\hat{u}_i = \mu_i^R + \mu_i + q_i\psi/\rho_i + gz, \tag{2.66}$$

$$\hat{\mu}_w = \mu_w^R + \mu_w + gz, \tag{2.67}$$

respectively, where μ_i^R, μ_w^R, and μ_g^R are the chemical potentials in a stress-free reference state. Combining Prigogine's theorem and the definition of the effective chemical potentials yields a new form of the dissipation function,

$$D = -\sum_{i=1}^{N}\mathbf{J}_i\nabla\hat{\mu}_i - \mathbf{J}_w\nabla\hat{\mu}_w - \mathbf{S}\nabla T + D^S. \tag{2.68}$$

We use now Curie's principle to set up the form of the linear constitutive equations for the fluxes; see Curie (1894). This principle states that for an isotropic system, the coupling of tensors, whose orders differ by an odd number, do not occur. From this principle, it follows that we must consider scalars, vectors and second-rank tensors in the dissipation function separately. We will focus below only on the vectorial fluxes entering the dissipation function. The vectorial contribution to the dissipation function is:

$$D = -\sum_{i=1}^{N} \mathbf{J}_i \nabla \hat{\mu}_i - \mathbf{J}_w \nabla \hat{\mu}_w - \mathbf{H} \nabla T / T. \tag{2.69}$$

Here we have replaced the entropy flux $\mathbf{S}$ by $\mathbf{H}/T$, where $\mathbf{H}$ is the heat flux including a convective contribution. The seepage (Darcy) velocity $\mathbf{U}$ is,

$$\mathbf{U} = \sum_{i=1}^{N} \mathbf{J}_i / \rho_i + \mathbf{J}_w / \rho_w, \tag{2.70}$$

$$\rho_w \mathbf{U} \approx \mathbf{J}_w. \tag{2.71}$$

Note that the definitions of the Darcy velocity and the heat flux are not unique because of the presence of the ions in the solvent (water). The gradients of the effective chemical potentials of the pore water and ions are:

$$\nabla \hat{\mu}_w = \frac{k_b T}{\rho_w} \nabla \ln \bar{C}_w + \frac{\nabla \bar{p}}{\rho_w} - \mathbf{g} - s_w \nabla T, \tag{2.72}$$

$$\nabla \hat{\mu}_i = \frac{k_b T}{\rho_i} \nabla \ln \bar{C}_i - \frac{q_i}{\rho_i} \mathbf{E} - s_i \nabla T, \tag{2.73}$$

$$s_i \equiv -\left(\frac{\partial \hat{\mu}_i}{\partial T} \right)_{Ci,p,\varphi}, \tag{2.74}$$

where p is the pore water pressure of a reservoir locally in equilibrium with the pore water of the porous medium, $\mathbf{E}$ is the electrical field, and s_i are the specific entropy values of pore fluid component i. The swelling pressure π is defined as $\pi \equiv -k_b T \ln \bar{C}_w$ and $\bar{p} \equiv p + \pi - \rho_f g z$ (in Pa) is the effective pore fluid pressure. Therefore the gradients of the effective chemical potentials of the water and ions per unit mass are given by

$$\nabla \hat{\mu}_w = \nabla p / \rho_w - s_w \nabla T, \tag{2.75}$$
$$\nabla \hat{\mu}_i = \nabla \tilde{\mu}_i / \rho_i - s_i \nabla T, \tag{2.76}$$

respectively, and $\tilde{\mu}_i$ represents the electrochemical potentials of species i per unit volume. The linear non-equilibrium thermodynamic theory implies linear relationships between the flows and the generalized driving forces that cause these flows. This yields a form of the constitutive equations that Revil (2007a) call "the N-form":

$$\begin{bmatrix} \mathbf{J}_i \\ \cdots \\ \mathbf{J}_i \\ \mathbf{H} \end{bmatrix} = -\overline{\overline{N}} \begin{bmatrix} \nabla \hat{\mu}_i \\ \cdots \\ \nabla \hat{\mu}_w \\ \nabla T / T \end{bmatrix}, \tag{2.77}$$

where $\overline{\overline{N}}$ is a $(N + 2)$ square matrix of material transport properties. This type of generalized constitutive set of equations has been discussed in the literature; see, for example, Mitchell (1993) and Degond *et al.* (1998). Inserting the linear phenomenological equations within the dissipation function shows that the entropy production is a homogeneous quadratic function of the independent forces entering Eq. (2.77). The second law of thermodynamics states that this expression is positive ($D \geq 0$). This yields $N_{ij} = N_{ji}$ (Onsager's reciprocity), $N_{ii} \geq 0$, and $N_{ij}^2 \leq N_{ii} N_{jj}$ ($i \neq j$). By volume-averaging the local Nernst–Planck and Stokes equations, Revil and Linde (2006) provided the necessary relationships between the material properties entering the matrix $\overline{\overline{N}}$ and both the constituent properties and two textural parameters, the formation factor and the permeability.

The continuity equations can be written as a function of volumetric fractions per unit initial volume and therefore involved the determinant J of the deformation gradient tensor. The $(N + 2)$ generalized continuity equations associated with the N-form of the constitutive equations are

$$J\nabla \cdot \begin{bmatrix} \mathbf{J}_i \\ \cdots \\ \mathbf{J}_w \\ \mathbf{H} \end{bmatrix} = -\frac{d}{dt} \begin{bmatrix} \rho_i \phi \\ \cdots \\ \rho_w \phi \\ \rho C_v T \end{bmatrix} + \begin{bmatrix} \rho_i R_i \\ \cdots \\ \rho_w R_w \\ Q_S \end{bmatrix}, \qquad (2.78)$$

where R_i and R_w are the production rates of ions i and water, respectively, within the controlled volume per unit mass of the porous porous medium (in mol m^{-3} s^{-1}) and Q_S is here the intrinsic heat source (bulk rate of heat production, expressed in J m^{-3} s^{-1}), see Eq. (2.50).

2.1.7 Generalized power balance equation

In order to discuss the influence of the reversible (elastic) component of the stress tensor in the present model, it is necessary first to derive a generalized power balance equation for the specific Helmholtz free energy of the porous composite. Taking the conservation equation for the internal energy and for the entropy of the system in a Lagrangian framework attached to the barycentric center of mass of the system yields:

$$\rho \left(\frac{Du}{Dt} - T\frac{Ds}{Dt} \right) = -\nabla \cdot (\mathbf{W}_u - T\mathbf{S}^m) + \overline{\overline{\sigma}} : \nabla \mathbf{v} - \mathbf{S}^m \nabla T$$
$$+ \rho T \Pi_T - D, \qquad (2.79)$$

$$\mathbf{W}_u - T\mathbf{S}^m = \sum_{i=1}^{N} \left(\Psi_i \overline{\overline{I}} - \frac{\overline{\overline{\sigma}}_i}{\rho_i} \right) \mathbf{J}_i^m = \sum_{i=1}^{N} \overline{u}_i \mathbf{J}_i^m, \qquad (2.80)$$

where the total Helmholtz free energy and the specific Helmholtz free energy per unit mass of the component i are defined by $\Psi = u - Ts$ and $\Psi_i = u_i - Ts_i$, respectively. This yields the following generalized power balance equation:

$$\rho \left(\frac{D\Psi}{Dt} + s \frac{DT}{Dt} \right) = -\nabla \cdot \left(\sum_{i=1}^{N} \bar{\mu}_i \mathbf{J}_i^m \right) + \bar{\bar{\sigma}} : \nabla \mathbf{v} - \mathbf{S}^m . \nabla T + \rho T \Pi_T - D.$$

(2.81)

which is the fundamental equation required to formulate the thermo-poroelastic constitutive equations. The term $d\Psi + s\,dT$ represents part of the energy externally supplied to the porous medium and stored (if $d\Psi + s\,dT > 0$) or extracted (if < 0) during the time interval dt and converted in heat. Introducing the expression of the dissipation function, Eq. (2.58) into Eq. (2.81), yields

$$\rho \left(\frac{D\Psi}{Dt} + s \frac{DT}{Dt} \right) = - \sum_{i=1}^{N} \bar{\mu}_i : \nabla \mathbf{J}_i^m + \bar{\bar{\sigma}}^e : \nabla \mathbf{v}.$$

(2.82)

The Helmholtz free energy corresponds to the potential associated with the entropy of the porous system.

2.1.8 Mechanical constitutive relationships

In the following, we assume that the porous medium has a thermo-poroelastic behavior. In addition, we assume small deformations for which the whole theory is linear and obeys the superposition principle. We denote Ω_0 (the volume bounded by the surface $\partial \Omega_0$) the Lagrangian reference configuration for the porous medium, $\mathbf{X}$ the vector locating an element of the porous grain skeleton in this reference configuration, and $\Omega(t)$ the Eulerian configuration at time t and $\mathbf{x}(\mathbf{X}, t)$ the position of the same material element at time t in the Eulerian configuration. The spatial position of the material element is given by $\mathbf{x}(\mathbf{X}, t) = \mathbf{X} + \mathbf{u}(\mathbf{X}, t)$, where $\mathbf{u}$ is the solid displacement. Deformation of the skeleton is characterized by the deformation gradient tensor $\overline{\overline{P}}$ by $d\mathbf{x}(\mathbf{X}, t \equiv \overline{\overline{P}} \cdot d\mathbf{X}$. The Green–Lagrange strain tensor $\bar{\bar{e}}$ and the second Piola–Kirchhoff stress tensor $\overline{\overline{T}}$ (both symmetric) are defined by (see Coussy (1995) and Levenston *et al.* (1999)),

$$\bar{\bar{e}} \equiv \frac{1}{2} (\overline{\overline{P}}^T \overline{\overline{P}} - \overline{\overline{I}}),$$

(2.83)

$$\overline{\overline{T}} \equiv J \overline{\overline{P}}^{-1} \bar{\bar{\sigma}} (\overline{\overline{P}}^T)^{-1},$$

(2.84)

$$J \equiv \det \overline{\overline{P}} = \frac{dV}{dV_0},$$

(2.85)

where $\overline{\overline{P}}^T$ represents the transposed matrix of $\overline{\overline{P}}$, $\overline{\overline{A}} = \overline{\overline{P}}^T \overline{\overline{P}}$ is the right Cauchy–Green tensor, $\overline{\overline{B}} = \overline{\overline{A}}^{-1} = \overline{\overline{P}}^{-1} \overline{\overline{P}}^{-T}$ is the Piola deformation tensor, J denotes the Jacobian of the deformation gradient tensor ($J > 0$) and is also equal to the ratio between the volume of the elementary volume in the current configuration dV divided by the volume this element occupied in the reference state dV_0. Tension is considered positive. In addition, Euler's formula is (see Huygue and Janssen (1997))

$$\frac{dJ}{dt} = J\nabla \cdot \mathbf{v}_s. \tag{2.86}$$

From Eq. (2.81) written now in the Lagrangian framework, we have

$$\frac{d(\rho\Psi)}{dt} + \rho\Psi\nabla \cdot \mathbf{v}_s = \text{Tr}(\bar{\bar{\sigma}}^e \nabla \mathbf{v}_s) - \sum_{i-1}^{N+1} \mu_i \nabla \cdot \mathbf{J}_i - \rho s \frac{dT}{dt}, \tag{2.87}$$

where the sum over i is extended to the $N + 1$ components of the pore water (N components plus the solvent, i.e. the water molecules). Using Eq. (2.86), we obtain

$$\frac{d(J\rho\Psi)}{dt} = J \left[\frac{d(\rho\Psi)}{dt} + d\Psi\nabla \cdot \mathbf{v}_s \right]. \tag{2.88}$$

The mass balance equation for species i is,

$$\frac{dm_i}{dt} = -J\nabla \cdot \mathbf{J}_i, \tag{2.89}$$

where $m_i = J\phi\rho_i$. Equations (2.87) to (2.89) yield

$$\frac{d(J\rho\Psi)}{dt} = \text{Tr}\left(\overline{\overline{T}}\frac{d\bar{\bar{e}}}{dt}\right) + \sum_{i=1}^{N+1} \mu_i \frac{dm_i}{dt} = J\rho s \frac{dT}{dt}. \tag{2.90}$$

The time derivative of the Helmholtz free energy per unit mass of the pore fluid, $\rho_f\Psi_f$, is obtained using the following chain of algebraic manipulations,

$$\Psi_f = -pv_f + \sum_{i=1}^{N+1} \mu_i \bar{C}_i^f, \tag{2.91}$$

$$\rho_f\Psi_f = -p + \sum_{i=1}^{N+1} \mu_i\rho_i, \tag{2.92}$$

$$\frac{d(\rho_f\Psi_f)}{dt} = -\frac{dp}{dt} + \sum_{i=1}^{N+1} u_i \frac{d\rho_i}{dt} + \sum_{i=1}^{N+1} \rho_i \frac{d\mu_i}{dt}, \tag{2.93}$$

where υ_f is the specific volume of the pore fluid in the porous medium and $\bar{C}_i^f = \rho_i/\rho_f$ the mass fraction of pore fluid component i. The Gibbs–Duhem relationship in the pore fluid is,

$$\rho_f s_f \delta T - \delta p + \sum_{i=1}^{N+1} \rho_i \delta \mu_i + \bar{Q}_V \delta \varphi + \rho_f g \delta z = 0, \qquad (2.94)$$

where s_f is the entropy per unit mass of the pore fluid, $\bar{Q}_V$ is the net excess of electrical charge in the connected porosity of the porous medium and $\bar{\varphi}$ is the local electrical potential (position dependent) in the pore fluid of the body. The osmotic pressure in the connected porosity of the porous continuum is given by

$$\pi = \int_0^{\bar{\varphi}} \bar{Q}_V d\bar{\varphi}'. \qquad (2.95)$$

This effective pressure includes gravitational and osmotic effects. In addition, state equations for the density and entropy per unit mass of the pore fluid, according to Coussy (1995), are

$$\frac{1}{\rho_f} = \frac{1}{\rho_f^0}(1 - \beta_f(\bar{p} - \bar{p}_0) + \alpha_f \theta), \qquad (2.96)$$

$$s_f - s_f^0 = \frac{C_p^f}{T_0}\theta - \alpha_f\left(\frac{\bar{p} - \bar{p}_0}{\rho_f^0}\right), \qquad (2.97)$$

$$\beta_f \equiv \frac{1}{K_f} \equiv \frac{1}{\rho_f}\left(\frac{\partial \rho_f}{\partial \bar{p}}\right)_T, \qquad (2.98)$$

$$\alpha_f \equiv -\frac{1}{\rho_f}\left(\frac{\partial \rho_f}{\partial T}\right)_p, \qquad (2.99)$$

$$C_\upsilon^f - C_p^f = \frac{T_0 \alpha_f^2}{\rho_f^0 \beta_f}, \qquad (2.100)$$

where $\theta = T - T_0$, C_p^f and C_υ^f are the fluid mass heat at constant pressure and volume, respectively, β_f is the isothermal bulk compressibility of the pore fluid, and α_f is the cubic thermal dilatation coefficient of the pore fluid. We write now the Gibbs–Duhem relationship, Eq. (2.94), in a more customary form;

$$\rho_f s_f \delta T - \delta \bar{p} + \sum_{i=1}^{N+1} \rho_i \delta \mu_i = 0, \qquad (2.101)$$

$$\rho_f s_f \frac{dT}{dt} - \frac{d\bar{p}}{dt} + \sum_{i=1}^{N+1} \rho_i \frac{d\mu_i}{dt} = 0. \qquad (2.102)$$

This yields

$$\frac{d(\rho_f \Psi_f)}{dt} = \sum_{i=1}^{N+1} \mu_i \frac{d\rho_i}{dt} + \rho_f s_f \frac{dT}{dt}. \tag{2.103}$$

We denote by $\upsilon_p = J\phi$ the specific pore volume of the porous medium in the deformed state per unit referential volume. We obtain

$$\frac{d(J\phi\rho_f \Psi_f)}{dt} = \upsilon_p \frac{d(\rho_f \Psi_f)}{dt} + \rho_f \Psi_f \frac{d\upsilon_p}{dt}, \tag{2.104}$$

$$\frac{d(J\phi\rho_f \Psi_f)}{dt} = \upsilon_p \left(\sum_{i=1}^{N+1} \mu_i \frac{d\rho_i}{dt} + \rho_f s_f \frac{dT}{dt} \right) + \left(-\bar{p} \sum_{i=1}^{N+1} \mu_i \rho_i \right) \frac{d\upsilon_p}{dt}, \tag{2.105}$$

and, in addition, we have,

$$\sum_{i=1}^{N+1} \mu_i \frac{dm_i}{dt} = \sum_{i=1}^{N+1} \mu_i \frac{d(\upsilon_p \rho_i)}{dt}, \tag{2.106}$$

$$\sum_{i=1}^{N+1} \mu_i \frac{dm_i}{dt} = \sum_{i=1}^{N+1} \mu_i \rho_i \frac{d\upsilon_p}{dt} + \sum_{i=1}^{N+1} \mu_i \upsilon_p \frac{d\rho_i}{dt}. \tag{2.107}$$

We now determine the free energy of the mineral matrix per unit initial volume of the porous medium, W. This quantity is obtained by removing from the free energy of the saturated porous medium, the free energy of the pore fluid. It corresponds to the elastic energy of the porous body. This yields $W \equiv J(\rho\Psi - \phi\rho_f \Psi_f)$ and, therefore,

$$\frac{dW}{dt} = \text{Tr}\left(\overline{\overline{T}} \frac{d\bar{\bar{e}}}{dt} \right) + \bar{p}\frac{d\upsilon_p}{dt} - J(\rho s - \phi\rho_f s_f)\frac{dT}{dt}. \tag{2.108}$$

We assume that the potential W depends only on the external variables $\bar{\bar{e}}$, υ_p, and T, i.e., $W = W(\bar{\bar{e}}, \upsilon_p, T)$. The differentiation of W yields

$$\frac{dW}{dt} = \left(\frac{\partial W}{\partial e_{ij}} \right)_{\upsilon_p,T} \frac{de_{ij}}{dt} + \left(\frac{\partial W}{\partial \upsilon_p} \right)_{e_{ij},T} \frac{d\upsilon_p}{dt} + \left(\frac{\partial W}{\partial T} \right)_{e_{ij},\upsilon_p} \frac{dT}{dt}, \tag{2.109}$$

$$\overline{\overline{T}} = \left(\frac{\partial W}{\partial \bar{\bar{e}}} \right)_{\upsilon_p,T}, \tag{2.110}$$

$$\bar{p} = \left(\frac{\partial W}{\partial \upsilon_p} \right)_{e_{ij},T}, \tag{2.111}$$

$$S - m_f s_f = -\left(\frac{\partial W}{\partial T} \right)_{e_{ij},\upsilon_p}, \tag{2.112}$$

where $S \equiv J\rho s$ is the specific entropy per unit referential volume and $m_f \equiv J\rho_f \phi = \upsilon_p \rho_f$ is the specific mass of water per unit referential volume. In linear thermo-poroelasticity, this yields the following constitutive equations for charged porous materials:

$$\overline{\overline{T}} - \overline{\overline{T}}^0 = {}^4\bar{C} : \bar{\bar{e}} - M\overline{\overline{B}}\frac{m_f}{\rho_f^0} - \overline{\overline{A}}\theta, \tag{2.113}$$

$$\bar{p} - \bar{p}_0 = -M\overline{\overline{B}} : \bar{\bar{e}} + M\frac{m_f}{\rho_f^0} + \rho_f^0 L\theta, \tag{2.114}$$

$$S - S_0 - m_f s_f^0 = \overline{\overline{A}} : \bar{\bar{e}} - Lm_f + C_v\theta/T_0. \tag{2.115}$$

Owing to the global electroneutrality of the porous medium, there are no electrostatic terms in the constitutive equations, except for the osmotic pressure, and therefore we recover the classical form of the mechanical constitutive equations given by Coussy (1995). The terms $\overline{\overline{T}}^0$, $\bar{p}_0$ and S_0 characterize the thermostatic state. If we consider a stress-free thermostatic configuration, $\overline{\overline{T}}^0 = 0$. We denote $^4\bar{C}$ the (fourth-order) stiffness tensor of undrained isothermic elastic moduli, $\overline{\overline{A}}$ and $\overline{\overline{B}}$ are symmetric second-order tensors, M is the Biot modulus, and C_v is the volume heat capacity per unit of initial volume in an isodeformation undrained experiment. They are defined by

$$C_{ijkl} \equiv \left(\frac{\partial T_{ij}}{\partial e_{kl}}\right)_{m_f,T} = \left(\frac{\partial T_{kl}}{\partial e_{ij}}\right)_{m_f,T}, \tag{2.116}$$

$$M \equiv \rho_f^0 \left(\frac{\partial \bar{p}}{\partial m_f}\right)_{e_{ij},T}, \tag{2.117}$$

$$B_{ij} \equiv -\left(\frac{\partial T_{ij}}{\partial \bar{p}}\right)_{e_{ij},T}, \tag{2.118}$$

$$A_{ij} \equiv -\left(\frac{\partial T_{ij}}{\partial T}\right)_{e_{ij},m_f}, \tag{2.119}$$

$$L \equiv s_f^0 - \left(\frac{\partial S}{\partial T}\right)_{e_{ij},m_f}, \tag{2.120}$$

$$C_v \equiv T_0 \left(\frac{\partial(S - m_f s_f^0)}{\partial T}\right)_{e_{ij}m_f}. \tag{2.121}$$

The tensor $T_0\overline{\overline{A}}$ is the tensor of undrained strain latent heats and $T_0 L$ represents the isodeformation latent heat of variation in fluid mass content. The tensor $\overline{\overline{B}}$ is the Biot stress tensor, which reduces to the Biot coefficient α in the isotropic case.

In the infinitesimal transformation assumption and for isotropic cases, the Green strain tensor $\bar{\bar{e}}$ is replaced by the infinitesimal strain tensor $\bar{\bar{\varepsilon}}$ and the Piola–Kirchhoff stress tensor $\overline{\overline{T}}$ can be replaced by the Cauchy stress tensor $\overline{\overline{\sigma}}$. Then, the constitutive equations of linear thermo-poroelasticity of charged porous materials are

$$\sigma_{ij} = \left(K_u - \frac{2}{3}G \right) \varepsilon_{kk}\delta_{ij} + 2G\varepsilon_{ij} - \alpha M \frac{m_f}{\rho_f^0}\delta_{ij} - K\alpha_b\theta\delta_{ij}, \quad (2.122)$$

$$p - p_0 = -\alpha M \varepsilon_{kk} + M\frac{m_f}{\rho_f^0} + C_v\theta, \quad (2.123)$$

$$S - S_0 = \rho_f \phi s_f^0 + K_u \alpha_b \varepsilon_{kk} - \alpha_m M\frac{m_f}{\rho_f^0}\delta_{ij} + \frac{C_v}{T_0}\theta, \quad (2.124)$$

where K_u is the undrained isothermal bulk modulus (in Pa), G is the shear modulus (in Pa), α_b is the cubic undrained thermal dilatation coefficient (K^{-1}), α denotes the Biot coefficient for the effective stress law (dimensionless), and $\alpha_m = (\alpha - \phi_0)\alpha_g + \phi_0\alpha_f$, where α_g and α_f denote the cubic thermal dilatation coefficients for the grains and fluid, respectively, and ϕ_0 denotes the porosity of the porous material in the reference state (see Coussy (1995), p. 91), and $\alpha M = C$, which denotes another Biot poroelastic modulus that will be discussed further in Chapter 8. The stress tensor obeys to the mechanical equilibrium condition. Appropriate boundary conditions can be used to close the system of equations and examples will be discussed later in Chapters 4 to 8. Equations (2.122) and (2.125), in isothermal conditions, will be used also in Chapter 8 to develop the seismoelectric theory and are important to understand and describe the propagation of seismic waves in porous media. The shear modulus remains the same for a drained or an undrained experiment as long as the pore fluid does not sustain shear stresses. A generalization of Biot theory to the case of a porous material saturated by a generalized Maxwell type fluid has also been given recently by Revil and Jardani (2010b). The extension makes the Biot's theory completely symmetric in terms of relationships between the drained and undrained moduli as the pore fluid can sustain shear stresses.

In summary, the previous set of equations provides a pretty general framework to analyze the occurrence of the self-potential signals in deformable porous media. We will see in several sections of this book how this framework can be applied to a variey of problems including the occurrence of electromagnetic signals associated with hydraulic fracturing and seismic wave propagation (Chapter 8) or the evolution of self-potential signals associated with pumping tests (Chapter 6). In the next section (Section 2.2), we provide a way to upscale the local equations (valid in each phase of the porous composite) to the scale of the representative elementary volume. This will provide the missing relationship between the material properties and (i) the textural properties of the porous material, (ii) the properties of each phase, and (iii) the properties of the electrical double layer (see Section 1.2 of Chapter 1).

2.2 Upscaling: from local to macroscopic equations

In this section, we provide one way to upscale the local equation (especially the Stokes and Nernst–Planck equations) from a single capillary to the scale of a represensative elementary volume of a porous material. This approach allows us to derive relationships between the material properties entering the macroscopic constitutive equations developed in the previous section and the properties of each phase as well as the textural properties of the porous material.

2.2.1 Upscaling using a single capillary

We consider first a capillary of radius R and length L ($L \gg R$ to avoid edge effects). This capillary is saturated by a binary symmetric 1:1 electrolyte like a NaCl or a KCl solution. We use below polar coordinates with r as the radial coordinate in the plane normal to the capillary and z in the direction along the capillary. The capillary is in contact with two neutral reservoirs of ions, one upstream at $z = 0$ (reservoir 1) and one downstream at $z = L$ (reservoir 2, see Figure 2.1a). The macroscopic potentials controlling transport phenomena in the capillary are controlled by the difference of potential between these two macroscopic reservoirs. In the case where there are no such reservoirs, a fictitious reservoir locally in equilibrium with the porous material should be considered.

We consider that the surface of the capillary is negatively charged. The opposite case of a positively charged mineral surface is just the symmetric case of the one considered below. The present model could also be easily coupled to electrical double layer theory, as explained below. As discussed in Chapter 1, the negative surface charge density Q_S includes the true mineral surface charge density Q_0 plus the surface charge density of the Stern layer of adsorbed counterions Q_β; (Figures 2.1b and 2.1c); see Leroy *et al.* (2007). In theory, Q_S may depend slightly on the capillary size, because of the overlapping of the electrical double layer and difference in salinity in the different capillaries; see Gonçalvès *et al.* (2007) and Wang and Revil (2010). In the present approach, we will consider that this dependence is so small, that it can be safely neglected. The charge density $\bar{q}_V$ (>0, in C m^{-3}) denotes the volumetric charge density in the capillary (the pore water is therefore not neutral because of the presence of the diffuse layer). The local electroneutrality equation of the capillary, as a whole, is therefore

$$\bar{q}_V + Q_S \frac{S}{V_p} = 0, \tag{2.125}$$

where $S = 2\pi R L$ is the internal surface of the capillary and $V_p = \pi R^2 L$ represents its volume.

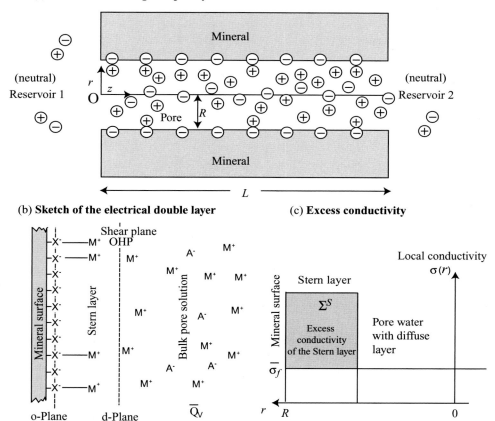

Figure 2.1. Description of a charged capillary in a porous material. (a) Sketch of a cylindrical pore separating two reservoirs of ions at two distinct pressures, salinities and electrical potentials. The electrolyte in the two reservoirs is the same (neutral) binary symmetric (1:1) electrolyte. The model corresponds here to the thick double layer assumption, e.g., there is no neutral pore water. (b) Sketch of the electrical double layer. The double layer comprises a layer of counterions sorbed onto the mineral surface and a diffuse layer with coulombic interaction only. (c) Sketch of the electrical conductivity distribution showing the excess surface conductivity Σ_S (in S) of the Stern layer and the effective conductivity of the pore water $\bar{\sigma}_f$ with an excess of counterions from the diffuse layer. The position $r = 0$ corresponds to the centerline of the capillary while the position $r = R$ characterizes the interface between the solid and fluid phases.

We use the Teorell–Meyer–Sievers (TMS) model (also called the Donnan model in the literature; see Teorell (1935) and Meyer and Sievers (1936)) to describe the mean concentrations in the pore space of the capillary (see solution of Exercise 2.2). This approximation is used to avoid involving the more rigorous solution of the

Poisson–Boltzmann equation; see, for example, Pride (1994) and Furini *et al.* (2006). As demonstrated by Westermann-Clark and Christoforou (1986), the use of the TMS approach is valid in the case where the thickness of the diffuse layer is on the same order of magnitude as the size of the pores. If we consider clay materials, this assumption is therefore valid. As for a lot of porous media, this assumption is incorrect (see discussion in Exercise 2.1 below), the charge density that is transported by advection is generally different and much smaller than the total charge density of the diffuse layer at the scale of a representative elementary volume. This difference will be considered in more details in Chapter 3.

From Eq. (2.125), the relationship between the volumetric charge density and the surface charge density Q_S (which is constant for all capillaries) is

$$\bar{q}_V = -Q_S \frac{S}{V_p} = -Q_S \frac{2}{R}. \tag{2.126}$$

Therefore $\bar{q}_V$ is inversely proportional to the pore radius.

The pore solution is a mixture of water molecules (the solvent) and ions. The Stokes equation for the transport of the salty water in a capillary should, therefore, be written in term of differences (or gradients) in the chemical potentials between the two reservoirs. This yields

$$-\bar{c}_w \nabla \mu_w - \bar{c}_{(+)} \nabla \mu_{(+)} - \bar{c}_{(-)} \nabla \mu_{(-)} + \mathbf{F} + \eta_f \nabla^2 \mathbf{v} = 0, \tag{2.127}$$

$$-\bar{c}_w \Omega_w \nabla(p - \pi) - \bar{c}_{(+)} \nabla \mu_{(+)} - \bar{c}_{(-)} \nabla \mu_{(-)} + \mathbf{F} + \eta_f \nabla^2 \mathbf{v} = 0, \tag{2.128}$$

$$-\nabla(p - \pi) + \eta_f \nabla^2 \mathbf{v} + \mathbf{F} - \bar{c}_{(+)} \nabla \mu_{(+)} - \bar{c}_{(-)} \nabla \mu_{(-)} = 0, \tag{2.129}$$

as $\bar{c}_w \Omega_w \approx 1$ where $\bar{c}_w$ is the concentration of water molecules per unit volume of the pore fluid in the capillary and Ω_w represents the molecular volume of water (in m^3 mol^{-1}) and where p is the mechanical pore water pressure (in Pa, not related to concentrations), π is the osmotic pressure (see the solution of Exercise 2.2), η_f is the dynamic viscosity of the pore water (in Pa s), $\mathbf{v}$ is the velocity of the pore fluid in the capillary (in m s^{-1}), $\mathbf{F}$ is the external body force applied to the pore fluid (in N m^{-3}), $\bar{c}_{(\pm)}$ are the average concentrations of the cations and anions in the capillary, and $\mu_{(\pm)} = \mu_{(\pm)}^0 + k_b T \ln C_{(\pm)}$ are the macroscopic chemical potentials of the cations and anions in the reservoirs (in J) where k_b is the Boltzmann constant (1.381×10^{-23} J K^{-1}), T is the absolute temperature (in K), $\mu_{(\pm)}^0$ is the chemical potential in a reference state, and $C_{(\pm)}$ are the macroscopic concentrations in the reservoirs in contact with the capillary (in m^{-3}). The pressure difference $p^* = p - \pi$ represents the total water potential in the reservoirs, omitting the gravitational term if this term is accounted for in the body force. There is a gradient in the osmotic pressure between the two reservoirs, because of the salinity gradient between these reservoirs.

The macroscopic boundary conditions for the hydrodynamic fluid pressure are:

$$p = p_1, \text{at } z = 0, \tag{2.130}$$
$$p = p_1 - \delta p, \text{at } z = L, \tag{2.131}$$

in reservoirs 1 and 2, respectively, with $\delta p > 0$. The unit vector $\hat{z}$ is in the flow direction. Neglecting gravity (or alternatively keeping the gravity in the total hydraulic head), the body force entering Eq. (2.127) is given by the Coulomb force,

$$\mathbf{F} = \bar{q}_V \mathbf{E}, \tag{2.132}$$

where $\mathbf{E}$ is the external electrical field (in V m^{-1}). In the quasi-static limit of the Maxwell equations (no electromagnetic induction), we have $\nabla \times \mathbf{E} = 0$ and therefore the electrical field can be derived from the gradient of a scalar potential,

$$\mathbf{E} = -\nabla \psi, \tag{2.133}$$

where ψ is the electrical potential (in V) (electromotive potential) defined in the two reservoirs (only the difference of potential is measurable). The boundary conditions for the macroscopic potential ψ are

$$\psi = \psi_1, \text{at } z = 0, \tag{2.134}$$
$$\psi = \psi_1 - \delta \psi, \text{at } z = L, \tag{2.135}$$

in reservoirs 1 and 2, respectively. The chemical potentials $\mu_{(\pm)}$ obey similar macroscopic boundary conditions between the upstream and the downstream reservoirs. Then, the Stokes equation can be written as

$$\nabla p^* + \bar{q}_V \nabla \psi + \bar{c}_{(+)} \nabla \mu_{(+)} + \bar{c}_{(-)} \nabla \mu_{(-)} = \eta_f \nabla^2 \mathbf{v}, \tag{2.136}$$

or, equivalently, in terms of electrochemical potential $\tilde{\mu}_{(\pm)} = \mu_{(\pm)}^0 + \mu_{(\pm)} \pm e\psi$ (see Newman (1991)),

$$\nabla p^* + \bar{c}_{(+)} \nabla \tilde{\mu}_{(+)} + \bar{c}_{(-)} \nabla \tilde{\mu}_{(-)} = \eta_f \nabla^2 \mathbf{v} \tag{2.137}$$

We can integrate the Stokes equation for the pore fluid in cylindrical coordinates. The standard potentials $\mu_{(\pm)}^0$ depend on the choice of the standard values for the concentrations, electrostatic potential, and fluid pressure. We first see from Eq. (2.136) that there are three forcing terms in the Stokes equation: (i) the effective fluid pressure gradient, (ii) the electrical field associated with external sources $\bar{q}_V \nabla \psi$, and (iii) the chemical potential gradients associated with the ionic species. Therefore, the velocity can be divided into three components including the mechanical, the electrical and the chemio-osmotic contributions,

$$\mathbf{v} = \mathbf{v}_m + \mathbf{v}_e + \mathbf{v}_c \tag{2.138}$$

The boundary condition for the mechanical contribution is $v_m(r = R) = 0$ on the surface of the capillary. We use identical boundary conditions for the two other velocity fields. This implies that the migration of the counterions in the Stern layer does not significantly influence this boundary condition, an assumption that may be not valid if the velocity of the pore water is too small. In addition, we will show that the contribution of the Stern layer can be neglected for d.c. conditions (because of the discontinuity of the solid phase); therefore, the boundary conditions for the velocity used above is correct.

In the capillary model shown in Figure 2.1, the effective fluid pressure, the electromotive potential ψ, and the chemical potentials $\mu_{(\pm)}$ depend only on the distance z along the capillary;

$$p^*(z) = -\frac{\delta p^*}{L}z + p_1^*, \tag{2.139}$$

$$\psi(z) = -\frac{\delta \psi}{L}z + \psi_1, \tag{2.140}$$

$$\mu_{(\pm)}(z) = -\frac{\delta \mu_{(\pm)}}{L}z + \mu_{(\pm)}^1. \tag{2.141}$$

The average concentrations in the pore space of the capillary can be determined from the TMS model using the equality between the electrochemical potentials in the pore space and with a fictitious reservoir of ions in local equilibrium with the pore space of the capillary (see the solution of Exercise 2.2). The excess of charge and the concentrations in the capillary are related by

$$\bar{q}_V = e\left(\bar{c}_{(+)} - \bar{c}_{(-)}\right), \tag{2.142}$$

$$\bar{c}_{(\pm)} = C_f\left(\sqrt{\frac{\bar{q}_V^2}{4C_f^2 e^2} + 1} \pm \frac{\bar{q}_V}{2eC_f}\right), \tag{2.143}$$

$$\bar{c}_{(\pm)} = C_f\left(\sqrt{\frac{Q_S^2}{e^2 R^2 C_f^2} + 1} \mp \frac{Q_S}{eRC_f}\right), \tag{2.144}$$

where Eq. (2.126) has been used, and e is the elementary charge (1.6×10^{-19} C). The three contributions to the velocity of the pore water are

$$v_m(r) = -\left(\frac{r^2 - R^2}{4\eta_f}\right)\frac{\delta p^*}{L}, \tag{2.145}$$

$$v_e(r) = -\left(\frac{r^2 - R^2}{4\eta_f}\right)\bar{q}_V\frac{\delta \psi}{L}, \tag{2.146}$$

$$v_c(r) = -\left(\frac{r^2 - R^2}{4\eta_f}\right)\left(\bar{c}_{(+)}\frac{\delta \mu_{(+)}}{L} + \bar{c}_{(-)}\frac{\delta \mu_{(-)}}{L}\right). \tag{2.147}$$

The total velocity can be integrated to get the flux of the pore water through the capillary. By definition, the flux is given by

$$u = \int_0^R v(r) 2\pi r \, dr. \tag{2.148}$$

After integration of the different contributions, we obtain

$$u = -\left(\frac{\pi R^4}{8\eta}\right)\left(\frac{\delta p^*}{L} + \bar{q}_v \frac{\delta \psi}{L} + \bar{c}_{(+)}\frac{\delta \mu_{(+)}}{L} + \bar{c}_{(-)}\frac{\delta \mu_{(-)}}{L}\right). \tag{2.149}$$

We are now in the position to upscale to a representative elementary volume made of a set of n capillaries of the same radius R and parallel to each other. We consider that the porous material is a jacketed cylindrical core sample (jacket on the external surface with the exception of the two end-faces). We note A the surface area of a cross-section of the cylinder. The connected porosity is defined as

$$\phi = \frac{n\pi R^2}{A}. \tag{2.150}$$

The macroscopic flux density over a cross section of surface area A of the porous material is

$$U = \frac{nu}{A}, \tag{2.151}$$

$$U = -\frac{n}{A}\left(\frac{\pi R^4}{8\eta}\right)\left(\frac{\delta p^*}{L} + \bar{q}_v \frac{\delta \psi}{L} + \bar{c}_{(+)}\frac{\delta \mu_{(+)}}{L} + \bar{c}_{(-)}\frac{\delta \mu_{(-)}}{L}\right), \tag{2.152}$$

$$U = -\phi\left(\frac{R^2}{8\eta}\right)\left(\frac{\delta p^*}{L} + \bar{q}_v \frac{\delta \psi}{L} + \bar{c}_{(+)}\frac{\delta \mu_{(+)}}{L} + \bar{c}_{(-)}\frac{\delta \mu_{(-)}}{L}\right). \tag{2.153}$$

If we consider the effect of the tortuosity of the capillaries, we need to replace the porosity by an effective porosity, which is the ratio of the porosity to the tortuosity of the pore space α. This ratio is equivalent to the inverse of the formation factor F used to describe the in-phase conductivity of the porous material (see Chapter 3 and Revil and Cathles (1999));

$$F = \frac{\alpha}{\phi}. \tag{2.154}$$

In porous media, F is usually correlated to the connected porosity via a power-law relationship named Archie's law,

$$F = \phi^{-m}, \tag{2.155}$$

where m is loosely called the cementation exponent; see Archie (1942). In a lot of papers, we find the relationship between the formation factor and the porosity written as $F = a\phi^{-m}$, where a is an empirical parameter. This equation cannot actually be found in the original work by Archie and should not be called Archie's law. It was introduced much later in the literature and works only for a colleciton of samples from the same formation reporting their formation factors as a function of their porosities.

Darcy's law is defined as a linear macroscopic relationship between the volumetric flux of water and the pressure gradient,

$$\mathbf{U} = -\left(\frac{k}{\eta}\right)\nabla p, \tag{2.156}$$

when the other potential gradients are equal to zero ($\delta\psi = \delta\mu_{(\pm)} = 0$), and in absence of ions in the pore water. Darcy's law can be seen either as a macroscopic constitutive equation for the flux of the pore water or as a macroscopic momentum conservation equation for the pore fluid. A comparison between Eqs. (2.153) and (2.156) reveals a well-known expression of the permeability for a set of capillaries of radius R; see Dullien (1992) and Jackson (2008, 2010),

$$k = \frac{\phi}{\alpha}\left(\frac{R^2}{8}\right) = \frac{R^2}{8F}. \tag{2.157}$$

In this expression, R is often called the hydraulic radius when Eq. (2.157) is applied to a real porous material and is eventually determined from specific surface area measurements. Therefore, the generalized macroscopic constitutive equations for the Darcy velocity is,

$$\mathbf{U} = -\left(\frac{k}{\eta}\right)\left(\nabla p^* - \bar{q}_V\mathbf{E} + \bar{c}_{(+)}\nabla\mu_{(+)} + \bar{c}_{(-)}\nabla\mu_{(-)}\right), \tag{2.158}$$

or alternatively,

$$\mathbf{U} = -\left(\frac{k}{\eta}\right)\left(\nabla p^* + \bar{c}_{(+)}\nabla\tilde{\mu}_{(+)} + \bar{c}_{(-)}\nabla\tilde{\mu}_{(-)}\right), \tag{2.159}$$

where $\mathbf{E} = -\nabla\psi$ is the macroscopic (low-frequency) electrical field and $\tilde{\mu}_{(\pm)} = \mu^0_{(\pm)} + k_bT\ln C_{(\pm)} \pm e\psi$ are the macroscopic electrochemical potentials of the two charge carriers (+ for cations and − for anions). If we use the equality of the macroscopic chemical potential gradients through the porous material, we obtain:

$$\nabla\mu_{(+)} = \nabla\mu_{(-)} = \nabla\mu_f = k_bT\nabla\ln C_f. \tag{2.160}$$

Equation (2.160) is only valid for a 1:1 electrolyte. Combining Eqs. (2.159) and (2.160), we can write the macroscopic equation for the Darcy's velocity as

$$\mathbf{U} = -\left(\frac{k}{\eta}\right)\left[\nabla p^* - \bar{q}_V\mathbf{E} + \left(\bar{c}_{(+)} + \bar{c}_{(-)}\right)\nabla\mu_f\right]. \tag{2.161}$$

We now turn our attention to the flux densities of the cations and anions. These flux densities are related to diffusion, electromigration, and advective transport of the ionic species. The Nernst–Planck equation is written locally as (see Appendix A)

$$\mathbf{j}_{(\pm)} = -b_{(\pm)}\bar{c}_{(\pm)}\nabla\tilde{\mu}_{(\pm)} + \bar{c}_{(\pm)}\mathbf{v}_m, \qquad (2.162)$$

$$\mathbf{j}_{(\pm)} = -b_{(\pm)}\bar{c}_{(\pm)}\nabla\left((\pm e)\,\psi + k_b T C_{(\pm)}\right) + \bar{c}_{(\pm)}\mathbf{v}_m, \qquad (2.163)$$

where $b_{(\pm)}$ represents the mobility of the ionic species in the pore water. Equation (2.162) is correct in the pore water (including the effect of the diffuse layer), but it does not take into account transport along the mineral surface (in the so-called Stern layer) by diffusion or by electromigration. The last term of Eqs. (2.162) and (2.163) is related to the mechanical velocity. Indeed, the electrical and endo-osmotic velocities are already captured by the gradient of the electrochemical potential of the ionic species in the first term of the right-hand side of Eqs. (2.162) and (2.163). In reference to the Stern layer contribution, the diffusion and the electromigration components should comprise two terms: one contribution from the bulk of the capillary, and one contribution along the mineral surface in the Stern layer; see Revil and Leroy (2001). The Stern layer contribution has been suggested by Zukoski and Saville (1986a, b) and is modeled in Exercise 2.3. We add this contribution for completeness, but we will show below that it is not required to explain the transport properties of a clay-rock, for which the solid phase is discontinuous. However, adding the Stern layer contribution would be helpful in order to model induced polarization, that is, the frequency dependence of electrical conductivity; see Chapter 3 and Revil and Florsch (2010), Vaudelet et al. (2011a, b) and Revil (2012).

Using these two contributions (in the bulk pore water and along the mineral surface), the flux densities of cations and anions are given by

$$\mathbf{j}_{(\pm)} = -b_{(\pm)}\bar{c}_{(\pm)}\nabla\tilde{\mu}_{(\pm)} - b_{(\pm)}^S\Gamma_{(\pm)}\nabla\tilde{\mu}_{(\pm)}^S\delta(r - R) + \bar{c}_{(\pm)}\mathbf{v}_m, \qquad (2.164)$$

$$\mathbf{j}_{(\pm)} = -k_b T \left(b_{(\pm)}\bar{c}_{(\pm)}\nabla\ln C_{(\pm)} + b_{(\pm)}^S\Gamma_{(\pm)}\nabla\ln\Gamma_{(\pm)}\delta(r - R)\right)$$
$$- (\pm e)\left[b_{(\pm)}\bar{c}_{(\pm)} + b_{(\pm)}^S\Gamma_{(\pm)}\delta(r - R)\right]\nabla\psi + \bar{c}_{(\pm)}\mathbf{v}_m, \qquad (2.165)$$

where $b_{(\pm)}^S$ represents the mobilities of the ionic species along the mineral surface, $\Gamma_{(\pm)}$ are the surface concentration densities of cations and anions adsorbed on the mineral surface, $\tilde{\mu}_{(\pm)}^S$ represents the electrochemical potential along the mineral surface, and $\delta(r - R) = 0$ if $r \neq R$ and $\delta(r - R) = 1$ if $r = R$ ($0 \leq r \leq R$). In Exercise 2.3, we demonstrate that

$$\nabla\tilde{\mu}_{(\pm)}^S = \nabla\tilde{\mu}_{(\pm)}, \qquad (2.166)$$

the gradient of the electrochemical potential along the mineral surface is equal to the gradient of the electrochemical potential in the pore water. This result is valid only if the solid phase is continuous at the scale of the representative elementary volume. The macroscopic ionic fluxes are defined by averaging the previous expression over a set of capillaries of same radius R. After straightforward algebraic manipulations, this yields

$$\mathbf{J}_{(\pm)} = \frac{n\pi R^2}{A}\left\{-b_{(\pm)}\bar{c}_{(\pm)}\nabla\tilde{\mu}_{(\pm)} + \bar{c}_{(\pm)}\mathbf{v}_m\right\} + \frac{n2\pi R}{A}\left[-b_{(\pm)}^S\Gamma_{(\pm)}\nabla\tilde{\mu}_{(\pm)}\right], \quad (2.167)$$

$$J_{(\pm)} = -\frac{n\pi R^2}{A}\left[k_bT\left(b_{(\pm)}\bar{c}_{(\pm)} + \frac{2}{R}b_{(\pm)}^S\Gamma_{(\pm)}\right)\frac{\delta\ln C_{(\pm)}}{L}\right.$$
$$\left. + (\pm e)\left(b_{(\pm)}\bar{c}_{(\pm)} + \frac{2}{R}b_{(\pm)}^S\Gamma_{(\pm)}\right)\frac{\delta\psi}{L}\right] + \bar{c}_{(\pm)}U_m, \quad (2.168)$$

where U_m is the mechanical contribution to the Darcy velocity $\mathbf{U}_m = -(k/\eta)\nabla p^*$ and $\mathbf{J}_{(\pm)} = J_{(\pm)}\mathbf{z}$. Using the expression for the porosity, scaling the porosity by the tortuosity, and replacing the ratio of the porosity to the tortuosity by the formation factor (see above), we obtain (see Revil *et al.* (2011a)),

$$J_{(\pm)} = -\frac{k_bT}{e^2}\sigma_{(\pm)}\frac{\delta\ln C_{(\pm)}}{L} - \frac{(\pm 1)}{e^2}\sigma_{(\pm)}\frac{\delta\psi}{L} + \bar{c}_{(\pm)}U_m, \quad (2.169)$$

$$\sigma_{(\pm)} = \frac{1}{F}\left(e\beta_{(\pm)}\bar{c}_{(\pm)} + \frac{2}{R}e\beta_{(\pm)}^S\Gamma_{(\pm)}\right), \quad (2.170)$$

where we have used an alternative definition of the ionic mobility, $\beta_{(\pm)} = b_{(\pm)}e$. In vectorial form, the macroscopic ionic flux densities are given by

$$\mathbf{J}_{(\pm)} = -\frac{1}{e^2}\sigma_{(\pm)}\nabla\tilde{\mu}_{(\pm)} + \bar{c}_{(\pm)}\mathbf{U}_m \quad (2.171)$$

The electrical current density is defined as the total amount of charges passing through the porous material per unit time and per unit surface area. The diffusion flux is defined as the mean flux of cations and anions passing through the porous material per unit surface area and per unit time. These two quantities are therefore related to the ionic flux densities by

$$J = e\left(J_{(+)} - J_{(-)}\right), \quad (2.172)$$

$$J_d = \frac{1}{2}\left(J_{(+)} + J_{(-)}\right), \quad (2.173)$$

where $\mathbf{J} = J\mathbf{z}$ and $\mathbf{J}_d = J_d\mathbf{z}$. Using the expression of the mechanical contribution to Darcy's velocity, we obtain, in vectorial form, the following macroscopic constitutive equations;

$$\mathbf{J} = -\frac{1}{e}\left(\sigma_{(+)} - \sigma_{(-)}\right)\nabla\mu_f + \sigma\mathbf{E} - \frac{k}{\eta}\bar{q}_V\nabla p^*, \quad (2.174)$$

$$\mathbf{J}_d = -\frac{\sigma}{2e^2}\nabla\mu_f + \frac{1}{2e}\left(\sigma_{(+)} - \sigma_{(-)}\right)\mathbf{E} - \frac{1}{2}\left(\bar{c}_{(+)} + \bar{c}_{(-)}\right)\frac{k}{\eta}\nabla p^*, \quad (2.175)$$

where the electrical conductivity is given as

$$\sigma = \sigma_{(+)} + \sigma_{(-)}. \tag{2.176}$$

The conductivity can be also written as the sum of pore water and surface conductivity acting in parallel,

$$\sigma = \frac{1}{F} \left(\bar{\sigma}_f + \sigma_S \right), \tag{2.177}$$

where the effective pore water conductivity and the surface conductivity are given by

$$\bar{\sigma}_f = e \left(\beta_{(+)} \bar{c}_{(+)} + \beta_{(-)} \bar{c}_{(-)} \right), \tag{2.178}$$

$$\sigma_S = \frac{2}{R} e \left(\beta_{(+)}^S \Gamma_{(+)} + \beta_{(-)}^S \Gamma_{(-)} \right), \tag{2.179}$$

where $\beta_{(\pm)} = b_{(\pm)} e$. The macroscopic electrical conductivity of the porous material can be also expressed as

$$\sigma = \frac{1}{F} \left(\bar{\sigma}_f + \frac{2}{R} \Sigma_S \right), \tag{2.180}$$

where Σ_S is the specific surface conductivity of the Stern layer defined by Revil and Leroy (2001) and Leroy and Revil (2009) as

$$\Sigma_S = e \left(\beta_{(+)}^S \Gamma_{(+)} + \beta_{(-)}^S \Gamma_{(-)} \right). \tag{2.181}$$

This conductivity is expressed in siemens and is therefore often called the specific surface conductance. The next section generalizes the previous equations to the case of a pore size distribution defined through a probability density.

2.2.2 Generalization to a pore size distribution

We define $g(R)$ as the probability density to have a capillary with a radius comprised between R and $R + dR$ (therefore $g(R)$ denotes the pore size distribution). Because $g(R)$ is normalized, we have

$$\int_0^\infty g(R) dR = 1. \tag{2.182}$$

We write the different raw moments of this probability distribution as

$$\Pi_p = \int_0^\infty R^p g(R) dR, \tag{2.183}$$

where p is a negative or positive integer (Π_p is the expectation of the distribution R^p according to the probability distribution $g(R)$). From Eq. (2.183), we have

$$\Pi_0 = \int_0^\infty g(R)dR = 1, \tag{2.184}$$

$$\Pi_1 = \int_0^\infty Rg(R)dR, \tag{2.185}$$

$$\Pi_2 = \int_0^\infty R^2 g(R)dR, \tag{2.186}$$

and so on.

To proceed with this model, we come back to the description of the concentration of the cations and anions as a function of the radius of the pore. With the Donnan approximation, the average concentrations of cations and anions in the pore space of a single capillary of radius R are (see the solution of Exercise 2.2),

$$\bar{c}_{(\pm)} = C_f \left(\sqrt{1 + \xi^2} \pm \xi \right), \tag{2.187}$$

$$\xi = \frac{Q_s}{eRC_f}. \tag{2.188}$$

For a distribution of pore radii described by $g(R)$, the (average) porosity, the macroscopic volumetric charge density of the diffuse layer $\bar{Q}_V$ (in C m^{-3}), and the macroscopic concentrations $\bar{C}_{(\pm)}$ (C m^{-3}) in the pore space are related to the raw moments of the pore size distribution by (see the solution of Exercise 2.4)

$$\phi = \frac{n\pi}{A} \Pi_2, \tag{2.189}$$

$$\bar{Q}_V = -2Q_s \frac{\Pi_1}{\Pi_2}, \tag{2.190}$$

$$\bar{C}_{(\pm)} = C_f \left(\sqrt{1 + \Theta^2} \pm \Theta \right), \tag{2.191}$$

respectively, where the dimensionless parameter Θ is defined by

$$\Theta \equiv \frac{Q_s}{eC_f} \frac{\Pi_1}{\Pi_2}. \tag{2.192}$$

This coefficient will be the main dimensionless coefficient controlling the dependence with salinity of some of the material properties entering the constitutive equations. The osmotic pressure is given by Eq. (2.135) with $\bar{Q}_V$ replacing $\bar{q}_V$. The parameters $\bar{Q}_V$, $\bar{C}_{(\pm)}$, and Θ are the macroscopic equivalents of the capillary-scale properties $\bar{q}_V$, $\bar{c}_{(\pm)}$, and ξ, respectively, as discussed by Revil *et al.* (2011a).

As in the previous section, the flux density (Darcy velocity) over a cross-section of the porous material is written as

$$U = \frac{nu}{A}, \qquad (2.193)$$

$$u = -\left(\frac{\pi \Pi_4}{8\eta}\right)\left(\frac{\delta p^*}{L} + \bar{Q}_V \frac{\delta \psi}{L} + \bar{C}_{(+)}\frac{\delta \mu_{(+)}}{L} + \bar{C}_{(-)}\frac{\delta \mu_{(-)}}{L}\right). \qquad (2.194)$$

Carrying out the convolution product, adding the tortuosity effect (and therefore introducing the formation factor), using the definitions above, the Darcy velocity is given by

$$U = -\frac{k}{\eta}\left[\frac{\delta p^*}{L} + \bar{Q}_V \frac{\delta \psi}{L} + \bar{C}_{(+)}\frac{\delta \mu_{(+)}}{L} + \bar{C}_{(-)}\frac{\delta \mu_{(-)}}{L}\right], \qquad (2.195)$$

where the (hydrodynamic) permeability is related to two raw moments of the pore size distribution by

$$k = \frac{1}{8F}\frac{\Pi_4}{\Pi_2}. \qquad (2.196)$$

Equation (2.196) has been derived by Dullien (1992) and Jackson (2008). In vectorial notations, Eq. (2.195) is written as

$$\mathbf{U} = -\frac{k}{\eta}\left[\nabla p^* + \bar{Q}_V \nabla \psi + \bar{C}_{(+)}\nabla \mu_{(+)} + \bar{C}_{(-)}\nabla \mu_{(-)}\right]. \qquad (2.197)$$

The Darcy velocity comprises the hydromechanical contribution including an osmotic pressure term (the first term in the brackets), the electro-osmotic contribution (the second term in the brackets), and the chemio-osmotic contributions (the two last terms in the brackets). The relationship between the macroscopic excess of electrical charges (from the diffuse layer) in the pore space and the concentrations of cations and anions in the pore space of the material is

$$\bar{Q}_V = e\left(\bar{C}_{(+)} - \bar{C}_{(-)}\right). \qquad (2.198)$$

Combining Eqs. (2.197) and (2.198), the constitutive equation for the Darcy velocity is given by

$$\mathbf{U} = -\frac{k}{\eta}\left[\nabla p^* + \bar{C}_{(+)}\nabla \tilde{\mu}_{(+)} + \bar{C}_{(-)}\nabla \tilde{\mu}_{(-)}\right], \qquad (2.199)$$

Applying the same rules to the ionic densities, we obtain the same macroscopic equation as in the case of capillaries of the same pore radius. We can write the final macroscopic constitutive equations as:

$$
\begin{bmatrix} 2\mathbf{J}_d \\ \mathbf{J} \\ \mathbf{U} \end{bmatrix} = -\overline{\overline{\mathbf{M}}} \begin{bmatrix} \nabla\mu_f \\ \nabla\psi \\ \nabla p^* \end{bmatrix},
\tag{2.200}
$$

$$
\overline{\overline{\mathbf{M}}} = \begin{bmatrix} \dfrac{\sigma}{e^2} & \dfrac{1}{e}\left(\sigma_{(+)} - \sigma_{(-)}\right) & \dfrac{k}{\eta}\left(\bar{C}_{(+)} + \bar{C}_{(-)}\right) \\[2mm] \dfrac{1}{e}\left(\sigma_{(+)} - \sigma_{(-)}\right) & \sigma & \dfrac{k}{\eta}\bar{Q}_V \\[2mm] \dfrac{k}{\eta}\left(\bar{C}_{(+)} + \bar{C}_{(-)}\right) & \dfrac{k}{\eta}\bar{Q}_V & \dfrac{k}{\eta} \end{bmatrix},
\tag{2.201}
$$

where $\overline{\overline{\mathbf{M}}}$ is a matrix of material properties (note that for an anisotropic material, each element of this matrix would be a second-order symmetric tensor). This matrix also obeys the Onsager reciprocity $M_{ij} = M_{ji}$; see Prigogine (1947).

The material properties entering $\overline{\overline{\mathbf{M}}}$ are defined below. The concentrations $\bar{C}_{(\pm)}$ are given by the TMS model (see the solution of Exercise 2.2). The electrical conductivity and the contributions to the total electrical conductivity are defined by

$$
\sigma = \sigma_{(+)} + \sigma_{(-)},
\tag{2.202}
$$

$$
\sigma_{(\pm)} = \frac{1}{F}\left(e\beta_{(\pm)}\bar{C}_{(\pm)} + 2\frac{\Pi_1}{\Pi_2}e\beta_{(\pm)}^S\Gamma_{(\pm)}\right),
\tag{2.203}
$$

respectively. The conductivity can, therefore, be written as the sum of pore water and surface conductivity acting in parallel as proposed in Waxman and Smits (1968),

$$
\sigma = \frac{1}{F}\left(\bar{\sigma}_f + \sigma_S\right),
\tag{2.204}
$$

where the effective pore water conductivity and the surface conductivity are given by

$$
\bar{\sigma}_f = e\left(\beta_{(+)}\bar{C}_{(+)} + \beta_{(-)}\bar{C}_{(-)}\right),
\tag{2.205}
$$

$$
\sigma_S = 2\frac{\Pi_1}{\Pi_2}e\left(\beta_{(+)}^S\Gamma_{(+)} + \beta_{(-)}^S\Gamma_{(-)}\right).
\tag{2.206}
$$

Equation (2.204) is a Waxman-and-Smits-type equation, except that the conductivity of the pore water is distinct from the conductivity of a reservoir in local equilibrium with the porous material. Bussian (1983) and Revil *et al.* (1998) have shown that this equation works well as long as the connectivity problem is ignored.

Indeed, in a complex porous material, the surface and bulk tortuosities are different; see also Schwartz *et al.* (1989), Bernabé and Revil (1995) and Bernabé (1998). This equation will be used in Chapter 3 below and extended to the frequency domain in this chapter.

For a bundle of capillaries, the electrical conductivity can be written as (solution of Exercise 2.4)

$$\sigma = \frac{1}{F}\left(\bar{\sigma}_f + 2\frac{\Pi_1}{\Pi_2}\Sigma_S\right). \tag{2.207}$$

This equation is distinct from the equation derived by previous authors (see, for instance, Pride (1994)), because the conductivity of the diffuse layer is encapsulated in the effective pore water conductivity, and the surface conductance is only due to the contribution of the Stern layer. Pride ignored the consequences of the existence of the Stern layer in his model.

2.2.3 *Introduction of hydrodynamic dispersion*

We should keep in mind that there are limitations in the use of the bundle of non-connecting capillaries. The assumptions made in this model are: (i) the connectivity of the pores is ignored, (ii) the solid phase is assumed continuous through the two reservoirs, i.e., at the scale of a representative elementary volume, and (iii) the thickness of the diffuse layer is on the same order of magnitude as the size of the pores. In addition, the previous transport model does not account for hydrodynamic dispersion. In fact because the pores do not intersect in the capillary model, they cannot explain transversal dispersivity. However, if we consider the migration of a salt plume in a porous material, we have to include the dispersion effect associated with mixing. We use below a classical Fickian approach where the diffusion coefficient appearing in the first Fick's law is replaced by an effective dispersion tensor accounting for both diffusion and hydrodynamic longitudinal and transversal dispersion coefficients. From Eq. (2.200), the electrical field can be written as a function of the current density $\mathbf{J}$ as

$$\mathbf{E} = \frac{1}{\sigma}\mathbf{J} + \frac{1}{e}\left(T_{(+)} - T_{(-)}\right)\nabla\mu_f + \frac{\bar{Q}_V k}{\sigma\eta}\nabla p^*, \tag{2.208}$$

where the macroscopic Hittorf numbers, $T_{(\pm)}$, are defined by

$$T_{(\pm)} = \frac{\sigma_{(\pm)}}{\sigma}. \tag{2.209}$$

Taking the expression of the flux of salt and after algebraic manipulations, we obtain (see Revil *et al.* (1996) and Revil (1999))

$$\mathbf{J}_d = -D\nabla C_f + \frac{1}{2e}\left(2T_{(+)} - 1\right)\left(\mathbf{J} + \frac{\bar{Q}_V k}{\eta}\nabla p^*\right)$$
$$- \frac{k}{2\eta}\left(\bar{C}_{(+)} + \bar{C}_{(-)}\right)\nabla p^*, \tag{2.210}$$

$$D = \left(\frac{2k_b T}{e^2 C_f}\right)\left(\frac{\sigma_{(+)}\sigma_{(-)}}{\sigma_{(+)} + \sigma_{(-)}}\right). \tag{2.211}$$

Here, D is the diffusion coefficient of the salt through the porous material. This approach generalizes the Nernst–Hartley equation describing the mutual diffusion coefficient of a binary salt in an electrolyte. Now we can replace D in Eq. (2.210) by the Fickian hydrodynamic dispersion tensor $\mathbf{D}$ defined as

$$\bar{\bar{\mathbf{D}}} = [D + \alpha_T U]\mathbf{I}_3 + \frac{\alpha_L - \alpha_T}{U}\mathbf{U} \otimes \mathbf{U}, \tag{2.212}$$
$$U = |\mathbf{U}|, \tag{2.213}$$

where $\mathbf{a} \otimes \mathbf{b}$ represents the tensorial product between vectors $\mathbf{a}$ and $\mathbf{b}$, and α_L and α_T are the longitudinal (along $\mathbf{U}$) and transverse (normal to $\mathbf{U}$) dispersivities (in m). The final constitutive equations for the salt flux density, the Darcy velocity, and the current density are, therefore,

$$\mathbf{J}_d = -\bar{\bar{\mathbf{D}}} \cdot \nabla C_f + \frac{1}{2e}\left(2T_{(+)} - 1\right)\left(\mathbf{J} + \frac{\bar{Q}_V k}{\eta}\nabla p^*\right)$$
$$- \frac{k}{2\eta}\left(\bar{C}_{(+)} + \bar{C}_{(-)}\right)\nabla p^*, \tag{2.214}$$

$$\mathbf{U} = -\left(\frac{k}{\eta}\right)\left[\nabla p^* - \bar{Q}_V\mathbf{E} + \left(\bar{C}_{(+)} + \bar{C}_{(-)}\right)\nabla\mu_f\right], \tag{2.215}$$

$$\mathbf{J} = -\frac{1}{e}\left(\sigma_{(+)} - \sigma_{(-)}\right)\nabla\mu_f + \sigma\mathbf{E} - \frac{\bar{Q}_V k}{\eta}\nabla p^*. \tag{2.216}$$

These equations are valid for any type of isotropic porous materials. They could be used as well to describe the transport properties of an anisotropic porous material if the inverse of the formation factor time the porosity is replaced by a second-order tortuosity tensor.

2.2.4 Application to transport properties of clay materials

One application of the present theory is to predict the coupled flow in clayey porous materials. Revil *et al.* (2011a) considered the Callovo-Oxfordian (COx) clay-rock to test this type of model. This well-described clay-rock is indeed under

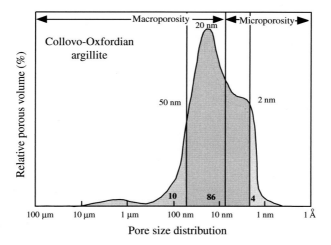

Figure 2.2. Pore size distribution of the Callovo-Oxfordian clay-rock. The peak of the distribution corresponds to a pore size of about 20 nm. The distinction between the macroporosity and the microporosity is based on the thickness of the diffuse layer computed from the Debye length (in first approximation, the thickness of the diffuse layer is equal to twice the Debye length) at the salinity of the ionic neutral reservoir in local equilibrium with the pores.

consideration for the long-term storage of nuclear waste in the Paris Basin (France). The Callovo-Oxfordian clay-rock can be conceptualized as a clay matrix with some imbedded grains of silica and carbonates (calcite and a minor fraction of dolomite) (see Figures 2.2 and 2.3a). The clay matrix represents 20% to 50% of the rock volume in the total formation; see Gaucher *et al.* (2004). This clay fraction is mainly composed of illite and interstratified illite-smectite clays. The size of the silica and carbonate grains is in the range 10 to 20 μm. The volume of the silica and carbonate grains represent between 20% and 40% of the rock assemblage. In addition, there are small amounts of potassic feldspar, plagioclase and pyrite. The pore size distribution of this clay-rock is shown in Figure 2.2 and the assumption that most of the connected porosity is occupied by the diffuse layer is valid.

We are looking, first, for an expression for the osmotic efficiency ε. Assuming that the total current density is zero (typically, the case of a cylindrical core sample between two reservoirs) and using $T_{(+)} + T_{(-)} = 1$ and Eq. (2.216), the electrical field associated with a gradient of the chemical potential of the solution and the flow of the pore water is given by

$$\mathbf{E}|_{J=0} = \frac{1}{e}\left(2T_{(+)} - 1\right)\nabla\mu_f - \frac{\bar{Q}_V}{\sigma}\mathbf{U}_m, \qquad (2.217)$$

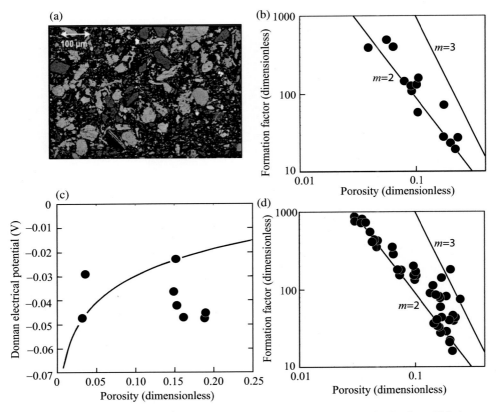

Figure 2.3. Mean potential and formation factor of the Callovo-Oxfordian (COx) clay-rock. (a) Thin section of a COx sample analyzed by scanning electron microscopy (credit: J. C. Robinet). (b) Formation factor versus porosity from electrical conductivity measurement; see Revil *et al.* (2005). (c) The mean electrical potential models are computed from the pore water chemistry used for the diffusion experiments in Melkior *et al.* (2007) (solid line). The experimental points are from Jougnot *et al.* (2009) and Jougnot *et al.* (2010b). (d) Formation factor obtained by the ratio $F = D^f_{HTO}/D_{HTO}$, where D^f_{HTO} denotes the diffusion coefficient of HTO in water and D_{HTO} denotes the diffusion coefficient of HTO in the core samples (experimental data from Descostes *et al.* (2008)). The parameter m denotes the cementation exponent of Archie's law obtained by $m = -\ln F/\ln \phi$.

where $\mathbf{U}_m$ denotes the mechanical contribution to the Darcy velocity $\mathbf{U}_m = -(k/\eta)\nabla p^*$. The streaming potential coupling coefficient is defined as

$$C \equiv \left. \frac{\Delta \psi}{\Delta p} \right|_{\mathbf{J}=0, \nabla \mu_f=0} = -\frac{\bar{Q}_V k}{\sigma \eta} = -\frac{\bar{Q}_V k F}{\eta \left(\bar{\sigma}_f + 2\frac{\Pi_1}{\Pi_2}\Sigma_S \right)}. \qquad (2.218)$$

Replacing the electrical field $\mathbf{E}$ in Eq. (2.215) by the expression obtained in Eq. (2.217), we obtain the following expressions for the Darcy velocity

$$\mathbf{U} = -\left(\frac{k}{\eta}\right)\left[\nabla p^* - \bar{Q}_V\left(\frac{1}{e}\left(2T_{(+)} - 1\right)\nabla\mu_f - C\nabla p^*\right) + \left(\bar{C}_{(+)} + \bar{C}_{(-)}\right)\nabla\mu_f\right], \tag{2.219}$$

$$\mathbf{U} = -\left(\frac{k}{\eta}\right)\left[(1+a)\nabla(p-\pi) + \left(-\bar{Q}_V\frac{1}{e}\left(2T_{(+)} - 1\right) + \left(\bar{C}_{(+)} + \bar{C}_{(-)}\right)\right)\nabla\mu_f\right], \tag{2.220}$$

where the dimensionless coefficient a is defined as

$$a = C\bar{Q}_V. \tag{2.221}$$

The parameter a is $\sim 2 \times 10^2$ and can therefore be neglected with respect to one in Eq. (2.220). We use now the Donnan model (see the solution of Exercise 2.2) to obtain the following expressions for the concentrations in the pore space of the porous material

$$\bar{C}_{(+)} + \bar{C}_{(-)} = 2C_f\sqrt{1 + \Theta^2}, \tag{2.222}$$

$$\Theta \equiv \frac{\bar{Q}_V}{2eC_f} = -\frac{Q_S}{eC_f}\frac{\Pi_1}{\Pi_2}. \tag{2.223}$$

In addition, the gradient of the chemical potential of the salt between the two reservoirs can be related to the gradients in salinity C_f and osmotic pressure between the two reservoirs by

$$\nabla\mu_f = k_bT\nabla\ln C_f = \left(\frac{1}{2C_f}\right)\nabla\pi. \tag{2.224}$$

Using Eqs. (2.222) and (2.224) in Eq. (2.220) yields the following expressions, for the Darcy velocity:

$$\mathbf{U} = -\left(\frac{k}{\eta}\right)\left[\nabla p - \nabla\pi + \left(\Theta\left(1 - 2T_{(+)}\right) + \sqrt{1 + \Theta^2}\right)\nabla\pi\right], \tag{2.225}$$

$$\mathbf{U} = -\left(\frac{k}{\eta}\right)[\nabla p - \varepsilon\nabla\pi], \tag{2.226}$$

$$\varepsilon = 1 - \Theta\left(1 - 2T_{(+)}\right) - \sqrt{1 + \Theta^2}, \tag{2.227}$$

$$k^* = k/(1+a). \tag{2.228}$$

In these equations k^* denotes an apparent permeability corresponding to the intrinsic hydrodynamic permeability corrected for electro-osmosis. For the Callovo-Oxfordian clay-rock, we have $k^* \approx k$ because $a \ll 1$. The coefficient ε is the osmotic efficiency defined by

$$\varepsilon \equiv \left(\frac{\partial p}{\partial \pi} \right)_{J,U=0}. \tag{2.229}$$

Rousseau-Gueutin *et al.* (2010) also use the following expression for the Darcy velocity

$$\mathbf{U} = -\frac{k^*}{\eta} \nabla p + \frac{k_c}{\eta} \nabla \pi, \tag{2.230}$$

where $k_c = k^* \varepsilon \approx k \varepsilon$ is termed the osmotic permeability (in m^2); see Revil and Leroy (2004).

2.2.5 Comparison between experimental data and theory

Below, we compare some material properties of the COx clay-rock that we have published in various papers with our model. We first determine some of the raw moments of the pore size distribution for the COx argillite. Using Eq. (2.196) for the permeability and the mean value of the porosity and mean value cementation exponent, we obtain $\Pi_4 / \Pi_2 = 1 \times 10^{-17}$ m^2. If the pore size distribution is described by a delta function $\Pi_4 / \Pi_2 = R^2$, this yields a mean pore size $R = 32 \pm 8$ nm. This value is slightly above the peak of the pore size distribution (20 nm) shown in Figure 2.2.

Using membrane diffusion potential measurements, the mean volumetric charge density of the diffuse layer is given by $\bar{Q}_V = 8 \times 10^6$ C m^{-3}; see Revil *et al.* (2005) and Leroy *et al.* (2007). The surface charge density on the mineral surface Q_0 can be obtained from the ratio of the cation exchange capacity (CEC) and the specific surface area. In the present case, this yields $Q_0 \approx 2$ elementary charge nm^{-2} (0.32 C m^{-2}), a value that agrees with the mean surface charge of clay minerals; see Patchett (1975), Revil *et al.* (1998), and Chapter 3, Figure 3.2. Typically, 90% of this surface charge is compensated in the Stern layer; see the electrical double layer modeling by Leroy and Revil (2009). This yields a surface charge density of $Q_S = -0.032$ C m^{-2}.

We now compare our model to measurements of the streaming potential coupling coefficient. The relationship between the conductivity of the pore water $\bar{\sigma}_f$ and the conductivity of the reservoir in contact with the charged porous material σ_f is given approximately by Revil and Leroy (2004) as

$$\bar{\sigma}_f \approx \sigma_f \sqrt{1 + \Theta^2}. \tag{2.231}$$

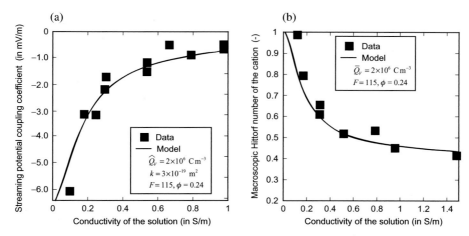

Figure 2.4. Streaming potential coupling coefficient (in $\mathrm{mV\,m^{-1}}$) and macroscopic Hittorf number of the cation (dimensionless) versus the electrical conductivity of the solution used to perform the experiments. Only the high porosity samples (samples #438, #480V and #512) have been selected here with the same porosity range (mean 0.24, s.d. 0.02). (a) Streaming potential coupling coefficient. (b) Macroscopic Hittorf number of the cations. The theoretical curves do not account for Stern layer transport. The charge density $\hat{Q}_V$ denotes the charge density of the diffuse layer dragged by the flow of the pore water while the charge density $\bar{Q}_V$ denotes the total charge density of the diffuse layer (for thick diffuse layer in relatively homogeneous materials, we expect $\hat{Q}_V = \bar{Q}_V$).

Therefore, using Eq. (2.218), the coupling coefficient can be expressed as

$$C = -\frac{\bar{Q}_V k F}{\eta \left(\sigma_f \sqrt{1+\Theta} + 2\frac{\Pi_1}{\Pi_2}\Sigma_S \right)}. \tag{2.232}$$

If the experiments are performed at a time-scale that is not compatible with the sorption/desorption of ionic species in the Stern layer, the Stern layer contribution (the second term in the paraenthesis of the denominator of Eq. 2.232) should be neglected. The streaming potential experiments performed by Revil *et al.* (2005) are usually done over few minutes, a time-scale that is much shorter than the time required for the kinetics of sorption to be efficient. Therefore, these experiments need to be compared with Eq. (2.232) without accounting for the Stern layer contribution, which leads to

$$C \approx -\frac{\bar{Q}_V k F}{\eta \sigma_f \sqrt{1+\Theta}}. \tag{2.233}$$

A comparison between this equation and the experimental data reported by Revil *et al.* (2005) is reported in Figure 2.4. There is a good agreement between the theory and these experimental data. When the dimensionless number Θ is very small (at high salinities), the coupling coefficient is given by

$$\lim_{\Theta \ll 1} C \approx -\frac{\bar{Q}_V k F}{\eta \sigma_f}. \tag{2.234}$$

In this high salinity limit ($\Theta \ll 1$), the streaming potential coupling coefficient is also given by the Helmholtz–Smoluchowski equation $C \approx \varepsilon_f \zeta / \eta \sigma_f$, where ζ (in V) denotes the zeta potential at the interface between the Stern and diffuse layers and ε_f is the dielectric constant of the pore water. The Helmholtz–Smoluchowski equation indicates that the coupling coefficient is independent of the microstructure at high salinities, implying that the volumetric charge density of the diffuse layer $\bar{Q}_V$ should scale as $1/kF$.

When the dimensionless number Θ is very large (at low salinities), we obtain

$$C \approx -\frac{2kF}{\eta(\beta_{(+)} + \beta_{(-)})}; \tag{2.235}$$

therefore, the coupling coefficient reaches a limiting value that depends only on the product of the formation factor by the permeability and that is independent on the salinity and the volumetric charge density. So, both the volumetric charge density and the product (kF) are uniquely determined by the streaming potential versus salinity curve (Figure 2.4a).

We now focus on the value of the macroscopic Hittorf number. Using Eq. (2.209) with Eqs. (2.203) and (2.191), the Hittorf number of the cations can be related to the dimensionless number Θ as

$$T_{(+)} = \frac{\sigma_{(+)}}{\sigma_{(+)} + \sigma_{(-)}}, \tag{2.236}$$

$$T_{(+)} = \frac{\beta_{(+)}(\sqrt{1 + \Theta^2} + \Theta) + \beta_{(+)}^S \Psi}{\beta_{(+)}(\sqrt{1 + \Theta^2} + \Theta) + \beta_{(+)}^S \Psi + \beta_{(-)}(\sqrt{1 + \Theta^2} - \Theta)}, \tag{2.237}$$

where we assumed that the Stern layer is populated by counterions and $\Psi = 2\Pi_{-1} \Gamma_{(+)}/C_f$ is a second dimensionless number related to the density of counterions in the Stern layer with respect to the salinity of the reservoir. We need to estimate the Stern layer conductivity $2e\Pi_{-1}\beta_{(+)}^S \Gamma_{(+)}$. We already know that $e\Gamma_{(+)} = Q_\beta = |Q_0 - Q_S| = 0.29$ C m^{-2} (the surface charge density of the Stern layer). The value of the surface mobility in the Stern layer for the counterions must also be known. Taking $\beta_{(+)}^S = 5 \times 10^{-8}$ m^2 V^{-1}s^{-1} and therefore $2e(\Pi_1/\Pi_2)\beta_{(+)}^S \Gamma_{(+)} \approx 4$ S m^{-1}. With this value, the Hittorf numbers are strongly over-predicted (not shown). A comparison between the theory (without transport in the Stern layer) and the experimental data reported by Revil *et al.* (2005) is

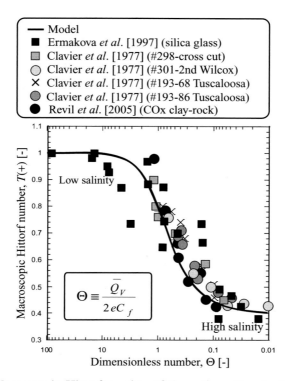

Figure 2.5. Macroscopic Hittorf number of the cations $T_{(+)}$ (scale of a porous core sample) versus the dimensionless parameter Θ corresponding to the ratio of the diffuse layer counterion volumetric density divided by the salinity of a reservoir locally in equilibrium with the pore space. Experimental data from Revil et al. (2005) (COx clay-rock, NaCl), Clavier et al. (1984) (shaly sands using the measured CEC and $f = 0.90$, NaCl) and Ermakova et al. (1997) (silica glass using the measured pore size and a surface charge density of 0.002 C m^{-2}, NaCl). The solid line represents the prediction from our theory. Correlation coefficient $R^2 = 0.87$.

shown in Figure 2.4b (using the same samples used in Figure 2.4a). The plain line represents the prediction of the theory without the Stern layer. Therefore, it seems that the transport in the Stern layer does not play a role for the COx argillite and can be neglected. Consequently, the macroscopic Hittorf number of the cations can be determined by the following relationship;

$$T_{(+)} = \frac{\beta_{(+)}(\sqrt{1 + \Theta^2} + \Theta)}{\beta_{(+)}(\sqrt{1 + \Theta^2} + \Theta) + \beta_{(-)}(\sqrt{1 + \Theta^2} - \Theta)}. \tag{2.238}$$

In Figure 2.5, we plot the Hittorf number versus the dimensionless parameter Θ for the COx clay-rock and for literature data for both shaley sands and porous glasses with known pore sizes. There is quite good agreement between the theory and the experimental data. In the case where the cations and the anions have roughly the same mobilities (e.g., KCl), the Hittorf number is given by the simplified equation,

$$T_{(+)} \approx \frac{\sqrt{1 + \Theta^2} + \Theta}{2\sqrt{1 + \Theta^2}}. \tag{2.239}$$

We can now find an expression for the osmotic efficiency. Using Eq. (2.227) and Eq. (2.238), we find

$$\varepsilon = 1 + \Theta \left(\frac{\beta_{(+)}(\sqrt{1 + \Theta^2} + \Theta) - \beta_{(-)}(\sqrt{1 + \Theta^2} - \Theta)}{\beta_{(+)}(\sqrt{1 + \Theta^2} + \Theta) + \beta_{(-)}(\sqrt{1 + \Theta^2} - \Theta)} \right) - \sqrt{1 + \Theta^2}. \tag{2.240}$$

If the mobility of the cations and anions are roughly the same, the osmotic efficiency is given by the simplified formula,

$$\varepsilon = \frac{\sqrt{1 + \Theta^2} - 1}{\sqrt{1 + \Theta^2}}. \tag{2.241}$$

We can check that if $\Theta \ll 1$ (high salinities) we have $\varepsilon = 0$ (no membrane behavior), while if $\Theta \gg 1$ (low salinities) we have $\varepsilon = 1$ (perfect membrane behavior). The behavior of ε versus Θ is shown in Figure 2.6. Our theory indicates that the osmotic efficiency should scale as $R\, C_f$. In order to compare our theory to experimental data, we first look for a simplified way to compute the dimensionless parameter Θ. Starting with the definition of this parameter, see Eq. (2.223), and using the relationship between the volumetric charge density of the diffuse layer and the cation exchange capacity, CEC (expressed in C kg^{-1}), we obtain

$$\Theta \equiv \frac{(1 - f)}{2eC_f} \rho_g \left(\frac{1 - \phi}{\phi} \right) \text{CEC}. \tag{2.242}$$

This equation can be expressed as follows,

$$\Theta \equiv \frac{10^{-3}(1 - f)}{2C_f} \rho_g \left(\frac{1 - \phi}{\phi} \right) \text{CEC}, \tag{2.243}$$

with C_f now expressed in mol l^{-1} and the CEC in meq g^{-1} (we have used the following conversion: 1 meq g^{-1} = $N\, e$ C kg = 96 320 C kg^{-1}, where N is Avogadro's constant), and f represents the fraction of counterions in the Stern layer. A comparison between our model, Eq. (2.240), and various experimental data is reported in Figure 2.6. There is good agreement between the model and the experimental data. In each case, the CEC values are taken from published data or estimated depending on the composition of the clay fraction and the the fraction of

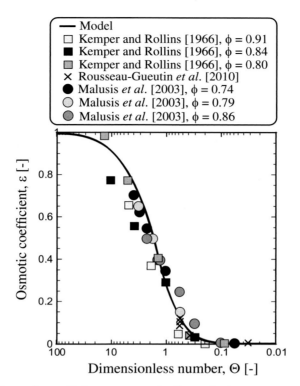

Figure 2.6. Osmotic coefficient ε versus the dimensionless parameter Θ. Experimental data from Malusis *et al.* (2003) (geosynthetic clay liner, KCl, $f_Q = 0.94$, measured CEC $= 0.48$ meq g^{-1}, $\rho_g = 2650$ kg m^{-3}), from Rousseau-Gueutin *et al.* (2010) (COx clayrock, NaCl) and from Kemper and Rollins (1966) (bentonite, $f = 0.90$, NaCl, CEC $= 0.8$ meq g^{-1}, $\rho_g = 2650$ kg m^{-3}). The solid line represents the prediction from our theory for KCl. Correlation coefficient $R^2 = 0.97$.

counterions in the Stern layer f is determined from the model of Leroy and Revil (2009).

The last parameter we want to investigate is the mutual diffusion coefficient of a salt, like NaCl or KCl, through the COx clay-rock in a salinity gradient. Using Eq. (2.211) with Eq. (2.203) for the expression of the conductivity contributions (without accounting for the Stern layer contribution), the effective diffusion coefficient D (in m^2 s^{-1}) is given by

$$D = \frac{D_f}{F}\gamma. \tag{2.244}$$

In Eq. (2.244), D_f is the mutual diffusion coefficient of the salt in the reservoir (in m^2 s^{-1}). This mutual diffusivity is given by the Nernst–Hartley equation (see

Newman (1991))

$$D_f = \frac{2D_{(+)}^f D_{(-)}^f}{D_{(+)}^f + D_{(-)}^f}, \tag{2.245}$$

where $D_{(\pm)}^f = (k_b T/e)\beta_{(\pm)}$ ($D_{(\pm)}^f$ are the diffusion coefficients of the cations and anions in the reservoirs) and where the correction coefficient γ (dimensionless) is given by

$$\gamma = \frac{\left(\beta_{(+)} + \beta_{(-)}\right)}{\beta_{(+)}\left(\sqrt{1 + \Theta^2} + \Theta\right) + \beta_{(-)}\left(\sqrt{1 + \Theta^2} - \Theta\right)}, \tag{2.246}$$

$$\gamma \approx \frac{1}{\sqrt{1 + \Theta^2}}. \tag{2.247}$$

Because the ratio D_f/F represents the effective diffusion coefficient when the material is uncharged (or at very high salinities $\Theta \ll 1$), the correction coefficient γ can be seen as a normalized diffusion coefficient pointing out the effect of the diffuse layer upon the effective diffusion coefficient D. A comparison between our theory and various experimental data is shown in Figure 2.7. Note that the experimental data reported in this figure have rather large error bars, which may explain some of the discrepancy between the model predictions and the data. The corrections factor γ explains the discrepancy between the apparent diffusion coefficient and electric formation factors for small porosities for clay and montmorillonite observed, for example, by Rosanne *et al.* (2003). Therefore, our approach shows that the macroscopic mutual diffusion coefficient is controlled by the electrical formation factor, but a correction term needs to be accounted for. This correction term depends on the ratio Θ.

The approach considered in this section works well at low frequency. However, as discussed in Chapter 3 below, the conductivity is frequency dependent. Revil and co-workers have pointed out that the Stern layer polarizes only above a certain frequency range associated with the grain size distribution; see Leroy *et al.* (2008), Revil and Florsch (2010) and Vaudelet *et al.* (2011a, b). A model of this polarization mechanism will be extensively described in Chapter 3.

2.3 The geobattery and biogeobattery concepts

2.3.1 Introduction to the concept of geobattery

The occurrence of strong negative (in few cases positive) self-potential anomalies associated with the presence of ore deposits has been known since the nineteenth

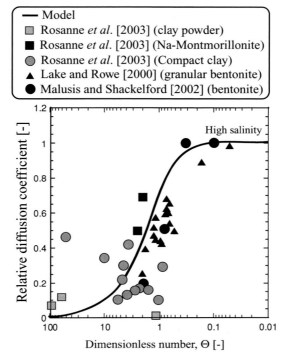

Figure 2.7. Relative diffusion coefficient γ versus the dimensionless parameter Θ. Experimental data from Malusis and Shackelford (2004) (geosynthetic clay liner, NaCl, $f = 0.94$, CEC $= 0.48$ meq g^{-1}, $\rho_g = 2650$ kg m^{-3}, $D_f = 1.60 \times 10^{-9}$ m^2 s^{-1}), Lake and Rowe (2000) (granular bentonite, NaCl, $f_Q = 0.85$, CEC $= 1.0$ meq g^{-1}, $\rho_g = 2650$ kg m^{-3}, $D_f = 1.60 \times 10^{-9}$ m^2 s^{-1}), Rosanne *et al.* (2003) (compact clay from the COx formation, NaCl, $f = 0.90$, CEC $= 0.18$ meq g^{-1}, $\rho_g = 2700$ kg m^{-3}, $m = 2.5$, $D_f = 1.60 \times 10^{-9}$ m^2 s^{-1}), Rosanne *et al.* (2003) (Na-Montmorillonite samples not free to swell, NaCl, $f = 0.90$, CEC $= 0.8$ meq g^{-1}, $\rho_g = 2650$ kg m^{-3}, $m = 2.5$, $D_f = 1.60 \times 10^{-9}$ m^2 s^{-1}) and Rosanne *et al.* (2003), (clay powder from the COx formation, NaCl, $f = 0.90$, CEC $= 0.18$ meq g^{-1}, $\rho_g = 2700$ kg m^{-3}, $m = 2.5$, $D_f = 1.60 \times 10^{-9}$ m^2 s^{-1}). The solid line represents the prediction from our theory. Correlation coefficient $R^2 = 0.55$.

century; see, for example, Fox (1830), Sato and Mooney (1960), Nourbehecht (1963), Bigalke and Grabner (1997) and Bigalke *et al.* (2004). The amplitude of these anomalies can reach few hundred of millivolts. Goldie (2002) reported a gigantic self-potential anomaly amounting to -10.2 V associated with the Yana-cocha high sulfidation gold deposit in Peru (see also Mendonça (2008)) and such high amplitude self-potential anomalies can be explained only through the super-position of multiple source contributions.

The occurrence of this type of self-potential anomalies is associated with redox half-reactions involving electron donors and electron acceptors and working together with an electronic conductor allowing the long-range transport of electrons (see Corwin and Hoover (1979), Bigalke and Grabner (1997), Maineult *et al.* (2006) and Revil and Linde (2006)). Note that faults with vein mineralizations (e.g., graphite) along the fault plane (see Bigalke and Grabner (1997)) are also associated with strong self-potential anomalies (see Bigalke and Grabner (1997)).

In this section, we will discuss the occurrence of self-potential anomalies in the case where the electronic conductor is abiotic (e.g., vein mineralization) or biotic in presence of conductive nanowires associated with bacteria (see also Section 6.4 of Chapter 6). The nature of the relationship between the distribution of the self-potential signals and the distribution of the redox potential in presence of biotic conductors at depth (biogeobattery) has been recently debated by several authors; (see, for example, Nyquist and Corey (2002) and Arora *et al.* (2007)). To explain the occurrence of self-potential anomalies observed with organic-rich contaminant plumes, Revil and co-workers (see Naudet *et al.* (2003), Arora *et al.* (2007) and Linde and Revil (2007)) have introduced a linear relationship between the source current density and the gradient of the redox potential. The assumption made by Naudet and Revil (2005) and Arora *et al.* (2007) was that biofilms of bacteria could transmit electrons and bridge areas characterized by different values of the redox potential (oxidized vadose zone and reduced contaminant plume). This idea was validated in the laboratory by Ntarlagiannis *et al.* (2007) and Risgaard-Petersen *et al.* (2012). This model was successfully applied to invert the distribution of the redox potential over the contaminant plume of Entressen in the South of France (see Linde and Revil (2007) and Chapter 6 below). In this section, we test simply the (abiotic) geobattery model in the case of a corrosion of a vertical bar in a sandbox experiment.

As self-potential anomalies include an electrical signature of on-going redox reaction processes occurring at depth, it should be possible to invert self-potential signals to obtain information related to these redox processes. The possibility to invert self-potential data in terms of the distribution of the redox potentials is mportant in ore prospection and in environmental applications where the self-potential method can be used as a non-intrusive sensor of the distribution of the redox potential over contaminant plumes after removal of the streaming potential component (see Naudet *et al.* (2003, 2004), Naudet and Revil (2005), Maineult *et al.* (2006) and Arora *et al.* (2007)). It can be also used to locate metallic pipes in the ground and abandoned boreholes because of the corrosion of their metallic casing. In contaminated shallow aquifers, the redox potential is usually measured in a set of boreholes. This is both time-consuming and expensive, and does not allow a dense sampling of subsurface. Furthermore, the physical meaning of redox potential estimates from in situ measurements is considered to be uncertain because of the introduction of oxygen in the system and perturbations of the redox reactions in the vicinity of the boreholes; see Christensen *et al.* (2000). Furthermore, it is not clear that geostatistical analysis of a few redox potential data collected locally is indicative of electrochemical conditions at larger scales (see Stoll *et al.* (1995)).

Laboratory experiments have been conducted by Bigalke and Grabner (1997) to investigate ore bodies. Timm and Möller (2001) and Maineult *et al.* (2006) developed laboratory experiments to characterize liquid–liquid redox reactions and their impact on self-potential signals. Naudet and Revil (2005) performed an experiment to study bacteria-mediated redox processes associated with contaminant plumes. A comprehensive validation of the inverted redox potential was performed by Castermant *et al.* (2008). This experiment will be described in more details in Chapter 3.

The relative electron activity of water is defined as $p\varepsilon = -\log\{e^-\}$, where $\{e^-\}$ represents the potential electron activity in the pore water phase; see Hostetler (1984) and Thorstenson (1984). In a reducing system, the tendency to donate electrons, or electron activity, is relatively large and $p\varepsilon$ is low. The opposite holds true in oxidizing systems. In a reduction reaction, an oxidized species Ox reacts with n electrons to form a reduced species Red. The half-reaction $Ox + ne^- \rightarrow Red$ is characterized by a reaction constant $K = \{Red\}/\{Ox\}\{e^-\}^n$. Because there is no electron in the pore water, the previous reduction reaction has to be coupled with an oxidation reaction (typically for reference purpose the oxidation of hydrogen). This leads to

$$p\varepsilon = p\varepsilon^0 + \frac{1}{n}\log\left(\frac{\{Ox\}}{\{Red\}}\right), \qquad (2.248)$$

where $p\varepsilon^0$ is the standard electron activity of the actual reduction half-reaction when coupled to the oxidation of hydrogen under standard conditions (see Christensen *et al.* (2000)). The redox potential is defined through the Nernst equation by

$$E_H = 2.3\frac{k_b T}{e}p\varepsilon, \qquad (2.249)$$

where T is the absolute temperature in K, e is the elementary charge of the electron, and k_b is the Boltzmann constant. We consider a massive ore body embedded in the conductive ground. Electrons within the ore body have a high mobility but do not exist in the surrounding rock mass. The presence of the ions in the pore water controls the electrical conductivity in the surrounding rock mass. Because the fugacity of oxygen (the concentration/activity of oxygen dissolved in water) decreases with depth, the redox potential has, in the far field of the ore body, a strong dependence with the depth z. In the vicinity of the ore body, the distribution of the redox potential can be more complex because of contribution of interfacial

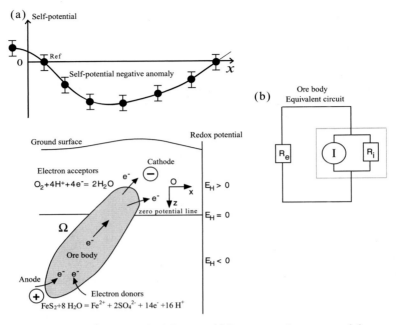

Figure 2.8. Sketch of the classical Sato and Mooney geobattery model proposed for ore bodies and its equivalent linear circuit. (a) Far from the ore body, the redox potential decreases with depth mainly because of the decrease with depth of the concentration of dissolved oxygen in the pore water. In the vicinity of the ore body, a redox potential disturbance is created because of the redox reactions at the surface of the ore body. Corrosion of the ore body can also be responsible for a resistive crust that adds an additional Ohmic resistance (and therefore an overvoltage) to the circuit. Typically, a self-potential anomaly associated with an ore body can amount to a few hundreds of millivolts at the ground surface but cannot be higher than the difference of the redox potential between the terminal points (anode and cathode) of the system. (b) The ore body is characterized by a source of current I and an internal resistance R_i. The resistance R_e stands for the external resistance of the conductive medium in which the ore body is embedded.

processes and possibly biological activity. The near-field distribution of the redox potential is influenced by anodic and cathodic reactions occurring at the surface of the ore metallic conductor.

Let us consider for example the corrosion of an ore body like pyrite (FeS_2, Figure 2.8). Reactions involve (i) the oxidation of the S(-II) and S(0) in pyrite coupled to release of SO_4^{2-} and Fe^{2+} at depth, coupled to (ii) the reduction of oxygen near the oxic–anoxic interface (typically the water table). The soluble Fe released during the anodic reaction at depth can then eventually react, through advective, dispersive and electromigration transport, with oxygen at the water table. These reactions impact the distribution of the redox potential in the vicinity of the ore body. This mechanism can be summarized by the following reactions. At depth at the surface of the ore body, the following half-reaction occurs;

$$FeS_2 + 8H_2O \Leftrightarrow Fe^{2+} + 2SO_4^{2-} + 14e^- + 16H^+, \qquad (2.250)$$

which is an abiotic half-reaction pulled along by sinks for electrons and Fe^{2+} at the oxic–anoxic interface. At the cathode, possibly in the vadose zone, we can have the following reactions:

$$14e^- + 3.5O_2 + 14H^+ \Leftrightarrow 7H_2O, \qquad (2.251)$$

$$4Fe^{2+} + O_2 + 4H^+ \Leftrightarrow 4Fe^{3+} + 2H_2O, \qquad (2.252)$$

$$4Fe^{3+} + 12H_2O \Leftrightarrow 4Fe(OH)_3 + 12H^+. \qquad (2.253)$$

Reaction (2.251) corresponds to the half-reaction associated with the electrons provided by the ore body. Reaction (2.252) is a redox reaction in the solution with the micro-organisms being potentially able to accelerate this reaction depending on the pH of the solution (low pH values favor the reaction). Reaction (2.253) is the Fe(III) oxide mineral precipitation, which is an abiotic reaction. It is important to realize that the vertical redox gradient in the vicinity of the ore body, in addition to being influenced by reactions associated with the corrosion of the ore body (Eqs (2.250) to (2.253)), could be also influenced by redox reactions (possibly microbially catalyzed) that are unrelated to ore body corrosion, e.g., reactions associated with degradation of organic matter in the aquifer sediments surrounding the ore body in the way envisioned below. In this case, the ore body would serve as a conductor for transfer of electrons released during these reactions from depth to the oxic–anoxic interface.

The distribution of the gradient of the redox potential should thus be viewed as a general thermodynamic force driving the flow of electrical charges inside the ore body. Two models of (geo)battery are possible. In the first, the ore body serves directly as a source of electrons, vis-à-vis Eq. (2.250), that flow from depth to the shallow subsurface through the ore body. Equation (2.250) describes a source of electrons originating from the oxidation of the ore body. This corresponds to an "active electrode" model. In the second model, called the "passive electrode" model, the ore body serves simply as a conductor for transfer of electrons that originate from redox reactions (potentially microbially catalyzed) that take place away from the ore body and have nothing to do with the corrosion of the ore body. Both models can co-exist. For example, the 14 electrons released in the half-reaction of Eq. (2.251) can pass directly through the ore body to oxygen at the aerobic–anaerobic interface, whilst the electrons released during the oxidation of Fe^{2+} in solution at or near the ore body–water interface can also be transmitted to the shallow subsurface through the ore body. Because of the existence of potential losses between the electron donors and the metallic body, for instance, the governing constitutive current/force equation is the non-linear Butler–Volmer equation; see Bockris and Reddy (1970) and Mendonça (2008).

Stoll *et al.* (1995) and Bigalke and Grabner (1997) derived an inert electrode model, which is based on the classical electrochemical model for metallic electrodes of Bockris and Reddy (1970). This model is based on the non-linear Butler–Volmer equation between the electrical potential and the current density generated at the surface of the metallic body. Once linearized, this equation yields

$$\mathbf{J}_S = \mathbf{J}_0 \frac{nF}{RT} (-E_H + E_m - \psi), \qquad (2.254)$$

where $\mathbf{J}_0$ is the exchange current density at the surface of the metallic body, n is the number of molar equivalents transferred for the given exchange reaction between the electronic conductor and the surrounding medium, $F = Ne$ is the Faraday constant (9.65×10^4 C mol^{-1}), R is the gas constant (8.31 J K^{-1}mol^{-1}), T is the temperature (in K), E_H is the redox potential (in V), and E_m is the redox potential of the metallic conductor (in V). The redox potential of the metallic conductor corresponds to the chemical potential of the electrons inside this body. Because the resistivity of the metallic conductor is very small, E_m has approximately a constant value. The current density $\mathbf{J}_S$ equals zero everywhere except at the surface of the iron bar. Mendonça (2008) presented an inverse procedure to determine the causative distribution of source currents in the ground responsible for the observed distribution of the self-potential at the ground surface and assuming a known electrical conductivity model.

An alternative model is considering a linear relationship between the source current density and the redox potential (see Arora *et al.* (2007) and Linde and Revil (2007)),

$$\mathbf{J}_S = -\sigma_0 \nabla E_H. \qquad (2.255)$$

In this equation, σ is the electrical conductivity of the volume characterized by a rapid change in the redox potential. This conductivity can be different from the conductivity of the host medium as discussed below.

In the quasi-static limit of the Maxwell equations, we have

$$\nabla \cdot (\sigma_0 \nabla \psi) = \nabla \cdot \mathbf{J}_S. \qquad (2.256)$$

For the two models described above, we have

$$\frac{1}{R_e} (-E_H + E_m - \psi) = \nabla \cdot \mathbf{J}_S, \qquad (2.257)$$

$$-\nabla \cdot (\sigma_0 \nabla E_H) = \nabla \cdot \mathbf{J}_S, \qquad (2.258)$$

respectively, where $R_e = RT/(gj_0nF)$ is the electrode resistance (expressed in Ω) in the inert electrode model and g the specific surface area of the surface of the metallic body.

2.3.2 Biogebattery

Negative self-potential signals of several hundred of mV have also been observed in association with contaminant plumes, rich in organic matter, associated with leakage from municipal landfills (see Nyquist and Corey (2002), Naudet *et al.* (2003) and Arora *et al.* (2007)). There are cases for which the relationship between the source current density and the redox potential gradient is linear, in presence of catalysts that lower the energy barriers between the electron donors and the metallic body or between the metallic body and electron acceptors. For instance, bacteria can be very efficient catalysts for redox reactions in the presence of a biofilm (Figure 2.9). In this case, the relationship between the current density and the gradient of the redox potential is found to be linear. The effects should not be mistaken with electrodic voltages associated with changes in the surface chemistry of the electrodes that are NOT self-potential signals (see Slater *et al.* (2008)).

2.4 Conclusions

In Section 2.1, we described a general thermodynamic model explaining the occurrence of source current densities in porous media. We have seen that several mechanisms are possible. In Section 2.2, we developed a petrophysical model showing that the current density (in absence of redox processses) is due to two contributions. In this case, there are two driving forces controlling the magnitude of the source current density. The first one is related to the gradient of the chemical potential of the pore water solution (e.g., associated with a gradient of salinity or ionic strength) and the second is related to the flow of the pore water, which is controlled by the gradient in the pore fluid pressure (accounting for an osmotic pressure term). The first contribution is called the diffusion current density and the second contribution is called the streaming current density. The associated voltage distributions are called the diffusion and streaming potentials, respectively. The second contribution is related to poromechanics through its influence to the pore fluid pressure. The model was developed for clayey materials with pores on the size of the electrical diffuse layer. We have therefore assumed that all the excess of electrical charges present in the pore space (due to the diffuse layer), $\bar{Q}_V$, is carried along by the flow of the pore water. Actually this is far from being true in most porous media with macropores and only a fraction of this charge density, $\hat{Q}_V$, is responsible for the

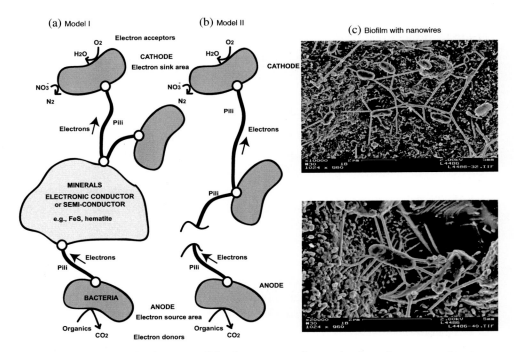

Figure 2.9. Sketch of two possible electron transfer mechanisms in a contaminant plume. (a) Model I, in which the presence of a conductive mineral facilitates electronic conduction. (b) Model II, in which only bacteria populations are connected by conductive pili. At the "bacterial anode," electrons are gained through the oxidation of the organic matter, iron oxides, or Fe-bearing phyllosilicates. The electrons are conveyed to the "bacterial cathode" through a network of conductive pili. At the "bacterial cathode," the reduction of oxygen and the nitrate prevails as electron acceptors. In this system, bacteria act as catalysts. The transport of electrons through the anode to the cathode of the microbattery may involve different bacterial communities and different electron transfer mechanisms including external electron shuttles. (c) *Shewanella oneidensis* with pili, showing the complexity of the 3D organization of pili. The picture shows how pili can be interconnected. The pictures show that several pili can start from a given bacterium and a pilum can split in several pili (pictures provided by Yuri Gorby and acquired in association with the study by Ntarlagiannis *et al.* (2007), using a Field Emission Electron Microscope; figure modified from Revil *et al.* (2010, 2012b).

streaming current density as discussed below in Chapter 3. From Eq. (2.216), the source current density can therefore be written as

$$\mathbf{J}_S = -\frac{1}{e}\left(\sigma_{(+)} - \sigma_{(-)}\right)\nabla\mu_f - \frac{\hat{Q}_V k}{\eta}\nabla p^*. \tag{2.259}$$

The second contribution (streaming current) is also related to the pore water chemistry through the osmotic component of the pore fluid pressure and to interfacial chemistry of the electrical double layer through the excess of charge $\hat{Q}_V$ that can be dragged by the flow of the pore water.

In the case of redox processes and in the presence of catalysts and biotic or abiotic electron pathways, another contribution to the source current density is given by

$$\mathbf{J}_S = -\sigma \nabla E_H. \tag{2.260}$$

This equation assumes that there are no potential losses between the electron donors and the electronic conductor, and between the electronic conductor and the electron acceptors. This contribution is called the electroredox current density and is directly related to the gradient of the chemical potential of the electrons. This contribution will be analyzed in more details in Chapters 3 and 6.

Exercises

Exercise 2.1. Prigogine thorem. As explained in the main text, Prigogine's theorem states that in mechanical equilibrium (inertial term neglected in the momentum conservation equation), the barycentric velocity $\mathbf{v}$ entering into the definition of the diffusion flow in the dissipation function can be replaced by another arbitrary velocity. The dissipation function D^m written in a Lagrangian framework attached to the barycentric center of mass motion is

$$D^m = -\mathbf{S}^m \nabla T - \sum_{i=1}^{N} (\nabla \mu_i - \mathbf{F}_i) \mathbf{J}_i^m - (\nabla \mu_w - \mathbf{F}_w) \mathbf{J}_w^m$$
$$- (\nabla \cdot \bar{\bar{\mu}}_s - \mathbf{F}_s) \mathbf{J}_s^m + \bar{\bar{\sigma}}^d : \nabla \mathbf{v}. \tag{2.261}$$

Using the force balance equation written under the assumption of mechanical equilibrium,

$$\sum_{i=1}^{N} \rho_i \phi \mathbf{F}_i + \rho_w \phi \mathbf{F}_w + \rho_s (1 - \phi) \mathbf{F}_s + \nabla_s + \bar{\bar{\sigma}} = 0, \tag{2.262}$$

demonstrate Prigogine theorem and show that the the barycentric velocity $\mathbf{v}$ can be replaced by the velocity of the grains

Exercise 2.2. Donnan model of osmotic pressure. In this exercise, we will first derive the expression of the equivalent salt concentration in the pore space of a porous material using the Donnan approach. In a second step, we will retrieve the classical expressions for the osmotic pressure in a fictitious reservoir in local equilibrium with the pore space of the porous material. The ionic concentration in the (neutral) fictitious reservoir of ions are $C_{(\pm)} = C_f$. The expression of the electrochemical potentials of the cations and anions in the fictitious reservoir and in the pore space are

$$\tilde{\mu}_{(\pm)} = \mu_{(\pm)}^0 + k_b T \ln C_{(\pm)}, \tag{2.263}$$

$$\bar{\tilde{\mu}}_{(\pm)} = \mu_{(\pm)}^0 + k_b T \ln \bar{C}_{(\pm)} \pm e \bar{\varphi}, \tag{2.264}$$

respectively. The bar is used to express that the quantity is considered in the pore space. In these expressions, $\bar{\varphi}$ denotes the mean electrical potential associated with the electrical diffuse layer in the capillary ($\bar{\varphi} = 0$ in the fictitious reservoir).

(1) Thermodynamic equilibrium between the fictitious reservoir and the pore space implies

$$\bar{\tilde{\mu}}_{(\pm)} = \tilde{\mu}_{(\pm)}. \tag{2.265}$$

Using this equilibrium assumption, demonstrate that the concentrations of the cations in the pore space follow Boltzmann distributions.

(2) The charge conservation equation in the pore space of the porous material is given by:

$$\bar{Q}_V = e \left(\bar{C}_{(+)} - \bar{C}_{(-)} \right). \tag{2.266}$$

Demonstrate that the ionic concentrations in the pore space of the material and the mean electrostatic potential are given by the following expressions:

$$\bar{C}_{(\pm)} = C_f \left(\sqrt{1 + \Theta^2} \pm \Theta \right), \tag{2.267}$$

$$\bar{\varphi} = -\frac{k_b T}{2e} \ln \left[\frac{\sqrt{1 + \Theta^2} + \Theta}{\sqrt{1 + \Theta^2} - \Theta} \right]. \tag{2.268}$$

Find the expression of Θ. These equations correspond to the Donnan model.

(3) The osmotic pressure in a reservoir and the chemical potential of the water in this reservoir are given by

$$\pi = -\frac{k_b T}{\Omega_w} \ln C_w = -\frac{k_b T}{\Omega_w} \ln \left(1 - 2\Omega_f C_f \right) \approx 2 k_b T C_f, \tag{2.269}$$

$$\mu_w = \mu_w^0 + \Omega_w p + k_b T \ln C_w, \tag{2.270}$$

respectively, where p is the mechanical pressure. Demonstrate that the gradient of the chemical potential of water is given by $\nabla \mu_w = \Omega_w \nabla p^*$, where $p^* = p - \pi$ denotes an effective fluid pressure.

(4) The chemical potential of the water in the pore space of the material is

$$\bar{\pi} = -\frac{k_b T}{\Omega_w} \ln \bar{C}_w, \tag{2.271}$$

$$\bar{\mu}_w = \mu_w^0 + \Omega_w \bar{p} + k_b T \ln \bar{C}_w, \tag{2.272}$$

where $\bar{C}_w$ and $\bar{p}$ are the water concentration and internal fluid pressure in the pore space of the porous material. We can also write the chemical potential of water in the pore space as $\bar{\mu}_w = \mu_w^0 + \Omega_w(\bar{p} - \bar{\pi})$. The conservation of mass in the reservoir and in the pore space can be described as

$$\Omega_w C_w + (\Omega_{(+)} + \Omega_{(-)})C_f = 1, \tag{2.273}$$

$$\Omega_w \bar{C}_w + \Omega_{(+)}\bar{C}_{(+)} + \Omega_{(-)}\bar{C}_{(-)} = 1, \tag{2.274}$$

respectively. Local thermodynamic equilibrium between the fictitious reservoir and the pore water implies $\bar{\mu}_w = \mu_w$. Demonstrate that the difference of osmotic pressure between the reservoir and the pore sapce is given by $\delta\pi \approx -k_bT\left(\bar{C}_{(+)} + \bar{C}_{(-)} - 2C_f\right)$ and that the internal pressure in the pore space is given by $\bar{p} = p - (\pi - \bar{\pi})$.

(5) Starting with the definitions of the osmotic pressures, Eqs. (2.269) and (2.271), and using the mass conservation equations, Eqs. (2.273) and (2.274), demonstrate the following expressions for the osmotic pressure in the reservoir and for the osmotic pressure in the pore space of the porous material;

$$\pi = k_bT(2C_f), \tag{2.275}$$

$$\bar{\pi} = k_bT(\bar{C}_{(+)} + \bar{C}_{(-)}). \tag{2.276}$$

Equation (2.275) is called the van 't Hoff equation in a neutral electrolyte while Eq. (2.276) is an extension of this equation for the pore space of a charge porous material.

(6) We can also express the difference in osmotic pressure between the reservoir and the local pore space as a function of the mean potential or as a function of the key dimensionless variable Θ. Demonstrate the following equation:

$$\bar{\pi} = 2C_f k_bT \cosh\left(\frac{e\bar{\varphi}}{k_bT}\right). \tag{2.277}$$

Demonstrate that the integration of the excess of charge with the mean potential is also equal to the difference of osmotic pressure between the reservoir and the pore space of the porous material:

$$\delta\pi = \int_0^{\bar{\varphi}} \bar{Q}_V d\bar{\varphi}', \tag{2.278}$$

(7) We now turn our attention to the local flux densities of ions. These fluxes are related to the ionic concentrations and to the drift velocity by $\mathbf{j}_{(\pm)} = \bar{C}_{(\pm)}\mathbf{v}$ (the Nernst–Planck equation). Demonstrate that the velocity $\mathbf{v}$ can be divided into three components including a mechanical, an electrical and a diffusional

contribution:

$$\mathbf{v} = \mathbf{v}_m + \mathbf{v}_e + \mathbf{v}_c$$

and find the expression of these three components.

Exercise 2.3. Ionic fluxes in the Stern layer in a capillary. In order to gain better insight into transport in the Stern layer, we introduce in this exercise the microscopic equations describing the physics of the transport of counterions and co-ions located in the Stern layer, which was described in Chapter 1. We consider a capillary coated by the Stern layer and located between two ionic reservoirs. To simplify our analysis, we consider the sorption of co-ions and co-counterions at the surface of silica:

$$>\text{SO}^- + \text{M}^+ \overset{K_{(+)}}{\Leftrightarrow} >\text{SO}^-\text{M}^+, \tag{2.279}$$

$$>\text{SOH}_2^+ + \text{A}^- \overset{K_{(-)}}{\Leftrightarrow} >\text{SOH}_2^+\text{A}^-, \tag{2.280}$$

where $>\text{SOH}$ represents a surface hydroxyl site where the $K_{(\pm)}$ are the sorption constants:

$$K_{(+)} = \frac{\Gamma_{(+)}}{\Gamma_{(-)}^0 C_{(+)} \exp\left(-\frac{e\varphi_d}{k_b T}\right)}, \tag{2.281}$$

$$K_{(-)} = \frac{\Gamma_{(-)}}{\Gamma_{(+)}^0 C_{(-)} \exp\left(\frac{e\varphi_d}{k_b T}\right)}, \tag{2.282}$$

where $\Gamma_{(+)}$ and $\Gamma_{(+)}$ denote the density of surface species $>\text{SO}^-\text{M}^*$ and $>\text{SOH}_2^+\text{A}^-$ (in m^{-2}), respectively, and $\Gamma_{(-)}^0$ and $\Gamma_{(-)}^0$ represent the density of surface species $>\text{SO}^-$ and $>\text{SOH}_2^+$ (in m^{-2}), respectively. The fluxes of counterions and co-ions along the mineral surface are given by an interfacial version of the Nernst–Planck local equation

$$\mathbf{j}_{(\pm)} = -b_{(\pm)}^S \Gamma_{(\pm)} \nabla_S \tilde{\mu}_{(\pm)}^S, \tag{2.283}$$

where $\tilde{\mu}_{(\pm)}^S$ denotes the electrochemical potential along the mineral surface (in J), ∇_S denotes a surface gradient along the mineral surface, and $(\pm 1)e$ is the charge of counterions and co-ions (in C). The potential tangential mobility of the weakly sorbed counterions at the surface of silica is confirmed by nuclear magnetic resonance (NMR) spectroscopy. Demonstrate that the fluxes along of the counterions in the Stern layer are given by

$$\mathbf{j}_{(\pm)} = -b_{(\pm)}^S \Gamma_{(\pm)} \nabla \tilde{\mu}_{(\pm)}, \tag{2.284}$$

where $\tilde{\mu}_{(\pm)}$ denotes the electrochemical potential in the two reservoirs.

Exercise 2.4. Electrical conductivity and effective charge density of a bundle of capillaries. We consider that the porous material is described by a bundle of capillaries between two reservoirs. We define $g(R)$ as the probability density to have a capillary with a radius comprised between R and $R + dR$. Because $g(R)$ is normalized, we have

$$\int_0^\infty g(R)dR = 1. \tag{2.285}$$

We write the different raw moments of this probability distribution as

$$\Pi_n = \int_0^\infty R^n g(R)dR, \tag{2.286}$$

where n is a negative or positive integer (Π_n is the expectation of the distribution R^n according to the probability distribution $g(R)$).

(1) Demonstrate that the conductivity of such a porous material is given by

$$\sigma = \frac{1}{F}\left(\sigma_w + 2\frac{\Pi_1}{\Pi_2}\Sigma_S\right). \tag{2.287}$$

(2) Demonstrate that the total charge density of the diffuse layer $\bar{Q}_V$ and the effective charge density dragged by the flow of the pore water $\hat{Q}_V$ (streaming current) are given by

$$\bar{Q}_V = -2Q_S\frac{\Pi_1}{\Pi_2}, \tag{2.288}$$

$$\hat{Q}_V = \bar{Q}_V\frac{\Pi_2\Pi_3}{\Pi_4\Pi_1} \leq \bar{Q}_V, \tag{2.289}$$

respectively.

3

Laboratory investigations

In Chapter 1, we showed how the occurrence of a natural source current density in a conductive porous material is responsible for an electrical field (the self-potential field). The electrical potential is solution of a Poisson-type equation with a source term given by the divergence of the source current density. Chapter 2 was dedicated to establishing a fundamental theory of the source current density in porous media, looking at the contributions of the different types of charge carriers (ionic species and electrons). The present chapter investigates two types of laboratory measurements, which are used to gain better insight into these processes: core sample measurements and sandbox experiments. In Section 3.1, we present electrical (complex) conductivity and streaming potential measurements, and show how these measurements can be considered in a unified framework of petrophysical properties. Such a unified framework is of paramount importance in considering the natural complementarity of d.c. resistivity, induced polarization, and self-potential in solving hydrogeophysical problems. Section 3.2 deals with a sandbox experiment investigating the geobattery concept and its predictions for ore bodies. We are especially interested in the occurrence of a dipolar anomaly and the role of the redox potential distribution in this behavior.

3.1 Analyzing low-frequency electrical properties

In this section, we are providing a unified approach for complex conductivity and streaming potential in porous materials, with particular emphasis on saprolitic core samples from the Oak Ridge Reservation Site in Tennessee.

3.1.1 Relevant properties of saprolites

A very important property in the description of the low frequency electrical phenomena in clayey materials is the cation exchange capacity, CEC, which is

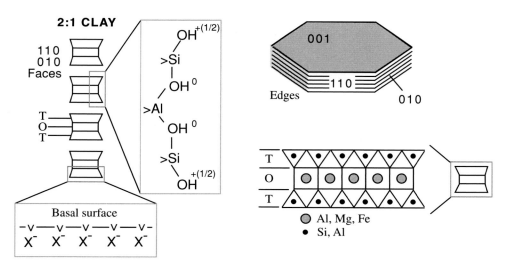

Figure 3.1. Electrical double layer at the surface of clay minerals. The surface charge of illite and smectite is heterogeneous with amphoteric sites (exchanging protons) located on the 110 and 010 crystalline planes. Isomorphic substitutions in the crystalline framework are responsible for negative charges on the basal surfaces (001 crystalline planes).

generally expressed in Coulomb (symbol C) per unit mass of solid (minerals). The CEC represents the number of active sites that exist on the mineral surface, per unit mass of solid of the material. The CEC is typically measured through potentiometric titrations (see Sumner and Miller (1996)): a solution with a known amount of a cation with a strong affinity for the mineral surface (e.g., copper Cu^{2+}, cobalt II Co^{2+}, or ammonium NH_4^+) is mixed with a certain amount of the solid phase (minerals) of the porous material. The metal ions in the fluid are sorbed to the mineral surface and are exchanged with the ions in the interlayer of the clay lattice, reducing their concentration in the electrolyte. The concentration of the cation is measured again in the solution, and the difference of their concentration provides a measure of the charge of the mineral surface, which is normalized by the mass of solid grains. For clay minerals, the CEC, at a given pH, is related both to the active amphoteric sites on the edges of the clay crystals, as well as (for illite and smectite) to the isomorphic substitutions in the crystalline framework (see Figure 3.1 for a description of the charges on the mineral surface of clays).

A number of publications describes the mineralogy of the saprolite from the Oak Ridge Reservation Site in Tennessee (see, e.g., McKay *et al.* (2005)), and some of their petrophysical, textural, and interfacial properties. The sand and silt fractions (by weight) are about 50% and 31%, respectively, and the clay fraction is about 19%; see Jardine *et al.* (1988, 1993a, b). The cation exchange capacity of saprolite soil samples collected at 1 m depth is typically 10.5 cmol kg^{-1} (0.105 meq g^{-1}); see Kim *et al.* (2009). Jardine *et al.* (1993a) reported a CEC in the range 0.07–0.16 meq g^{-1}. According to the CEC/specific area relationship developed recently by Woodruff and Revil (2011), taking 10.5 cmol kg^{-1} is consistent with a specific surface area of 30 m^2 g^{-1}. This value is consistent with the reactive surface area of the saprolite from Oak Ridge measured by Kooner *et al.* (1995), 40 m^2 g^{-1}. According to Kim *et al.* (2009), quartz, illite, and microcline (K-feldspar) make up 95% of the total mineral composition. The main component of the minor fraction (the remaining 5%) is vermiculite, a 2:1 clay mineral. Taking 5% weight fraction of vermiculite, with a CEC of 1 meq g^{-1}, implies that the weight fraction of illite, with a CEC of 0.20 meq g^{-1}, is 28%, on average. The grain density is about 2650 kg m^{-3}.

Quartz, potassium feldspar, plagioclase, and clay minerals were identified by X-ray diffraction (XRD) analyses of the whole rock samples. Illite and mixed clays consisting of 80%–90% illite and 10%–20% smectite dominated the clay fractions of all three samples. A trace amount of chlorite may be present in sample S23. An important point is that the CEC scales linearly with the specific surface area S_{Sp} measured by BET experiments as shown in Figure 3.2. The specific surface area measures the amount of surface area (in m^2) of the minerals per unit mass of the minerals. The ratio of the CEC by S_{Sp} represents the surface charge density of the mineral surfaces Q_S (in C m^{-2}), which was discussed in Chapter 2 (see Section 2.2). This surface charge density is, at first approximation and for near neutral pH values (5 to 8), constant for clay minerals. It is typically equal to 1 to 3 elementary charges ($e = 1.6 \times 10^{-19}$ C) per nm^2 (Figure 3.2). The data for the saprolite (mean CEC = 0.105 meq g^{-1} and mean S_{Sp} = 40 m^2 g^{-1}) are consistent with the trend shown in Figure 3.2.

3.1.2 Material and methods

Three saprolite core samples were collected from the Oak Ridge Reservation Site in Tennessee at depth of 2.7 m (9 feet), 4.9 m (16 feet), and 6.7 m (22 feet) below the ground surface (Samples S9, S16, and S22, respectively). The core samples were cut from the original plastic sleeves, homogenized and packed into 5 cm long portion of 4.4 cm diameter acrylic columns. The experimental setup used for the measurements described in this chapter is shown in Figure 3.3. This system is able to measure the permeability, the electrical conductivity and the streaming potential coupling coefficient during the same experiment, as described below. The end caps of the columns were slotted to distribute the flow evenly across the samples. At the mid-height of the samples, four Ag(s)/AgCl electrodes

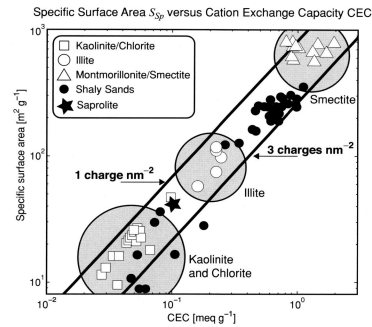

Figure 3.2. Specific surface area of clay minerals (in $m^2\ g^{-1}$) as a function of the (absolute) (in meq g^{-1} with 1 meq $g^{-1} = 96\,320$ C kg^{-1} in SI units) for various clay minerals. The ratio between the CEC and the specific surface area gives the equivalent total surface charge density of the mineral surface. Generalized regions corresponding to kaolinite, illite, and smectite are represented by shaded circles. The two lines correspond to 1 to 3 elementary charges per unit surface area. Data from: Patchett (1975), Lipsicas (1984), Zundel and Siffert (1985), Lockhart (1980), Sinitsyn *et al.* (2000), Avena and De Pauli (1998), Shainberg *et al.* (1988), Su *et al.* (2000), Ma and Eggleton (1999) and Revil *et al.* (unpublished work, 2012).

were installed circumferentially, at 60° angular spacing, for complex conductivity measurements and two additional Ag(s)/AgCl electrodes close to the end-faces of the core samples were used to measure the streaming potential in response to pore water flow. The experimental setup served as a constant head permeameter, and was used for all petrophysical measurements. The samples were saturated from the bottom up, with background ground water under vacuum for three days. The porosity of the packed samples was approximately 45%. The samples were initially flushed with 42–124 pore volumes of background groundwater to establish equilibrium conditions. Once the electrical conductivity of the influent and effluent solutions were within 20 μS cm^{-1}, and the pH differed less than approximately 0.5 pH units, the hydraulic conductivity, streaming potential coupling coefficient, and complex conductivity were measured. The samples were subsequently flushed

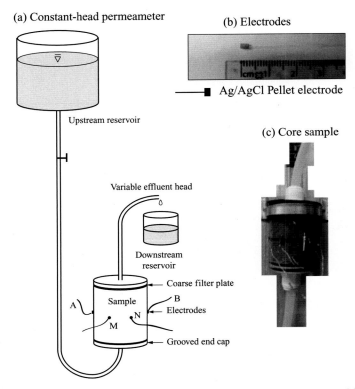

Figure 3.3. Sketch of the experimental setup used to estimate the permeability and the electrical properties of core samples. (a) Constant head permeameter. The non-polarizing Ag/AgCl electrodes for the measurement of the complex conductivity are organized in a Wenner array (A and B are the current electrodes, M and N denote the potential electrodes). The ABMN electrodes are in the middle of the sample holder. The geometrical factor for this array is modeled by solving the Laplace equation with Comsol Multiphysics 3.5a. The non-polarizing electrodes for the streaming potential measurements are placed at the two end-faces of the core sample (not shown here). (b) Non-polarizing Ag/AgCl electrodes. (c) Picture of the sampler holder.

with 3, 10, 30, 100 and 300 mM NaCl solutions near pH 5.7 and with 10 mM NaCl solutions with pH adjusted to various values comprised between 3 and 6 as discussed further below. Typically, 100 pore volumes were flushed through each sample before equilibrium was reached. The experiments took several months to complete.

The determination of the permeability of the core samples is based on Darcy's law, which can be seen as a macroscopic volume average of the fluid momentum equation for the pore water (the Stokes equation, see Chapter 2, Section 2.2). At the scale of a representative elementary volume, Darcy's law is a constitutive equation connecting the volumetric flux of water through a porous material $\mathbf{U}$ (the Darcy velocity, expressed in m s^{-1}) to the gradient of the pore fluid pressure p (in Pa) or hydraulic head h (in m). In saturated conditions, Darcy's law is written as

$$\mathbf{U} = -\frac{k}{\eta_f}\nabla p = -K\nabla h, \tag{3.1}$$

where k denotes the permeability, K the hydraulic conductivity, and η_f the dynamic viscosity of the pore water ($\eta_f = 10^{-3}$ Pa s at 25 °C for water). The sample holder shown in Figure 3.3 is used as a constant head permeameter to perform the permeability measurements. The flux can be measured for different heads and Darcy's law is used to estimate the permeability of the core samples.

The sample holder shown in Figure 3.3 was also used to determine the complex conductivity σ^* or its inverse, the complex resistivity $\rho^* = 1/\sigma^*$ (in Ω m), over several orders of magnitude of frequency (1 mHz to 45 kHz); see Zimmermann *et al.* (2008). We measured the complex impedance Z^* (in Ω) defined by

$$Z^*(\omega) = \frac{U}{I} = |Z^*(\omega)|e^{i\varphi(\omega)}, \tag{3.2}$$

where U is the voltage between the potential electrodes M and N (Figure 3.3), I the magnitude of the current injected between the current electrodes A and B (Figure 3.3), and $|Z^*(\omega)|$ and $\varphi(\omega)$ are the amplitude and the phase of the complex impedance, respectively. The complex resistivity ρ^* is related to Z^* by a geometrical factor G (in m) by: $\rho^*(\omega) = GZ^*(\omega)$. The geometrical factor takes into account the position of the electrodes on the sample and the size and the shape of the samples and the boundary conditions for the potential. The complex resistivity ρ^* can be recast into the complex conductivity as described in the previous section. The measurements were performed with the high precision impedance meter designed by Zimmerman *et al.* (2008). The accuracy of the instrument is approximately 0.1–0.3 mrad at frequencies below 1 kHz.

The experimental setup shown in Figure 3.3 was also used for the measurement of the streaming potential coupling coefficient. To measure the streaming potential, a given hydraulic head is imposed on the cylindrical sample inside the tube, by adding water to the water column in the tube in such a way that the hydraulic head is maintained constant (as done for the permeability measurements). The hydraulic head in the tube and the length of the porous pack control the gradient of the fluid pressure. The brine flows through the porous sample; the resulting electrical potential associated with the drag of the excess charge in the diffuse layer is measured with two non-polarizable Ag/AgCl$_2$ electrodes (Ref321/XR300, Radiometer Analytical) located in the vicinity of the end-faces of the sample.

The difference of the electrical potential measured between the end-faces of the porous pack, divided by the length of the sample, is the streaming electrical field associated with the flow of the brine through the porous core sample. The voltages were measured with a data logger (Easy Log, internal impedance of 10 MOhm, sensitivity of 0.1 mV) or with a voltmeter (Metrix MX-20, internal impedance 100 MOhm, sensitivity of 0.1 mV). Both provided consistent measurements.

3.1.3 Complex electrical conductivity

We begin our analysis with the description of complex conductivity σ^* of a porous material, written as

$$\sigma^* = |\sigma| \exp(i\varphi) = \sigma' + i\sigma'', \tag{3.3}$$

where $|\sigma|$ denotes the amplitude of the conductivity (in S m^{-1}), $|\varphi|$ denotes the phase lag (usually expressed in mrad), σ' and σ'' denote the real (in phase) and imaginary (quadrature) components of the conductivity (both in S m^{-1}), and $i^2 = -1$ denotes the pure imaginary number ($|\sigma| = \sqrt{\sigma' + \sigma''}$ and $\varphi = \mathrm{atan}(\sigma''/\sigma')$). Figure 3.4 provides a typical spectrum showing the dependence of the real (in phase) and imaginary (quadrature) conductivities on the frequency. Because the phase is small (smaller than 100 mrad in amplitude), the phase is given by

$$\varphi \approx \sigma''/\sigma'. \tag{3.4}$$

From a physical standpoint, the in-phase conductivity represents the ability of the porous material to transmit electrical current (conduction), while the quadrature conductivity describes the ability of the porous material to reversibly store electrical charges (polarization of mobile charges). In this context, the term polarization has nothing to do with the dielectric polarization of bound charges observed at high frequencies. Low-frequency polarization is always driven by an electrochemical potential and the existence of polarization length scales, along which accumulation or depletion of charge carriers are observed; see Vinegar and Waxman (1984), Revil and Florsch (2010) and Revil *et al.* (2012a, b). Alternatively, induced polarization can be seen as the induction of a source current density (not as in the classical meaning of electromagnetic induction, but in the sense of the movement of charges induced by the low-frequency polarization of the Stern layer); therefore, there is a strong relationship between self-potential and induced polarization in terms of fundamental physics, a point always missed in the literature. Details of this

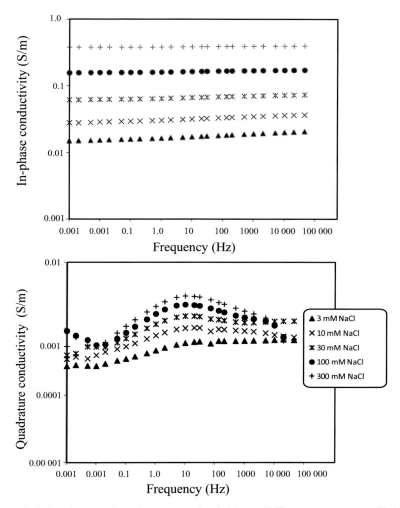

Figure 3.4. In-phase and quadrature conductivities at different pore water salinities (pH in the range 5–6) for a core sample of saprolite from Oak Ridge, Tennessee. The in-phase electrical codnuctivity increase slightly with the frequency, especially at high salinity. The quadrature conductivity shows a peak at a frequency of about 10 Hz.

connection will not be formally investigated in the present book and will be the subject of further research with the goal of unifying low-frequency electrical methods of geophysics.

In the following, we will also not explicitly consider the frequency dependence of the complex conductivity. Instead, we will take the values of the in-phase and quadrature conductivities at 1 Hz, which is (as shown later) an adequate frequency in probing the polarization of a clayey material. According to the model developed by Revil (2012) and Skold *et al.* (2012), the complex conductivity of fully water-saturated, porous clayey materials can be written as the sum of two contributions:

$$\sigma^* = \frac{1}{F}\sigma_f + \sigma_S^* \tag{3.5}$$

$$\sigma^* = \sigma_f \left(\frac{1}{F} + \mathrm{Du}^* \right), \tag{3.6}$$

where F denotes the formation factor (dimensionless), σ_f denotes the (real) electrical conductivity of the pore water (in S m^{-1}), σ_S^* denotes the complex-valued surface conductivity (components expressed in S m^{-1}), and $\mathrm{Du}^* = \sigma_S^*/\sigma_f$ denotes the complex-valued Dukhin number (the unitless ratio of the complex surface conductivity divided by the pore water conductivity). The formation factor F is related to the connected porosity ϕ by Archie's first law $F = \phi^{-m}$, with m denoting the cementation exponent; see Archie (1942). Figure 3.5 shows saprolite conductivity data as a function of the conductivity of the pore water. The data are very well predicted by Eq. (3.5), showing the existence of an isoconductivity point, below which the conductivity of the rock can be higher than the conductivity of the pore water.

According to the model developed by Revil (2012), the complex surface conductivity and the complex Dukhin number are related to the cation exchange capacity by

$$\sigma_S^* = \frac{2}{3}\rho_g \mathrm{CEC} \left[\beta_{(+)}(1 - f) - i\beta_{(+)}^S f \right] \tag{3.7}$$

$$\mathrm{Du}^* = \frac{2}{3} \left(\frac{\rho_g \mathrm{CEC}}{\sigma_w} \right) \left[\beta_{(+)}(1 - f) - i\beta_{(+)}^S f \right], \tag{3.8}$$

where f denotes the fraction of counterions in the Stern layer (dimensionless), ρ_g denotes the grain density (typically 2650 kg m^{-3}), $\beta_{(+)}$ denotes the mobility of the counterions in the diffuse layer (equal to the mobility of the same cations in the bulk pore water, $\beta_{(+)}(\mathrm{Na}^+, 25\,°\mathrm{C}) = 5.2 \times 10^{-8}$ m^{2}s^{-1}V^{-1}) and $\beta_{(+)}^S$ denotes the mobility of the counterions in the Stern layer ($\beta_{(+)}^S(25\,°\mathrm{C}, \mathrm{Na}^+) = 1.5 \times 10^{-10}$ m^{2}s^{-1}V^{-1}); according to Revil (2012), $f \approx 0.90$ for illite and smectite (the salinity dependence of f is discussed in Appendix A). The surface conductivity and the Dukhin number are defined as the real part of the complex surface conductivity and the real part of

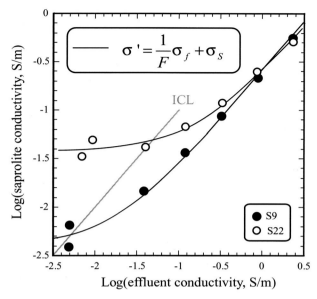

Figure 3.5. Analysis of the in-phase conductivity data of two saprolite core samples versus the conductivity of the effluent taken as a proxy for the pore water conductivity (measurement at 1 Hz). ICL stands for Iso-Conductivity Line (straight gray line). This line crosses the model line (black plain line) at a point for which the conductivity of the core sample is equal to the pore water conductivity. Below this point, the conductivity of the material can be higher than the conductivity of its pore water, a concept that is often not understood or forgotten by many hydrogeophysicists despite of its importance in interpreting electrical resistivity tomograms in low salinity environments.

the complex Dukhin number:

$$\sigma_S = \text{Re}[\sigma_S{}^*] \tag{3.9}$$

$$\text{Du} = \text{Re}[\text{Du}^*], \tag{3.10}$$

respectively, where Re[.] denotes the real part of a complex number.

The in-phase conductivity normalized by the pore water conductivity and the phase (quadrature conductivity normalized by the in-phase conductivity) are written as

$$\frac{\sigma}{\sigma_f} = \frac{1}{F} + \text{Du}, \tag{3.11}$$

$$\varphi \approx -\frac{\frac{2}{3} F \beta_{(+)}^S f \rho_g \text{CEC}}{\sigma_f + \frac{2}{3} \beta_{(+)} F(1-f) \rho_g \text{CEC}}. \tag{3.12}$$

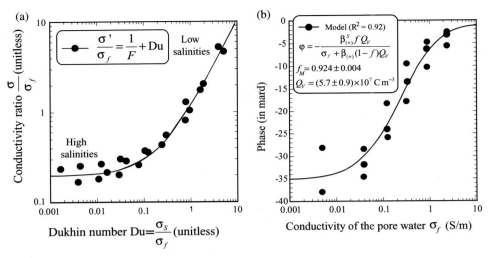

Figure 3.6. Complex conductivity properties of three core sample of saprolite from Oak Ridge saturated with NaCl solutions. (a) Normalized conductivity versus the Dukhin number for three core samples (pH ~6). The Dukhin number corresponds to the ratio between the surface conductivity and the pore water conductivity. Q_V denotes the total charge of the Stern plus diffuse layers per unit volume of pores. (b) Phase as a function of the conductivity of the pore water for three core samples of saprolite. The plain line corresponds to the best fit of the model with a constant partition coefficient f (the rationale for this coefficient is provided in Figure 1.9 and Appendix A).

Following Revil (2012), this last equation for the phase can also be approximated by

$$\varphi \approx -\frac{\beta_{(+)}^S \, f \, Q_V}{\sigma_f + \beta_{(+)}(1 - f)\, Q_V} \tag{3.13}$$

$$Q_V \equiv \rho_g \left(\frac{1-\phi}{\phi}\right) \mathrm{CEC}, \tag{3.14}$$

where Q_V denotes the total charge per pore volume (including the Stern and diffuse layer counterions and co-ions). Figures 3.6a and 3.6b show that Eqs. (3.11) and (3.13) provide a correct representation of the experimental data (conductivity and phase).

In Appendix A, following the model developed by Revil (2012), we show that the salinity of the quadrature conductivity of a clayey material can be expressed as

$$\sigma'' \approx \sigma_M'' \left(\frac{C_f K_{\mathrm{Na}}}{1 + C_f K_{\mathrm{Na}}}\right) \tag{3.15}$$

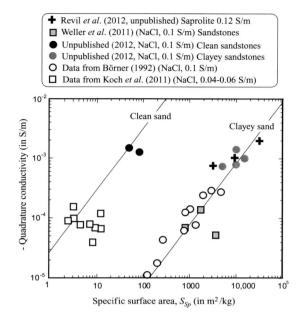

Figure 3.7. Influence of the specific surface area S_{Sp} upon the quadrature conductivity. We observe that the saprolite samples are characterized by relatively high specific surface areas and high quadrature conductivities in agreement with the trend determined for clayey sands by Revil (2012). Data from Weller *et al.* (2011), Börner (1992) and Koch *et al.* (2011). For the saprolite, the values are taken at 10 Hz and the CEC values are from surface conductivity measurements. The two lines (one for silicates and one for clayey materials) indicate that the mobility of the counterions in the Stern layer of silica is much higher than for clays.

$$\sigma''_M = -\frac{2}{3} \left(\rho_g \beta^S_{(+)} Q_S f_M \right) S_{sp} \tag{3.16}$$

$$\sigma''_M = -\frac{2}{3} \left(\rho_g \beta^S_{(+)} f_M \right) \text{CEC}, \tag{3.17}$$

where σ''_M denotes the maximum value of the quadrature conductivity reached at high pH and salinity values and K_{Na} denotes the apparent sorption constant of Na^+ in the Stern layer of the clay minerals. Taking $\beta^S_{(+)}(25\ °\text{C}, \text{Na}^+) = 1.5 \times 10^{-10}\ \text{m}^2\text{s}^{-1}\text{V}^{-1}$, $f = 0.90$, and $\rho_g = 2650\ \text{kg m}^{-3}$, we obtain $\sigma''_M = -a\,\text{CEC}$ with $a \approx (2/3)\beta^S_{(+)} f \rho_g = 2.4 \times 10^{-7}\ \text{S kg C}^{-1}\text{m}^{-1}$. Taking $\beta^S_{(+)}(25\ °\text{C}, \text{Na}^+) = 1.5 \times 10^{-10}\ \text{m}^2\text{s}^{-1}\text{V}^{-1}$, $f = 0.90$, $Q_S = 0.32\ \text{C m}^{-2}$ (see Revil (2012)), $\rho_g = 2650\ \text{kg m}^{-3}$, we have $\sigma'' \approx -a S_{Sp}$ with $a = 7.6 \times 10^{-8}\ \text{S kg m}^{-3}$. This trend, shown in Figure 3.7 for the shaly sands, is in agreement with the quadrature conductivity data of the saprolite.

We performed CEC measurements on three saprolite core samples from the Oak Ridge site, which is believed to be a representative background of the contaminated

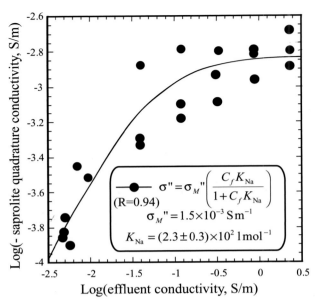

Figure 3.8. Dependence of the quadrature conductivity with the salinity (three saprolite core samples, NaCl, pH $\sim$ 5). The plain line corresponds to the prediction of the model described in the main text with a Langmuir-type isotherm for the partition coefficient f (see Appendix A).

portion of the site. The measurements were performed with a barium chloride solution to displace the cations sorbed to the clay mineral surfaces; see the protocol in Sumner and Miller (1996). All seven samples were analyzed in duplicate, with a relative standard deviation less than 17%. The average measured CEC value of the samples ranged from 5.0 to 8.6 cmol kg^{-1} or $CEC_M = (4.8 - 8.3) \times 10^3$ C kg^{-1}. The specific surface area is $S_{Sp} = 19\,000$ m^2 kg^{-1} (measured on three samples from the background site). Taking $CEC_M = 6 \times 10^3$ C kg^{-1}, the equivalent total charge per unit surface area is therefore $Q_S = CEC_M/S_{Sp} = 0.32$ C m^{-2} (two elementary charges per nm^2 as predicted by Revil *et al.* (1998)). The CEC range reported above, together with $\beta_{(+)}^S$(25 °C, Na$^+$) = 1.5 $\times$ 10^{-10} m^{2}s^{-1}V^{-1} (From Revil (2012)) and $f_M = 0.92$, yields $\sigma_M'' = -(1.1$–$1.9)\times10^{-3}$ S m^{-1}, in excellent agreement with the laboratory data (Figure 3.8).

3.1.4 The streaming potential coupling coefficient

The streaming current is the quasi-static source current density associated with the drag of the excess of electrical charge in the electrical diffuse layer of a porous material. There is an electrical charge in the pore water of any material, which compensates the charge density of the mineral surfaces (at near neutral pH, the surface charge is negative and the pore water exhibits a resulting positive charge excess, see Chapter 1). We write the macroscopic movable charge density (volumetric charge density per unit pore volume, in C m^{-3}) as

$$\hat{Q}_V \langle \mathbf{v}(\mathbf{x}) \rangle = \langle \rho(\mathbf{x})\mathbf{v}(\mathbf{x}) \rangle \tag{3.18}$$

$$\langle \cdot \rangle = \frac{1}{V_p} \int_{V_p} (\cdot)d\tau, \tag{3.19}$$

where the brackets denote a pore volume averaging, $\rho(\mathbf{x})$ denotes the local charge density in the pore space (in C m^{-3}), $\boldsymbol{v}(\mathbf{x})$ denotes the local instantaneous velocity of the pore water (in m s^{-1}), $\mathbf{x}$ denotes a local position in the pore space of the material, and $d\tau$ is an elementary volume around point $M(\mathbf{x})$. This equation is valid irrespective of the size of the diffuse layer relative to the pore size. In the case of a thin double layer (i.e., the thickness of the diffuse layer is much smaller than the equivalent pore radius), the macroscopic charge density $\hat{Q}_V \langle \boldsymbol{v}(\mathbf{x}) \rangle$ can be substantially smaller than the charge density associated with the diffuse layer (see Jougnot *et al.* (2012)), given by,

$$\bar{Q}_V = (1 - f)\rho_g \left(\frac{1 - \phi}{\phi} \right) \text{CEC}, \tag{3.20}$$

where f denotes the fraction of counterions in the Stern layer, ρ_g is the mass density of the grains (kg m^{-3}), ϕ denotes the porosity, and CEC denotes the cation exchange capacity of the material (in C kg^{-1}). The macroscopic source current density $\mathbf{J}_S$ (called the streaming current density in this case) associated with the drag of the excess of charge in the diffuse layer, is defined as

$$\mathbf{J}_S = \phi \langle \mathbf{j}_S \rangle \tag{3.21}$$

$$\mathbf{J}_S = \phi \langle \rho \mathbf{v} \rangle \tag{3.22}$$

$$\mathbf{J}_S = \hat{Q}_V \phi \langle \mathbf{v} \rangle \tag{3.23}$$

$$\mathbf{J}_S = \hat{Q}_V \mathbf{U}, \tag{3.24}$$

where $\mathbf{U}$ denotes the macroscopic Darcy velocity. The total (electromigration) current density is given for quasi-static conditions by Revil *et al.* (2011a) as

$$\mathbf{J} = \sigma \mathbf{E} + \hat{Q}_V \mathbf{U}, \tag{3.25}$$

where $\hat{Q}_V$ represents the effective (excess) moveable charge associated with pore water flow (see the conclusions of Chapter 2). This value is expected to be only a fraction of the local volumetric charge in the diffuse layer, which is itself a fraction of the total charge density associated with the cation exchange capacity and is given by $\hat{Q}_V = (1 - f)Q_V$ (see Eqs. (3.14) and (3.20) and Revil and Florsch (2010) and Jougnot *et al.* (2012)). Using $Q_V = 4.0 \times 10^7$ C m^{-3} and $f = 0.9$ for the saprolite core samples (see previous section), we obtain $\hat{Q}_V = 4 \times 10^6$ C m^{-3}.

Using Darcy's law, Eq. (3.1), the total current density is given by

$$\mathbf{J} = -\sigma \nabla \psi - K \hat{Q}_V \nabla h. \qquad (3.26)$$

The streaming potential coupling coefficient is given by Bolève *et al.* (2007b) and Revil *et al.* (2010) as

$$C \equiv \left(\frac{\partial \psi}{\partial h} \right)_{\mathbf{J}=0} = -\frac{K \hat{Q}_V}{\sigma}. \qquad (3.27)$$

This coupling coefficient characterizes the sensitivity of the electrical potential, in this case called the streaming potential, to the hydraulic head. Using Eqs. (3.26) and (3.27), the coupling coefficient is related to the pore water conductivity as

$$C = -\frac{FK\hat{Q}_V}{\sigma_f + F\sigma_S}. \qquad (3.28)$$

The streaming potential coupling coefficient is the fundamental petrophysical parameter characterizing streaming potentials in the field.

An alternative equation for the streaming potential coupling coefficient is written as a function of the zeta potential (see Merkler *et al.* (1989)). In this case the expression for the streaming potential coupling coefficient becomes (see Revil *et al.* (1999a))

$$C = -\frac{\varepsilon_f \rho_f g \zeta}{\eta_f(\sigma_f + F\sigma_S)}, \qquad (3.29)$$

where ρ_f is the mass density of the pore water, g is the gravitational acceleration (9.81 m s^{-2}), and ε_f denotes the dielectric constant of the pore water (81 × 8.84 × 10^{-12} F m^{-1}).

Streaming potential data from two typical experiments are shown at Figure 3.9. The differences in electrical potential, measured in the vicinity of the end-faces of the core sample, are proportional to the imposed hydraulic heads. The slope of the linear trend of streaming potential vs. head corresponds to the streaming potential coupling coefficient (see Eq. 3.28). The streaming potential coupling coefficient was determined for the three samples at NaCl concentrations ranging from 3 mM to 300 mM. In Figure 3.10, we report the value of the streaming potential coupling coefficient as a function of the conductivity of the pore water. We fit a function of the form

$$C = -\frac{a}{\sigma_f + b}. \qquad (3.30)$$

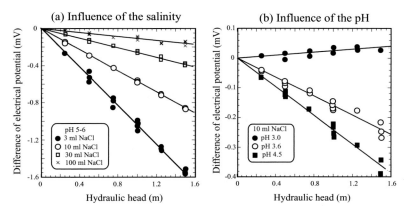

Figure 3.9. Example of the measurement of the streaming potential for various values of the hydraulic head using the sample holder shown in Figure 3.3. (a) Influence of the salinity (NaCl solutions, Sample S16). (b) Influence of the pH (NaCl solution 10 mM, Sample S9). We can see clearly that around pH 3, there is a change in the surface charge density of the mineral (change in the zeta potential). The point at which the zeta potential is zero is called the isoelectric point in pH.

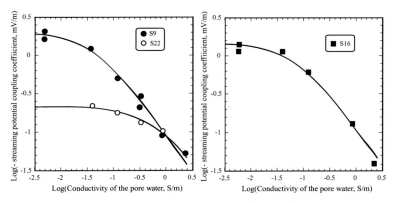

Figure 3.10. Value of the streaming potential coupling coefficient (negative) versus the pore water conductivity (NaCl solutions, pH $\approx$ 6). The plain lines correspond to the fit of the data with a function $C = a/(\sigma_f + b)$. Sample S22 is influenced by the presence of the smectite, which is generating a high surface conductivity and therefore a smaller value of the magnitude of the streaming potential coupling coefficient at low salinities.

Once a is determined, the value of the effective charge density dragged by the flow of the pore water is determined as $\hat{Q}_V = a/FK$. Equation (3.30) agrees with the experimental data shown in Figure 3.10 very well. The volumetric charge density $\hat{Q}_V$ is in the range 140–490 C m^{-3}, which represents only a small fraction of the total charge density of the diffuse layer ($\bar{Q}_V = 4 \times 10^6$ C m^{-3}), which is consistent

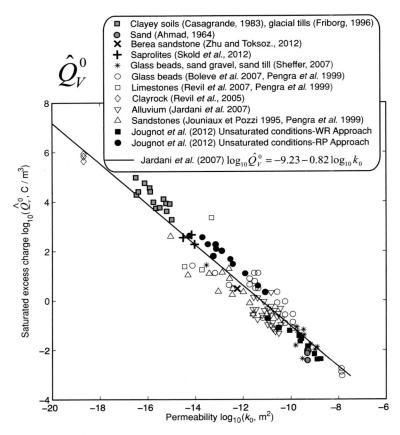

Figure 3.11. Quasi-static charge density (excess pore charge moveable by the quasi-static pore water flow) versus the quasi-static permeability for a broad collection of core samples and porous materials. This charge density is derived directly from the streaming potential coupling coefficient using Eq. (3.27). Data from Ahmad (1964), Bolève *et al.* (2007b), Casagrande (1983), Friborg (1996), Jougnot *et al.* (2012), Jardani *et al.* (2007a, b), Pengra *et al.* (1999), Revil *et al.* (2005, 2007), Sheffer (2007), Skold *et al.* (2012) and Zhu and Toksöz (2012). The subscript/superscript 0 will take its meaning in Chapter 8 in which seismoelectric signals are analyzed.

with the results found in Jougnot *et al.* (2012). The estimates for the effective charge per unit pore volume are compared with other literature data in Figure 3.11. As evidenced by these and other materials in the literature, the effective (moveable) charge per unit pore volume (subject to the drag of advective flow in porous media) can be directly correlated to the permeability of the porous material material. This key relationship is of fundamental importance in self-potential ground water flow modeling as discussed below in Chapters 4 to 8.

3.2 Investigating the geobattery concept in the laboratory

Sandbox experiments can be used to test conceptual models at an intermediate scale, between the core sample and field scales. In this section, we present an experiment designed to test the concept of geobattery introduced in Chapter 2.

3.2.1 Material and methods

The geobattery experiment was conducted in the 200 cm $\times$ 50 cm $\times$ 7 cm Plexiglas sandbox shown in Figure 3.12a. This tank was filled with well-calibrated sieved sand, characterized by a log-normal grain size distribution (mean grain size 132 μm). The porosity of the sand was 0.34 ± 0.01. X-ray analysis shows that it is composed of 95% silica, 4% orthoclase feldspar, and less than 1% albite. The sand was mixed with an aqueous solution (0.01 M KCl, 2×10^{-3} M NaOH) and was used to fill the tank entirely, taking care to avoid air entrapment. The use of NaOH was intended to accelerate the process of corrosion of an iron bar that was introduced into the sandbox as described below. Despite the initial pH of the solution of 10.5, the equilibrium pH of the pore water obtained from the slurry was equal to 7.0. Further information regarding this experiment can be found in Castermant *et al.* (2008).

The sand in the tank was left to compact and equilibrate with the pore water for 24 hours. A small amount of formaldehyde (135 μl per liter of solution) was also added to the pore water to impede the development of micro-organisms during the course of the experiment. Previous tests have shown that the presence of formaldehyde does not influence the measurement of the redox and self-potentials.

After the equilibration period, a clean iron bar was introduced vertically into the sand, offset to one side in the tank (Figure 3.12a). The iron bar was a rectangular piece of iron, with a thickness of 2 cm and a height of 50 cm. It was left in contact with the bottom of the tank. The upper boundary was exposed to the air, fixing the value of the redox potential value at 680 mV. The bar was not in contact with the front and back sides of the tank (Figure 3.12a). Thereafter, the bar was left to corrode for a period of six weeks at room temperature (24 ± 2 °C). The water level was kept constant during the experiment.

All self-potential signals were measured with reference to a non-polarizing electrode (termed the reference electrode) located on the surface at the opposite end of the tank (Figures 3.12a and 3.12b). Measurements were taken at the surface of the tank ($z = 0$) every 5 cm at time $t_0 = 0^+$ (defined as the time corresponding to the introduction of the iron bar), four weeks after the introduction of the bar and six weeks after the introduction of the bar. After six weeks, we also measured the distributions of the self-potential signals, the redox potential and the pH at different depths and distances from the iron bar (Figure 3.13). The self-potentials were

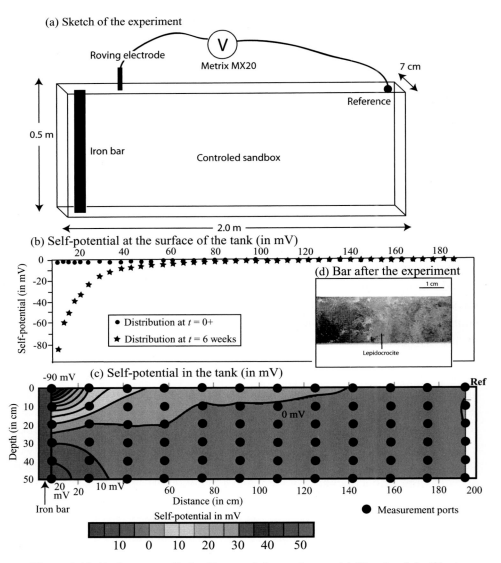

Figure 3.12. Redox-controlled self-potential experiment. (a) Sketch of the Plexi-glas controlled sandbox experiment. (b) Self-potential profiles (in mV) at the top surface of the tank at $t = 0^+$ and $t = 6$ weeks. (c) Side view of the saturated sandbox used for the experiment. Ref indicates the position of the reference elec-trode. Ref indicates the position of the reference electrode. (d) Picture showing the presence of lepidocrocite on the upper part of the iron bar from the tank at the end of the experiment after six weeks (the surface of the initial bar was totally clean).

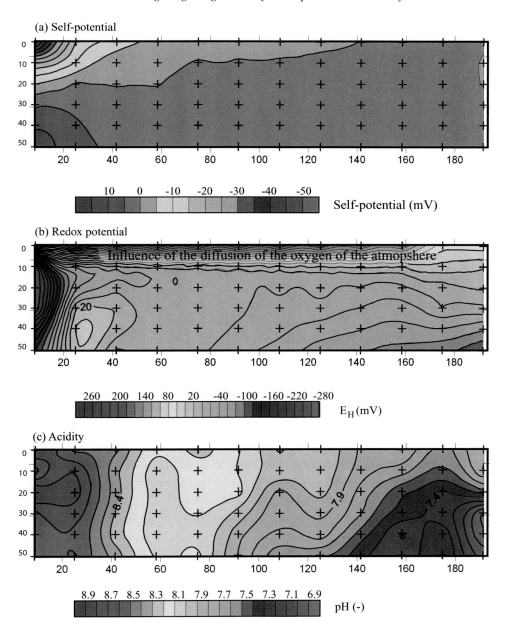

Figure 3.13. Distribution of the self-potential (a), distribution of the redox potential (b), and distribution of the pH (c) at the end of the experiment (distributions are shown six weeks after the introduction of the iron bar). Note that the self-potential signals exhibit a clear dipolar distribution with a negative pole located near the surface of the tank and a positive pole located at depth. See also color plate section.

measured with Ag/AgCl non-polarizing electrodes (REF321/XR300 from Radiometer Analytical, with a diameter of 5 mm) and a calibrated voltmeter (MX-20 from Metrix with a sensitivity of 0.1 mV and an internal impedance of 100 MΩ). The reference electrode for the self-potential measurements should be ideally placed at infinity, where, by definition, the electrical potential falls to zero. We placed the reference electrode as far as possible from the perturbed zone (Figures 3.12a and 3.13).

The pH was measured with a calibrated pH meter (pH 330/SET-1 WTW from Fisher Scientific) and an electrode PH/T SENTIX 41. The redox potential was measured with redox combination electrodes (InLab501 from Mettler Toledo). The measured values ($E_{Ag/AgCl}$) were converted to the normal hydrogen electrode (E_{NHE} or E_H), according to the relationship $E_{NHE} = E_{Ag/AgCl} + 208.56$ mV from Macaskill and Bates (1978).

We did not observe any self-potential anomalies immediately after the introduction of the iron bar (Figure 3.12b). A negative self-potential anomaly begins to develop at the surface of the tank in the days following the introduction of the bar; the peak of this anomaly is clearly associated with the presence of the iron. After six weeks, the amplitude of this anomaly reaches -82 mV in the vicinity of the iron bar. The polarity of the anomaly agrees with field observations by, for example, Sato and Mooney (1960) and the amplitude is similar to those reported in the laboratory by Bigalke and Grabner (1997). The self-potential distribution shown in Figure 3.13a has a dipolar character, with a small positive anomaly localized around the bottom portion of the iron bar; a stronger negative anomaly is located in the vicinity of the top part of the iron bar (Figure 3.13a). This observation is also consistent with the data reported by Bigalke and Grabner (1997), as well as the classical theory of the geobattery from Sato and Mooney (1960). After six weeks, the iron bar was removed from the tank, and evidence of corrosion was observed on its surface (Figure 3.12d).

The distribution of the redox potential (Figure 3.13b) appears to consist of two contributions. The far field contribution (at distances $x > 20$ cm from the iron bar) varies mainly with depth, and ranges from $+260$ mV at the top surface of the tank to -40 mV at the bottom of the tank. The strong gradient in redox potential within the upper 20 cm implies an abrupt change from oxidizing to reducing conditions (Figure 3.13b). This feature can be observed after six weeks, while the dissolved oxygen concentration was initially constant inside the tank. It is likely that the chemical reactions at the surface of the iron bar, and the degradation of a small amount of organic matter observed inside the sand, consumed the dissolved oxygen in the pore solution. Therefore, the relatively slow diffusion rate of oxygen in the slurry is responsible for a vertical gradient in the redox potential in the upper 20 cm of the tank.

In the vicinity of the iron bar (at $x < 20$ cm), there is a strong perturbation of the redox potential with respect to the far-field distribution. The redox potential is strongly negative in the immediate vicinity of the iron bar on the order of -200 to -250 mV. We observed that the color of the sand in the vicinity of the iron bar turns to a blueish-green color over time. If this sand was exposed to air, this color turns to ochre. This behavior is typical of the formation of fougerite, as this green rust is transformed to lepidocrocite when exposed to air, as reported by Trolard *et al.* (2007). Far from the iron bar, the pH is equal to the initial pH of the pore water solution (pH $= 7.0$) (Figure 3.13c). In the vicinity of the iron bar, the pH was alkaline, equal to 9.0. The pH gradient was mainly horizontal.

3.2.2 Geochemistry

We now describe the electrochemistry governing the corrosion of the iron bar. The corrosion of iron results in the formation of a green rust called fougerite; see, for example, Trolard *et al.* (2007). In our experiment, the values of the redox potential and the pH measured in the vicinity of the iron bar are compatible with the stability domain of fougerite. The first step in corroding iron in a neutral aqueous solution is the formation of $Fe(OH)_2$ according to the following sequence of redox reactions:

$$Fe \rightarrow Fe^{2+} + 2e^- \tag{3.31}$$

$$Fe^{2+} + 2\,OH^- \rightarrow Fe(OH)_2. \tag{3.32}$$

Then $Fe(OH)_2$, which is unstable, is oxidized to ferric oxyhydroxide depending on the anion in the solution (mainly lepidocrocite, goethite or magnetite). Fougerite is an intermediate component in these reactions that can be written (see Trolard *et al.* (2007)) as

$$Fe(OH)_2 + x A^- \rightarrow \left[Fe^{II}_{1-x} Fe^{III}_x (OH)_2 \right]^{x+} [x A^-]^{x-} + x e^- \tag{3.33}$$

$$\left[Fe^{II}_{1-x} Fe^{III}_x (OH)_2 \right]^{x+} [x A^-]^{x-} + OH^- \rightarrow FeOOH + x A^- + (1-x)e^- + H_2O, \tag{3.34}$$

where A^- is an interlayer anion that compensates for the excess of positive charge of the layer due to the partial oxidation of Fe^{II} to Fe^{III}. In this experiment, either Cl^- or OH^- are possible candidates for A^-; see Trolard *et al.* (2007). However, OH^- can be dismissed, because its concentration is too small (see the value of the pH in the vicinity of the bar on Figure 3.13).

Oxygen, O_2, plays the role of electron acceptor in the geobattery according to

$$1/4\,O_2 + 1/2\,H_2O + e^- \rightarrow OH^-. \tag{3.35}$$

Consequently the redox reaction for iron is,

$$Fe + 3/4\,O_2 + 1/2\,H_2O \rightarrow FeOOH. \tag{3.36}$$

Since atmospheric oxygen is constantly diffusing into the tank, chemical equilibrium is never established. The observed horizontal pH gradient in the tank (see Figure 3.13c) is likely related to the generation of OH^- in the vicinity of the iron bar and to the diffusion of OH^- down gradient, away from the increased concentrations around the iron, imposed by the reactions outlined above. To verify this, we consider the value of the diffusivity of OH^- in water: $D_f\,(OH^-) = 5.3 \times 10^{-9}\,m^2\,s^{-1}$ (see Samson *et al.* (2003). The tortuosity of the pore space in the sand is given by the product between the formation factor and the porosity, equal to 1.7 for the sand used in this sandbox; see Castermant *et al.* (2008). For a time constant τ of six weeks, the characteristic diffusion length of OH^- in the sandbox is $L \approx 2(D_f\tau)^{1/2} \approx 20$ cm. This order of magnitude is compatible with the distance between the pH front and the bar (about 40 cm, see Figure 3.13c).

In summary, the iron bar was originally placed in an equilibrated environment, in which there was no pre-existing gradient in the redox potential distribution; at the beginning of the experiment, the pH and the redox potential are constant over the entire volume of the tank. Chemical reactions at the surface of the iron bar are responsible for its corrosion. This corrosion perturbs the redox potential distribution in the vicinity of the iron bar, depletes the oxygen concentration inside the tank, generates a basic pH front diffusing away from the viscinity of the iron inside the tank, and causes the formation of a crust at the surface of the bar. This crust is probably responsible for an increase of the resistivity at the surface of the iron bar.

3.3 Conclusions

In this chapter, we have covered two types of laboratory methods, at two different scales, that are crucial to understanding the physics of self-potential signals. The first set of methods concerns core sample measurements. We measured the streaming potential coupling coefficient, which describes the sensitivity of the electrical field to a gradient in the pore fluid pressure. We have shown that the electrical conductivity, the permeability and the streaming potential coupling coefficient can be all measured with the same sample holder; therefore enabling the measurement of the petrophysical parameters of the porous material under exactly the same conditions. The second laboratory method involved a sandbox experiment designed to test the concept of geobattery at an intermediate scale. These types of core and sandbox experiments are extremely complementary to field measurements, allowing us to test the physics of these problems, and providing relative orders of magnitude for the material properties. We will see the importance of petrophysics in the following chapters, both in the forward and inverse modeling of self-potential signals. We believe that petrophysics is presently underused in some hydrogeophysical work because most geophysicists are not experienced enough with the physics of porous media. We have seen too many papers, for instance, where Archie's law is abusively used and misused to interpret resistivity measurements in low salinity environments or in clayey materials. We will demonstrate in the next chapters that petrophysics is really required to interpret quantitatively self-potential measurements in the field.

Exercises

Exercise 3.1. Electrical conductivity tensor of a pile of N planar sedimentary layers. In this chapter, electrical conductivity was considered as a scalar. This scalar is a constant, in the case of a homogeneous conductive space, or it depends on the position, in the case of a heterogeneous conductive material. In an anisotropic material, the value of conductivity depends on the direction and electrical conductivity, a second-order symmetric tensor that can be represented by a matrix. In the present exercise, we examine the case of a pile of N isotropic sedimentary layers. This symmetry is called transverse isotropy. Each layer is characterized by a conductivity σ_i (i from 1 to N), and a thickness m_i. The length and width of the geological pile is L. In this case, the electrical conductivity is not a scalar, but a second-order symmetric tensor. In a Cartesian coordinate system, with x in the horizontal direction and z in the vertical direction, Ohm's law is written as $\mathbf{J} = -\bar{\bar{\sigma}} \cdot \nabla \psi$ where the conductivity tensor is given by

$$\bar{\bar{\sigma}} = \begin{bmatrix} \sigma_x & 0 & 0 \\ 0 & \sigma_x & 0 \\ 0 & 0 & \sigma_z \end{bmatrix}. \tag{3.37}$$

In this exercise we are going to get expressions for the horizontal and the vertical components of the electrical conductivity tensor.

(1) Using Ohm's law, demonstrate that the effective conductivity of the block along the x- or y-directions is given by

$$\sigma_x = \frac{\sum_{i=1}^{N} m_i \sigma_i}{\sum_{i=1}^{N} m_i}. \tag{3.38}$$

(2) Using Ohm's law, demonstrate that the effective conductivity of the block along the z-direction is given by

$$\sigma_z = \frac{\sum_{i=1}^{N} m_i}{\sum_{i=1}^{N} m_i / \sigma_i}. \tag{3.39}$$

(3) We consider a 300 m sequence of interbedded sandstone and shale with 75% sandstone. We consider the case in which the conductivity of the sandstone is 0.01 S m^{-1}, and that of the shale is 1 S m^{-1}. Determine the two components of the electrical conductivity tensor.

(4) Do the same exercise for the permeability using Darcy's law and for the streaming current upscaling the coefficient L (in A m^{-2}) in the expression of the source current density $\mathbf{J}_S = -L\nabla h$, where h denotes the hydraulic head.

Exercise 3.2. Complex conductivity. For a clayey material, the in-phase conductivity of the material is given by

$$\sigma' = \frac{1}{F}\sigma_f + \sigma_S, \tag{3.40}$$

where F denotes the formation factor and σ_f denotes the conductivity of the pore water. The surface conductivity σ_S is given by:

$$\sigma_S = \frac{1}{F}\beta_{(+)}(1-f)Q_V \tag{3.41}$$

$$Q_V = \rho_g\left(\frac{1-\phi}{\phi}\right)\text{CEC}, \tag{3.42}$$

where f is the fraction of counterions in the Stern layer (also called the partition coefficient). Equation (3.41) means that the surface conductivity is controlled by the diffuse layer with a fraction of counterions $(1-f)$ and a mobility of the counterions $\beta_{(+)}$, which is equal to the mobility of the counterions in the bulk pore water $\beta_{(+)}(\text{Na}^+, 25\text{ °C}) = 5.2\times10^{-8}$ m^{2}s^{-1}V^{-1}. The quadrature conductivity is given by $\sigma'' = (1/F)\beta_{(+)}^S f Q_V$ with $\beta_{(+)}^S(25\text{ °C, Na}^+) = 1.5 \times 10^{-10}$ m^{2}s^{-1}V^{-1} the mobility of the counterions in the Stern layer. Use the data in Table 3.1 (the in-phase conductivity and the phase) to determine S9, S16 and S22 for the three core samples, the value of the formation factor, the surface conductivity, the cementation exponent m, the fraction of counterions in the Stern layer f, the total charge per unit pore volume Q_V, and the cation exchange capacity (CEC).

Table 3.1. *Conductivity* (σ') *and phase* (φ) *data at 1 Hz for different pore fluid salinities. The conductivity* σ_w *denotes the pore water conductivity.*

Sample	Effluent σ_w (S m^{-1})	σ' (S m^{-1})	φ (mrad)
S9	4.95e − 3	3.98e − 3	−37.9
S9	4.95e − 3	6.45e − 3	−28.1
S9	3.82e − 2	1.47e − 2	−34.7
S9	1.21e − 1	3.63e − 2	−18.2
S9	3.23e − 1	8.66e − 2	−9.4
S9	8.73e − 1	2.20e − 1	−4.8
S9	2.33	5.5e − 1	−2.5
S16	4.95e − 3	1.02e − 2	−
S16	5.80e − 3	1.03e − 2	−
S16	3.88e − 2	1.67e − 2	−28.4
S16	1.21e − 1	3.07e − 2	−25.9
S16	3.22e − 1	6.49e − 2	−17.9
S16	8.83e − 1	1.63e − 1	−10.2
S16	2.35	3.9e − 1	−5.6
S22	9.53e − 3	4.94e − 2	−
S22	7.09e − 3	3.34e − 2	−
S22	4.01e − 2	4.15e − 2	−31.8
S22	1.22e − 1	6.73e − 2	−24.2
S22	3.27e − 1	1.18e − 1	−13.5
S22	8.79e − 1	2.55e − 1	−6.1
S22	2.34	5.1e − 1	−3.1

The porosity data are Sample S9: 0.48, Sample S16: 0.49, and Sample S22: 0.43. Complete this table. Note: to interpret the conductivity data, it is better to analyze the conductivity equation in a log-log plot using $Y = \log \sigma'$, $X = \log \sigma_w$; therefore, using a non-linear fit $Y = \log(a\,10^X + b)$ rather than $Y = aX + b$. Explain why.

Exercise 3.3. Interpretation of a geobattery experiment. We selected two ISO (Industry Standard Object, Part number 44615K466 of the McMaster-Carr catalog) to serve as proxies for UXO as recommended by SERDP (Strategic Environmental Research and Development Program). These two ISOs were buried in a sandbox, partially filled with a uniform sand (Figure 3.14a). The level of water was adjusted in such a way that the two ISOs were inside the capillary fringe, one vertical and one horizontal. The self-potential measurements were performed with two non-polarizing Pb/PbCl electrodes and a high-input impedance voltmeter (MX20,

108   *Laboratory investigations*

(a) Geometry of the sandbox

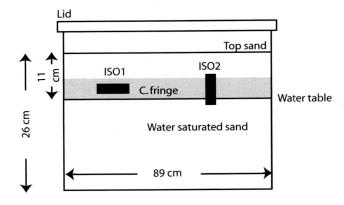

(b) Self-potential map (in mV, top surface)

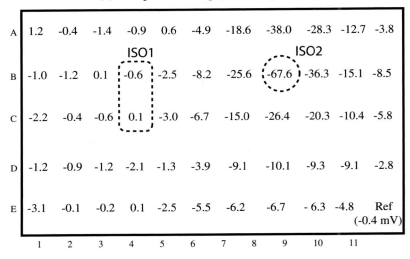

	1	2	3	4	5	6	7	8	9	10	11
A	1.2	-0.4	-1.4	-0.9	0.6	-4.9	-18.6	-38.0	-28.3	-12.7	-3.8
B	-1.0	-1.2	0.1	-0.6	-2.5	-8.2	-25.6	-67.6	-36.3	-15.1	-8.5
C	-2.2	-0.4	-0.6	0.1	-3.0	-6.7	-15.0	-26.4	-20.3	-10.4	-5.8
D	-1.2	-0.9	-1.2	-2.1	-1.3	-3.9	-9.1	-10.1	-9.3	-9.1	-2.8
E	-3.1	-0.1	-0.2	0.1	-2.5	-5.5	-6.2	-6.7	-6.3	-4.8	Ref (-0.4 mV)

Figure 3.14. Sandbox experiment used to investigate the redox potential contribution to self-potential signals. (a) Two ISO are buried in the capillary fringe, one completely inside the capillary fringe (ISO1) and the second is crossing entirely the capillary fringe. (b) These self-potential data were collected 37 days after burying the ISOs. The distance between the measurement stations is 7.8 cm along x and y. The potential value at the reference station (-0.4 mV) indicates the difference of potential between the scanning electrode and the reference electrode (connected to the COM of the voltmeter) with the electrodes face-to-face. The ISOs are tubes 10.2 cm long and 3.2 cm in diameter.

100 MOhm of input impedance, sensitivity 0.1 mV). A self-potential map was performed right after the burial of the two ISOs. No signals could be detected. The self-potential map was performed again after 37 days, and the measurements are shown in Figure 3.14b. Interpret this self-potential map. The water used in the box was tap water (conductivity 262 μS cm^{-1}, pH $= 7.8$).

4

Forward and inverse modeling

In this chapter, we introduce two very distinct ways of inverting self-potential data. One approach is to invert for the source current density distribution and then to interpret this source current density in terms of parameters that are relevant to hydrogeochemistry (e.g., hydraulic head, redox potential, salinity). The second approach is to fully couple the self-potential inverse problem with the physics of the primary flow problem, solving the non-reactive or reactive transport equations and performing the inverse problem with either in a stochastic or determinsitic gradient-based type approach. Several routines are provided to the readers on the website associated with this book. In this chapter, we will focus mostly with the self-potential anomalies associated with the flow of the ground water.

4.1 Position of the problem

In early works, the generation of self-potential signals associated with ground water flow (the streaming potential) has been numerically interpreted using linear semi-empirical models connecting the electrical potential at the ground surface, and the thickness of the vadose zone or the pizometric level. Aubert and Atangana (1996) proposed a first-order empirical method, termed the "SPS method," to linearly relate the self-potential signals to the thickness of the vadose zone. Fournier (1989) presented another model connecting the self-potential signals to the piezometric level for unconfined aquifers. Both approaches are applicable only for homogeneous unconfined aquifers and vadose zones. Sill (1983) introduced a rigorous approach to numerically simulate the occurrence of streaming potentials in heterogeneous media, taking into account the modeling of the ground water flow and the electrical conductivity distribution. His approach combined the numerical solution of ground water flow equation (primary flow problem) combined with the solution of the Poisson equation (see Chapter 1) for the electrical potential using the finite difference method. His approach was the first able to simulate self-potential signals in complex hydrogeosystems. Later, such hydroelectric semi-coupled modeling has been performed with the help of various numerical methods such as the finite element approach of Wurmstich *et al.* (1991) and Revil *et al.* (1999b), and more recently using the finite volume approach of Sheffer and Oldenburg (2007).

The inverse problem to localize the causative source of self-potential signals associated with ground water flow or the occurrence of ore bodies and associated redox processes also has a long history. Early works were based on analytical solutions for simple geometries (see, for example, Nourbehecht (1963), Paul (1965), Fitterman (1976) and Rao and Babu (1984)) or the use of the cross-correlation approach with monopoles (see Patella (1997)) and dipoles (see Revil *et al.* (2001)). We will not review these approaches in our book, because much better algorithms have been proposed in the past decade. For instance, Minsley *et al.* (2007a, b) and Jardani *et al.* (2008) developed gradient-based algorithms to invert self-potential signals in terms of the distribution of the source current density; this is the divergence of the source current density in the case of Minsley *et al.* (2007a) and the direct inversion of the source current density vector distribution, in the case of Jardani *et al.* (2008). Using the physical model developed by Arora *et al.* (2007), Linde and Revil (2007) solved the Poisson equation to determine the distribution of the redox potential versus depth over a contaminant plume associated with the presence of a municipal landfill. Mendonça (2008) developed an algorithm to invert self-potential signals to delineate the position of ore bodies at depth using the geobattery model developed by Stoll *et al.* (1995). These models generally include the distribution of the electrical resistivity as prior information in the inversion process. This is definitely important in the case of ore deposits, because of the strong contrast in the electrical resistivity between the ore body and the host material. We will show, below, that these approaches can be used to determine some information regarding the ground water flow pattern, as well.

We saw in Chapter 1 that the forward problem for the self-potential involves solving a Poisson equation. We show how to compute streaming potential signals in saturated flow conditions by solving the two semi-coupled primary flow problem (ground water flow) and the secondary problem (solving for the electrical potential) in transient conditions (Figure 4.1). In an anisotropic, heterogeneous and saturated porous material, the hydraulic problem can be expressed using the following continuity and constitutive equations:

$$S_s \frac{\partial h}{\partial t} + \nabla \cdot \mathbf{U} = Q_s, \tag{4.1}$$

$$\mathbf{U} = -\overline{\overline{K}} \nabla h, \tag{4.2}$$

$$\overline{\overline{K}} = \begin{bmatrix} K_x & 0 & 0 \\ 0 & K_y & 0 \\ 0 & 0 & K_z \end{bmatrix}. \tag{4.3}$$

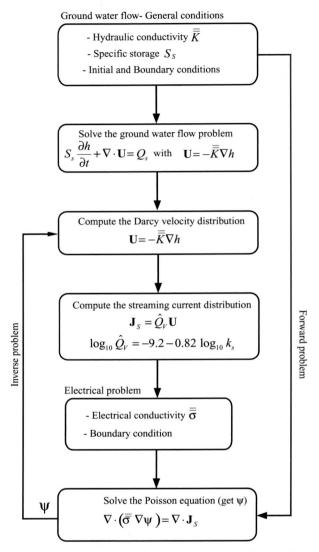

Figure 4.1. Flowchart for the simulation and inversion of the self-potential data associated with ground water flow. The vector field $\mathbf{U}$ and $\mathbf{J}_S$ denote the Darcy velocity and the source current density, respectively, while the potentials h and ψ denote the hydraulic head and the electrical potential, respectively. In saturated conditions, the material properties entering the forward modeling are the hydraulic conductivity and electrical conductivity tensors and the specific storage coefficient.

Equation (4.1) corresponds to the conservation equation for the mass of the pore water and Eq. (4.2) corresponds to Darcy's law. In these equations $\mathbf{U}$ denotes the Darcy velocity (in m s^{-1}) and h denotes the hydraulic head (in m), which is the main variable to solve in the primary flow problem. Hydraulic diffusivity depends on the ratio between the specific storage S_s (in m) and the hydraulic conductivity tensor $\overline{\overline{K}}$ (in m s^{-1}). Equations (4.1) and (4.2) lead to a diffusion equation for h, which is subject to the following initial and boundary conditions:

$$h = h_D \text{ at } \Gamma_D, \tag{4.4}$$

$$-\boldsymbol{n}.\overline{\overline{K}}\nabla h = q_0 \text{ at } \Gamma_N, \tag{4.5}$$

$$h = 0, \forall \mathbf{x} \text{ at } t = 0, \tag{4.6}$$

Equations (4.4) and (4.5) correspond to Dirichlet and Neumann boundary conditions, respectively. The hydraulic head h_0 denotes the head fixed at the boundary Γ_D, q_0 is the hydraulic flux (m^2 s^{-1}) assumed at the Neumann boundary Γ_N, and $\mathbf{n}$ is the unit vector normal to the boundary Γ_N. Equation (4.6) represents an initial condition.

The field equation to solve for the electrical problem is expressed (see Chapter 1) by

$$\nabla \cdot (\overline{\overline{\sigma}}\nabla\psi) = \nabla \cdot \mathbf{J}_S \tag{4.7}$$

$$\mathbf{J}_S = \hat{Q}_V \mathbf{U}, \tag{4.8}$$

where ψ denotes the self-potential field and $\overline{\overline{\sigma}}$ the d.c. electrical conductivity tensor. Equation (4.7) is solved with the following boundary conditions:

$$\psi = 0 \text{ at } \Gamma_D, \tag{4.9}$$

$$-\mathbf{n} \cdot [\overline{\overline{\sigma}}\nabla\psi - \mathbf{J}_S] = 0 \text{ at } \Gamma_N. \tag{4.10}$$

The Neumann boundary condition Γ_N is imposed at the insulating air–ground interface and a Dirichlet boundary condition Γ_D is imposed at the other boundaries. The physical model described above indicates that the source current density responsible for the occurrence of the self-potential signals is strongly related to the ground water flow pathways; see Revil and Jardani (2010a).

Similarly, there are several ways to invert self-potential data to determine the properties the ground, or to locate the source of the self-potential signals. The former has not yet attracted too much attention thus far. It consists of measuring the electric potential ψ, to determine $\Im = \nabla \cdot \mathbf{J}_S$ (by modeling the ground water flow and computing the streaming current density, which requires knowledge of the hydraulic conductivity field and the hydraulic boundary conditions), and to then determine the distribution of the electrical conductivity $\overline{\overline{\sigma}}$ by use of Eq. (4.7). This inverse problem is non-linear, and equivalent to the case of electrical resistivity tomography, except that internal sources of current are used, rather than imposed currents. Slob *et al.* (2010) recently proposed the use of a cross-correlation approach of self-potential fluctuations to determine the resistivity between a set of electrodes. Jardani and Revil (2009) used a stochastic approach to determine the probability density distributions of the resistivity values of a set of geological units. In both cases, self-potential signals and their fluctuations can be used to image the resistivity contrasts of the ground without injecting any current.

A more classical approach is to measure ψ, assume that the spatial or spatio-temporal distribution of the electrical conductivity is known (or independently measured through electrical resistivity tomography), and to then compute the volumetric source density $\Im = \nabla \cdot \mathbf{J}_S$ or the source current density $\mathbf{J}_S$. In this case, the corresponding inverse problem is linear and a variety of methods can be used to solve it. Like for any potential field problem (e.g., gravity, magnetism), the inverse problem is ill-posed and generally underdetermined. This self-potential inverse problem is fundamentally similar to that of electroencephalography (EEG), because in both cases, the same underlying linear (Poisson) elliptic equation is solved; see Grech *et al.* (2008).

The self-potential problem is known to be non-unique; see Pascal-Marquis *et al.* (2002) and Trujillo-Barreto *et al.* (2004) for EEG, and Minsley *et al.* (2007a) and Jardani *et al.* (2008) for the self-potential inverse problem. In other words, an infinite number of source configurations can produce the same measured self-potential field. Therefore it is important to add additional constraints or prior information regarding the number of sources or the spatial extension of the electrical sources to reduce the parameter space of the solution. Therefore, there is a strong interest in fully coupling the inversion of the self-potential data to the primary flow problem and other geophysical methods. In Section 4.3 we will present a general stochastic approach to address such a fully coupled inverse problem.

4.2 Gradient-based approaches and their limitations

4.2.1 The kernel

Using Green's method of integration, the electrical potential distribution can be written as

$$\psi(P) = \frac{1}{2\pi} \int_\Omega \rho(M) \frac{\nabla \cdot \mathbf{J}_S(M)}{MP} dV + \frac{1}{2\pi} \int_\Omega \frac{\nabla \rho(M)}{\rho(M)} \cdot \frac{\mathbf{E}(M)}{MP} dV, \quad (4.11)$$

where $x = MP$ is the distance from the source, located at position $M(r')$, to the electrode located at position $P(r)$ where the self-potential signal is recorded and Ω represents the conducting subsurface. The first term of the right-hand side of Eq. (4.11) corresponds to the primary source term while the second term corresponds to secondary sources associated with heterogeneities in the distribution of the electrical resistivity in the medium. Usually (but not necessarily), we know the resistivity and want to focus the inversion on the primary source current density.

It is also possible to rewrite Eq. (4.11) as the following convolution product, according to Spinelli (1999):

$$\psi(P) = \int_\Omega \mathbf{K}(P, M)\mathbf{J}_S(M)dV, \qquad (4.12)$$

where $\mathbf{K}(P, M)$ denotes a linear mapping function termed the kernel (see Grech *et al.* (2008)). The elements of the kernel are the Green's functions connecting the self-potential data at a set of measurement stations P located at the ground surface (possibly in boreholes) and the sources of current density at a set of source points M located in the conducting ground. The kernel $\mathbf{K}$ depends on the number of measurement stations at the ground surface, the number of discretized elements in which the source current density is going to be determined, and the resistivity distribution. For example, the resistivity distribution can be characterized from a number of geological strata with distinct resistivity values. Generally, the subsurface is discretized. That said, other strategies such as using wavelets are possible.

The self-potential is measured at only a few points (a few tens to a few thousands of self-potential stations are usually used in the field). We use the letter N to represent the number of these self-potential stations. The current density $\mathbf{J}_S$ is discretized over M elements. Each element of the discretized grid can be characterized by a small volume element (cell-based approach). It is possible to characterize the source by the distribution of the volumetric current density $\Im$ or by a vectorial current density. In our case, the forward problem is related to the determination of the kernel $\mathbf{K}$ in Eq. (4.12). The computation of the Green's functions entering $\mathbf{K}$ incorporates the resistivity distribution of the model by solving the Poisson equation numerically for $3M$ elementary dipoles using the finite element method. In our case, the 3D (three-dimensional) inversion corresponds to the determination of the three components of the current density for each discretized element of the medium (two components for 2D inversion; see Jardani *et al.* (2007b) and Revil *et al.* (2011b)). Because $3M \gg N$, the inverse problem is strongly underdetermined.

4.2.2 Finite element formulation

In this book, we use the finite element method to solve the field (partial differential) Eqs. (4.1)–(4.10). The unknown fields are the hydraulic head h (in m) and the electrical potential ψ (in V). In order to set up a variational formulation of both equations, we define the following functional spaces:

$$H^1(\Omega) := \{v \in L^2(\Omega), \nabla v \in L^2(\Omega)\}, \tag{4.13}$$

$$H^2(\Omega) := \left\{v \in H^1(\Omega), \frac{\partial^2 v}{\partial x^2}, \frac{\partial^2 v}{\partial y^2}, \frac{\partial^2 v}{\partial z^2}, \frac{\partial^2 v}{\partial x \partial y \partial z} \in L^2(\Omega)\right\}, \tag{4.14}$$

$$W(\Omega) := \{v \in H^1(\Omega), v|\Gamma_D = 0\}. \tag{4.15}$$

To write the variational formulation, we multiply Eq. (4.1) by a test function v and we integrate over domain Ω, giving

$$\int_\Omega S_s \frac{\partial h}{\partial t} v + \int_\Omega \nabla \cdot (-\overline{\overline{K}} \nabla h) v = \int_\Omega Q_s v. \tag{4.16}$$

In terms of Cartesian coordinates, we have:

$$\int_\Omega \nabla \cdot (-\overline{\overline{K}} \nabla h) v = \int_\Omega \nabla \cdot \left(-\begin{pmatrix} K_x & 0 & 0 \\ 0 & K_y & 0 \\ 0 & 0 & K_z \end{pmatrix} \begin{pmatrix} \frac{\partial h}{\partial x} \\ \frac{\partial h}{\partial y} \\ \frac{\partial h}{\partial z} \end{pmatrix}\right) v = \int_\Omega \nabla \cdot \begin{pmatrix} -K_x \frac{\partial h}{\partial x} \\ -K_y \frac{\partial h}{\partial y} \\ -K_z \frac{\partial h}{\partial z} \end{pmatrix} v$$

$$= \int_\Omega -\frac{\partial}{\partial x}\left(K_x \frac{\partial h}{\partial x}\right) v - \frac{\partial}{\partial y}\left(K_y \frac{\partial h}{\partial y}\right) v$$

$$-\frac{\partial}{\partial z}\left(K_z \frac{\partial h}{\partial z}\right) v. \tag{4.17}$$

By applying Green's theorem (see Hildebrand (1965)):

$$\int_\Omega \nabla \cdot (\overline{\overline{K}} \nabla h) v = -\int_\Omega K_x \frac{\partial h}{\partial x}\frac{\partial v}{\partial x} - \int_\Omega K_y \frac{\partial h}{\partial y}\frac{\partial v}{\partial y} - \int_\Omega K_z \frac{\partial h}{\partial z}\frac{\partial v}{\partial z}$$

$$+ \int_{\partial\Omega} K_x \frac{\partial h}{\partial x} v n_x + \int_{\partial\Omega} K_y \frac{\partial h}{\partial y} v n_y + \int_{\partial\Omega} K_z \frac{\partial h}{\partial z} v n_z$$

$$= -\int_\Omega \overline{\overline{K}} \nabla h \nabla v + \int_{\partial\Omega} \overline{\overline{K}} \nabla h . \mathbf{n} v. \tag{4.18}$$

Finally, Eq. (4.16) is equivalent to the following variational formulation, which consists in finding $h \in H^2(\Omega)$), with $h = h_D$ at Γ_D

$$\int_\Omega S_s \frac{\partial h}{\partial t} v + \int_\Omega \overline{\overline{K}} \nabla h \nabla v = \int_\Omega Q_s v + \int_{\Gamma_N} \overline{\overline{K}} \nabla h . \mathbf{n} v \quad \forall v \in W(\Omega). \tag{4.19}$$

In the discretization step, we replace the space $H^1(\Omega)$ by the discrete space $W^d(\Omega) \subset H^1(\Omega)$ with $W^d(\Omega) := \{v \in C^0(\Omega), v|T_i \in P_1\}, \forall T_i \in \Im_d$, where $\Im_d$

denotes a triangular mesh of Ω and $P_1(\Omega) := \{v, v \text{ polynomial of degree} \leq 1 \text{on } \Omega\}$ and $C^0(\Omega)$ represents the space of continuous functions on Ω. The functions $\varphi_{i=1,m}(m = \dim(W^d))$ are used as basis of P_1. A discrete formulation of h can be written as:

$$h^d = \sum_{i=1}^{m} h_i \varphi_i. \tag{4.20}$$

So, we can rewrite the discrete variational formulation of the hydraulic problem as:

$$\int_{\Omega} S_s \frac{\partial h^d}{\partial t} v^d + \int_{\Omega} \overline{\overline{K}} \nabla h^d \nabla v^d = \int_{\Omega} Q_s v^d + \int_{\Gamma_N} \underbrace{\overline{\overline{K}} \nabla h^d . \boldsymbol{n}}_{=U_n} v^d \; \forall v^d \in W^d(\Omega). \tag{4.21}$$

By rewriting the above discrete formulation with respect to Eq. (4.21), and by taking $v^d = \varphi_i$ we have

$$\frac{\partial}{\partial t} \sum_{j=1}^{m} h_j \int_{\Omega} S_s \varphi_i \varphi_j + \sum_{j=1}^{m} h_j \int_{\Omega} \overline{\overline{K}} \nabla \varphi_i \nabla \varphi_j = \int_{\Omega} Q_s \varphi_j + \int_{\Gamma_N} U_n \varphi_j. \tag{4.22}$$

In terms of matrices, we write:

$$\frac{\partial}{\partial t} \sum_{j=1}^{m} h_j \int_{\Omega} S_s \varphi_i \varphi_j + \sum_{j=1}^{m} h_j \int_{\Omega} \overline{\overline{K}} \nabla \varphi_i \nabla \varphi_j = \int_{\Omega} Q_s \varphi_j + \int_{\Gamma_N} U_n \varphi_j, \tag{4.23}$$

$$\frac{\partial \mathbf{M}}{\partial t} h + \mathbf{K}^h h = \overline{\mathbf{Q}}, \tag{4.24}$$

$$\mathbf{M}_{ij} = \int_{\Omega} S_s \varphi_i \varphi_j, \tag{4.25}$$

$$\mathbf{K}_{ij}^h = \int_{\Omega} \overline{\overline{K}} \nabla \varphi_i \nabla \varphi_j, \tag{4.26}$$

$$\overline{\mathbf{Q}} = \int_{\Omega} Q_s \varphi_j + \int_{\Gamma_N} U_n \varphi_j, \tag{4.27}$$

$$h = (h_1, h_2 \ldots h_m)^T. \tag{4.28}$$

We use a θ-scheme to solve Eq. (4.24):

$$\left[\mathbf{M} + \theta \Delta t \mathbf{K}_{ij}^h\right] h^{n+1} = \left[\mathbf{M} - (1-\theta)\Delta t \mathbf{K}_{ij}^h\right] h^n + \Delta t \overline{\mathbf{Q}}^{n+\theta}, \tag{4.29}$$

$$\overline{\mathbf{Q}}^{n+\theta} = \theta \overline{\mathbf{Q}}^{n+1} + (1-\theta)\overline{\mathbf{Q}}^n, \qquad 0 \leq \theta \leq 1, \tag{4.30}$$

where $h(0) = h^0$ is the given initial condition, and h^n corresponds to the hydraulic head distribution at time n. Computing the hydraulic head permits the determination of the electrokinetic source $q_s = \nabla \cdot \mathbf{J}_S$ (see Eq. (4.8)), which will be used as source term of Poisson's equation after rewriting its variational formulation under a discretization following the same method used above for the hydraulic problem:

$$\int_\Omega \overline{\overline{\sigma}} \nabla \left(\sum_{j=1}^m \psi_i \varphi_i \right) \nabla \varphi_j = \int_\Omega q_s \varphi_j + \int_{\Gamma_N} \underbrace{\overline{\overline{\sigma}} \nabla \left(\sum_{j=1}^m \psi_i \varphi_i \right) \mathbf{n} \varphi_j}_{=J n}, \quad (4.31)$$

$$\sum_{j=1}^m \psi_i \int_\Omega \overline{\overline{\sigma}} \nabla \varphi_i \nabla \varphi_j = \int_\Omega q_s \varphi_j + \int_{\Gamma_N} J_n \varphi_j. \quad (4.32)$$

In terms of matrices, we write:

$$\mathbf{K}^e \psi = \overline{\mathbf{Q}}^e, \quad (4.33)$$

$$K_{ij}^e = \int_\Omega \overline{\overline{\sigma}} \nabla \varphi_i \nabla \varphi_j, \quad (4.34)$$

$$\overline{\mathbf{Q}}^e = \int_\Omega q_s \varphi_j + \int_{\Gamma_N} J_n \varphi_j. \quad (4.35)$$

Equation (4.33) is solved by using Gaussian elimination with partial pivoting to determine the distribution of the self-potential. The book contains a code that has been implemented with Matlab® to solve the direct and inverse problems in the anistropic conditions of the electrical and hydraulic properties of the materials.

4.2.3 Classical Tikhonov regularization of the inverse problem

The first criterion for inversion requires that the predicted data must fit, or agree with, the observed data. This condition requires the use of a quantitative measure of the agreement between both data sets, known as the data misfit function C_d. From Eq. (4.12), this function is defined by

$$C_d = \| \mathbf{W}_d (\mathbf{Km} - \mathbf{\Psi}^{obs}) \|, \quad (4.36)$$

where $\|\mathbf{v}\| = (\mathbf{v}^T \mathbf{v})^{1/2}$ denotes the Euclidian (L_2) norm, $\mathbf{K} = (\mathbf{K}_{ij}^x, \mathbf{K}_{ij}^y, \mathbf{K}_{ij}^z)$ is the $N \times 3M$ kernel matrix corresponding to the self-potential, which can be measured by each component of the sources coordinates $\mathbf{m} = (\mathbf{J}_i^x, \mathbf{J}_i^y, \mathbf{J}_i^z)$ (N is the number of self-potential stations, while M is the disctretized elements composing the ground, $3M$ represents the number of elementary current sources to consider – two horizontal components and one vertical component), $\boldsymbol{\Psi}^{\mathrm{obs}}$ is vector of N elements corresponding to the self-potential data measured at the ground surface, or in boreholes, and $\mathbf{W}_d = \mathrm{diag}\{1/\varepsilon_1, \ldots, 1/\varepsilon_N\}$ is a square $N \times N$ diagonal weighting matrix. Elements along the diagonal of this matrix are the reciprocal of the standard deviation σ_ι squared, $\varepsilon_i = \sigma_\iota^2$. Linde *et al.* (2007b) have shown that the probability distribution of the self-potential measurements in the field corresponds to a Gaussian distribution. The inversion of self-potential data is ill-posed, and does not have a unique solution. This complication arises because there is a finite number of inaccurate data describing the model response, and the number of observations is much smaller than the number of unknowns, $3\,M \gg N$.

To reduce the number of solutions that equally reproduce the observed data, an additional criterion must be introduced to distinguish the model that most likely represents the subsurface source distribution among the range of possible models. Measuring the structural complexity of a model serves as a good criterion for this purpose. A model objective function C_m, is introduced,

$$C_m = \|\mathbf{W}_m(\mathbf{m} - \mathbf{m}_0)\|, \tag{4.37}$$

where $\mathbf{W}_m$ is a $3(M-2) \times 3M$ weighting matrix (e.g., the flatness matrix or the differential Laplacian operator), $\mathbf{m}$ is the vector of $3M$ model parameters, and $\boldsymbol{m}_0$ is a reference model. If we use a null distribution of prior information ($\boldsymbol{m}_0 = 0$), the previous model is similar to a damped weighted linear least squares or biased linear estimation problem. Using the differential Laplacian operator (second-order derivative), the matrix $\mathbf{W}_m$ is given by

$$\mathbf{W}_{\mathrm{m}}^2 = \begin{bmatrix} 1 & -2 & 1 & 0 & \ldots & 0 \\ 0 & 1 & -2 & 1 & \ldots & 0 \\ \vdots & & \ddots & \ddots & & \vdots \\ \vdots & & & \ddots & \ddots & \vdots \\ 0 & \ldots & 1 & -2 & 1 & 0 \\ 0 & \ldots & 0 & 1 & -2 & 1 \end{bmatrix}. \tag{4.38}$$

This operator smoothes the final result of the inversion. The criteria of data misfit and model objective function place different, and competing, requirements on the models. The best model will be one that minimizes the model objective function, C_m, while fitting the data within an acceptable range of data misfit, C_d. To harness the benefits of both criteria, it is necessary to determine how to control the contribution of each in order to obtain the best solution. The roles of data misfit and model objective function are balanced using Tikhonov regularization (see Tikhonov and Arsenin (1977)) through a global objective function, C, defined as,

$$C = \|\mathbf{W}_d(\mathbf{Km} - \boldsymbol{\Psi}^{\mathrm{obs}})\| + \lambda\|\mathbf{W}_m(\mathbf{m} - \mathbf{m}_0)\|, \tag{4.39}$$

where λ is a regularization parameter under the constraint that $0 < \lambda < \infty$. The value of the regularization parameter balances the data misfit term and the regularizer (last term of Eq. (4.39)).

The previous problem is a simple linear problem and the minimization of the cost function C yields the following solution:

$$m^*(\lambda) = \left[\mathbf{K}^T \left(\mathbf{W}_d^T \mathbf{W}_d \right) \mathbf{K} + \lambda \left(\mathbf{W}_m^T \mathbf{W}_m \right) \right]^{-1}$$
$$\times \left(\mathbf{K}^T \left(\mathbf{W}_d^T \mathbf{W}_d \right) \mathbf{\Psi}^{\text{obs}} + \lambda \left(\mathbf{W}_m^T \mathbf{W}_m \right) \mathbf{m}_0 \right). \qquad (4.40)$$

One of the important questions regarding the regularized inversion is the particular choice of the regularization parameter λ. A popular approach for choosing the regularization parameter is the *L*-curve criterion of Hansen (1998). The *L*-curve is a plot of the norm of the regularized smoothing solutions C_m, versus the norm of the residuals of data misfit function C_d. This dependence often has an L-shaped form, which reflects the heuristics that for large λs the residual increases without reducing the model norm of the solution much, while for small λs the norm of the solutions increases rapidly without much decrease in the data residual. Thus, the best regularization parameter should lie on the corner of the *L*-curve. We can also use a second approach called generalized cross-validation (GCV). This geostatistical method is based on the assumption that the data $\mathbf{\Psi}^{\text{obs}}$ is affected by normally distributed noise (see Wahba and Wang (1995)). The optimum λ for the GCV method corresponds to the minimum of the following function $V(\lambda)$:

$$V(\lambda) = \frac{\| \mathbf{K}\mathbf{m}^*(\lambda) - \mathbf{\Psi}^{\text{obs}} \|^2}{\left[\text{Trace} \left(\mathbf{I} - \mathbf{A}(\lambda)\mathbf{K} \right) \right]^2}, \qquad (4.41)$$

where $\mathbf{A}(\lambda) = \mathbf{K}\mathbf{W}_d[\mathbf{K}^T (\mathbf{W}_d^T \mathbf{W}_d)\mathbf{K} + \lambda(\mathbf{W}_m^T \mathbf{W}_m)]^{-1}\mathbf{K}^T \mathbf{W}_d^T$, $\mathbf{I}$ is the identity matrix, $m^*(\lambda)$ is the minimum of the cost function C, and "Trace" denotes the trace of the matrix. Numerator and denominator of generalized cross-validation criteria can be regarded as representing the variance of the estimated observation error, and the bias resulting from the regularization term, respectively, for the practicable reason that the singular value decomposition approach was used to compute the matrix $\mathbf{A}(\lambda)$.

As the self-potential sources are mainly dipolar in nature, according to Revil *et al.* (2001) and Minsley *et al.* (2007a), the Green's function decays as $1/r^2$ in an homogeneous volume, where r is the distance between the source and the location of the measurement stations. Therefore, because the sensibility of the self-potential field decays quickly with the distance, the inversion of surface self-potential data will generate a shallow source current density distribution. The resulting solution in terms of ground water flow, would not be hydrogeologically or hydromechanically meaningful. This is true, for example, when several aquifers contribute to the self-potential signals measured at the ground surface (see Titov *et al.* (2005) for a field example). The use of a depth-weighting function is then necessary to counteract this drawback. To provide cells at depth, with equal probability of obtaining non-zero values during inversion, a generalized depth-weighting function can be incorporated into the model objective function. The depth-weighting ($N \times 3M$) matrix is designed to match the overall sensitivity of the data set to a particular cell (see, for example, Spinelli (1999))

$$S = \text{diag}\left(\frac{1}{N} \sqrt{\sum_{j=1}^{N} (K_{ij})^2} \right). \tag{4.42}$$

The solution of the inverse problem is then given by Tikhonov and Arsenin, (1977) as

$$\mathbf{m}_w = \left[\mathbf{K}_w^T (\mathbf{W}_d^T \mathbf{W}_d) \mathbf{K}_w + \lambda (\mathbf{W}_m^T \mathbf{W}_m) \right]^{-1}$$
$$\times \left(\mathbf{K}_w^T (\mathbf{W}_d^T \mathbf{W}_d) \mathbf{\Psi}^{\text{obs}} + \lambda (\mathbf{W}_m^T \mathbf{W}_m) \mathbf{m}_0 \right), \tag{4.43}$$

where $\mathbf{K}_w = \mathbf{K} \mathbf{S}^{-1}$. The model is then given by $\mathbf{m}^* = \mathbf{S} \mathbf{m_w}$, as the objective function can be reformulated in the standard form

$$C = \|\overline{\mathbf{K}\mathbf{m}} - \overline{\psi}^{\text{obs}}\|_2^2 + \lambda^2 \|\overline{\mathbf{m}} - \overline{\mathbf{m}}_0\|_2, \tag{4.44}$$

with $\mathbf{m} = \boldsymbol{L_p} \overline{\mathbf{m}} + \mathbf{m}_0$, and where $\overline{\mathbf{K}}$, $\overline{\mathbf{m}}$, $\overline{\psi}^{\text{obs}}$ and $\boldsymbol{L_p}$ are the matrices deduced by using Elédn's algorithm, which is based on the QR factorizations. In this case, the solution of the problem corresponding to the minimum of the cost function C is given by Hansen (1998) as

$$\overline{\mathbf{m}}^* = \left[\overline{\mathbf{K}}^T \overline{\mathbf{K}} + \lambda^2 I \right]^{-1} \left(\overline{\mathbf{K}}^T \overline{\psi}^{\text{obs}} + \lambda \overline{\mathbf{m}}_0 \right). \tag{4.45}$$

In absence of a prior model ($\overline{\mathbf{m}}_0 = 0$), we obtain

$$\overline{\mathbf{m}}^* = \left[\overline{\mathbf{K}}^T \overline{\mathbf{K}} + \lambda^2 I \right]^{-1} \left(\overline{\mathbf{K}}^T \overline{\psi}^{\text{obs}} \right). \tag{4.46}$$

The computation of this minimum demands a lot of time, due to the inversion of a $N \times N$ matrix. To reduce the computation time, we used a singular value decomposition method, which consists of rewriting the solution $\overline{\mathbf{m}}^*$ under the following form:

$$\overline{\mathbf{m}}^* = \left(\frac{\sigma_i^2}{\sigma_i^2 + \lambda^2} \right) \frac{u_i^T \overline{\psi}^{\text{obs}}}{\sigma_i} v_i \tag{4.47}$$

$$\overline{K} = \sum_{i=1}^{Q} u_i \sigma_i v_i^T, \quad \text{where } Q = \min(N, M), \tag{4.48}$$

where, u_i and v_i denote the left and right singular vectors and σ_i are the singular values which are positive and appear in decreasing order. This Tikhonov solution is more efficient than Eq. (4.46) and is dependent on the choice of the regularization parameter.

4.2.4 Localization of compact self-potential sources

Hydraulic fracturing would correspond to a very compact electrokinetic source. In this case, compactness can be used as a criterion to locate more properly the extension of the source; see Minsley *et al.* (2007a). As a side note, compact sources can also be obtained if we use a first-order differential operator (see Jardani *et al.* (2008))

$$
\mathbf{W}_m^1 =
\begin{bmatrix}
1 & -1 & 0 & 0 & \cdots & 0 \\
0 & 1 & -1 & 0 & \cdots & 0 \\
\vdots & & \ddots & \ddots & & \vdots \\
\vdots & & & \ddots & \ddots & \vdots \\
0 & \cdots & 0 & 1 & -1 & 0 \\
0 & \cdots & 0 & 0 & 1 & -1
\end{bmatrix}
\tag{4.49}
$$

rather than the Laplacian operator given by Eq. (4.38). We now return to the development of an iterative compactness operator, following Minsley *et al.* (2007a). The matrix form of Eq. (4.33) is

$$
\mathbf{K\Psi} = \mathbf{s},
\tag{4.50}
$$

where $\mathbf{K}$ denotes the kernel matrix, $\mathbf{s}$ denotes a vector containing the M volumetric source current density terms $\Im(\mathbf{x}, t)$ (in A m^{-3}) and $\mathbf{\Psi}$ is the vector of electric potential observations at the N receivers locations. At the boundaries of the domain, we used the following boundary conditions: a Neumann condition is used at the interface between the model and the padding layer; at the outer edge of the padding layer, a Dirichlet's boundary condition is used. The electric potential at the outer edge of the padding layer tends to zero thus simulating an infinite domain.

The inverse problem involves reconstructing the spatial distribution of the volumetric source field (right-hand side of Eq. (4.50)) at each time t, through the optimization of the objective functions following the approach outlined by Minsley (2007):

$$
C = \|\mathbf{W}_d(\mathbf{Gs} - \mathbf{\Psi}^{obs})\| + \lambda\|\Lambda\mathbf{s}\|,
\tag{4.51}
$$

$$
\Lambda = \mathrm{diag}\sqrt{\sum_{i-1}^{N} G_{kj}^{T^2}},
\tag{4.52}
$$

where $\mathbf{G} = \mathbf{P}\mathbf{K}^{-1}$ denotes the Green's matrix ($N \times M$) computed as the product of the inverse kernel matrix times a sparse selector operator matrix $\mathbf{P}_{(N \times M)}$ that contains a single 1 on each row in the column corresponding to the location of that receiver. The rows of $\mathbf{G}$ can be computed effectively using reciprocity, which involves computing the forward response to a unit source located at each receiver. The vector $\boldsymbol{\Lambda}$ represents an inverse-sensitivity weighting function that accounts for distance from the receivers, as well as the resistivity structure; $\mathbf{s}$ is the vector containing the discretized source current density terms $\Im(\mathbf{r}, t)$ with dimension M, $\boldsymbol{\Psi}^{\mathrm{obs}}$ is the observed electric potential vector at the N sensors, and $\mathbf{W}_{\mathrm{d}}$ is a matrix that contains the information about the expected noise in the data.

In the following, we consider $\mathbf{W}_{\mathrm{d}}$ to be diagonal with each element on the diagonal being the inverse of the estimated variance of the measurement errors. In all the tests, we incorporate Gaussian noise with a standard deviation equal to 10% of the computed data mean. This is a realistic noise level for this type of experiment accounting for the amplitude of the signals that are measured. In addition, adding data weights helps to stabilize the inversion by eliminating artifacts that come from over-fitting the data.

The vector $\boldsymbol{\Lambda}$ represents the inverse-sensitivity weighting function. This function is needed because sensitivities decay quickly away from the receiver locations (typically as a power-law function of the distance from the receiver in a homogeneous medium, but is also affected by heterogeneous resistivity distributions). The weighting function is, therefore, needed to recover sources that are distant from the receivers.

Applying the following transform $\mathbf{s}_{\mathrm{w}} = \boldsymbol{\Lambda}\mathbf{s}$ and minimizing Eq. (4.51) gives the equation

$$\left[(\boldsymbol{\Lambda}^{-1}\mathbf{G}^{\mathrm{T}}\mathbf{W}_{\mathrm{d}}^{\mathrm{T}}\mathbf{W}_{\mathrm{d}}\mathbf{G}\boldsymbol{\Lambda}^{-1}) + \lambda\mathbf{I} \right]\mathbf{s}_{\mathrm{w}} = \boldsymbol{\Lambda}^{-1}\mathbf{G}^{\mathrm{T}}\mathbf{W}_{\mathrm{d}}^{\mathrm{T}}\mathbf{W}_{\mathrm{d}}\boldsymbol{\Psi}^{\mathrm{obs}}. \qquad (4.53)$$

The result of such an inversion is a smooth volumetric source current distribution. However, we know from the physics of the problem that the solution should be spatially compact.

Source compactness is a relatively classical technique that suits the nature of the electrical problem, because the volumetric source current densities associated with the seismoelectric conversion tend to be spatially localized. Compactness is based on the minimization of the spatial support of the source. The new global objective function is modified to include compactness as a regularization term

$$C = \|\mathbf{W}_{\mathrm{d}}(\mathbf{G}\mathbf{s} - \boldsymbol{\Psi}^{\mathrm{obs}})\|_2^2 + \lambda \sum_{k=1}^{M} \frac{s_k^2}{s_k^2 + \beta^2}, \qquad (4.54)$$

where β is the threshold term introduced to provide stability as $s_k \to 0$. This form of the objective function is now non-linear, since the compactness portion of the objective function is non-quadratic. The compactness term is effectively a measure of the number of source parameters that are greater than β, regardless of their magnitude. Minimization of this objective function, Eq. (4.54), results in the solution that uses the fewest number of source parameters that are still consistent with the measured data, which enforces sparseness of the source distribution. As model values fall below the threshold β, they no longer contribute to the sum in Eq. (4.54), and will be effectively masked from the solution.

In order to make this compact source problem linear, so that it can be solved in a least-squares framework, and to incorporate the inverse sensitivity scaling, the model weighting operator Λ in Eq. (4.51) is modified as

$$\mathbf{\Omega} = \mathrm{diag}\left\{\frac{\Lambda_{kk}^2}{s_{k_{(j-1)}}^2 + \beta^2}\right\}, \tag{4.55}$$

where diag(.) extracts the diagonal elements of the argument. Hence, the problem is linearized by making the objective function quadratic in s_k by fixing the denominator of the model objective function $\mathbf{\Omega}$, with respect to the previous solution at step $(j-1)$, using an iteratively reweighted least squares approach. The vector $\mathbf{s}_{j-1}$ is the initial model used to compute the first degree of compactness. A new vector $\mathbf{\Omega}$ is determined for every compactness degree, based on the previous model generated from the immediate previous compactness degree. Using the renormalization with $\mathbf{s}_w = \mathbf{\Omega}\mathbf{s}$ and minimizing the global model objective function in Eq. (4.51) gives us the iterative solution which utilizes compactness

$$\left(\mathbf{\Omega}_{j-1}^{-1}\mathbf{G}^{\mathrm{T}}\mathbf{W}_{\mathrm{d}}^{\mathrm{T}}\mathbf{W}_{\mathrm{d}}\mathbf{G}\mathbf{\Omega}_{j-1}^{-1} + \lambda\mathbf{I}\right)\mathbf{s}_{w,j} = \mathbf{\Omega}_{j-1}^{-1}\mathbf{G}^{\mathrm{T}}\mathbf{W}_{\mathrm{d}}^{\mathrm{T}}\mathbf{W}_{\mathrm{d}}\mathbf{\Psi}^{\mathrm{obs}}. \tag{4.56}$$

The process is halted after several iterations. Focusing the image is a subjective choice; we found that nine iterations offers a good compactness level to localize the sources responsible for the observed self-potential data.

4.2.5 Tests

To test our algorithm, we use the synthetic case shown in Figure 4.2. The result of the inversion is shown in Figure 4.3. The system is composed of two layers: the resistivity of the upper layer is 10 Ohm m, and its thickness is 10 m; the second layer has a resistivity of 100 Ohm m and a thickness of 30 m; the source is characterized by a current density of magnitude 10 mA m^{-2}. The polarized volume has the following coordinates: $x \in (50, 60)$ m, $y \in (40, 50)$ m, and $z \in (10, 20)$ m. The direction of the source current density is shown in Figure 4.4a, 4.4c, and 4.4e and is downward with two specified angles; see Jardani *et al.* (2008).

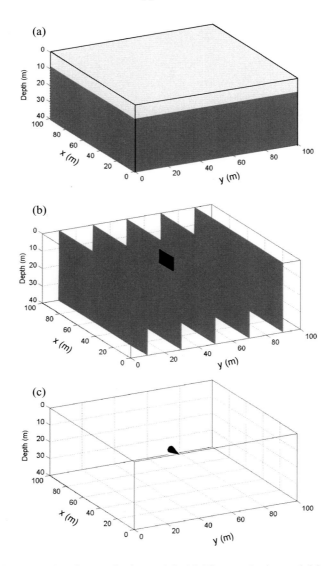

Figure 4.2. Example of a synthetic model. (a) The synthetic model is composed of two layers with distinct values of the electrical resistivity (upper layer: 10 Ohm m, thickness 10 m, deeper layer: resistivity 100 Ohm m, thickness of 30 m). (b) Position of the source current density. The magnitude of the source current density is 10 mA m^{-2}. (c) Direction of the current source density (modified from Jardani *et al.* (2008)).

The self-potential distribution is obtained by solving the Poisson's equation, Eq. (4.7), using the finite element commercial code Comsol Multiphysics 3.3. Forty-eight equally spaced stations are used at the top surface of the model domain, to simulate measurements stations for the self-potential signals (Figure 4.3a). We

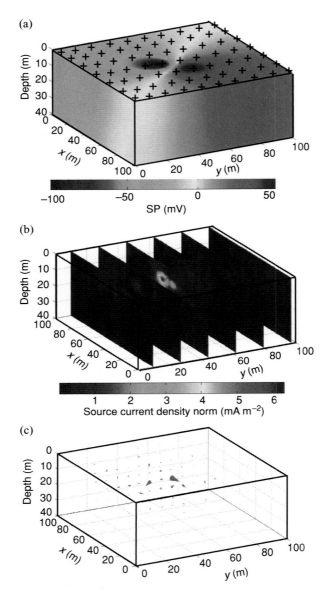

Figure 4.3. Result of the 3D-inversion of the source current density. (a) Distribution of the self-potential over the medium. The crosses corresponds to the
self-potential stations. The self-potential is regularly sampled at the top surface of
the system where the crosses are located. (b) Result of the inversion in terms of
distribution of the magnitude of the current density (the maximum of the recovered
distribution is 6.3 mA m^{-2}). (c) Result of the inversion in terms of direction of
the source current density using the first derivative operator for the regularization
of the inverse problem (no prior model used in the inversion) (see Jardani *et al.*
(2008)). See also color plate section.

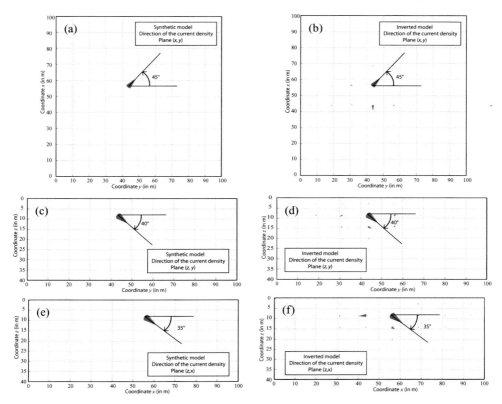

Figure 4.4. Comparison between the direction of the source current density in the synthetic model and the result of the inversion. The inversion is clearly able to give the correct directions of the source current density in addition to its magnitude (modified from Jardani *et al.* (2008)).

invert this set of discretized synthetic self-potential data for the source streaming current density. We use a null prior model ($\mathbf{m}_0 = 0$) and the first-order differential operator to regularize the inverse problem because the source is compact. The result of the inversion is shown on Figures 4.3 and 4.4. The choice of the regularization parameter is illustrated in Figure 4.5 by the L-shape method. The inversion result is in good agreement with the synthetic model (compare Figures 4.3c and 4.4c) both in terms of the intensity of the current density, the direction (Figure 4.4), and the position of the source. Note that the intensity of the inverted current density (6.3 mA m^{-2}) is only slightly smaller than the true current density (10 mA m^{-2}), showing the accuracy of the proposed method (the use of the smoothness operator yields an inverted current density smaller because the source is spreading over a larger volume).

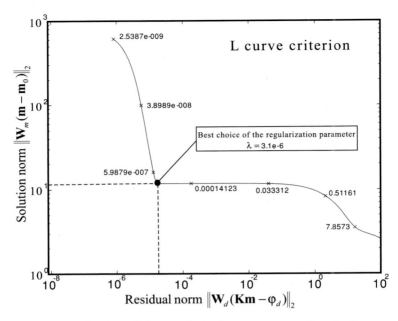

Figure 4.5. Determination of the regularization parameter using the L-curve criterion for the inversion of the synthetic case discussed in Figures 4.2 to 4.4. The optization is run for various value of the regularization parameter and the L-curve is built. The "best" value of the regularization parameter is chosen at the corner of the L-curve and the optimization algorithm is run a last time to obtain the "best" model shown in Figures 4.3 and 4.4 (see Jardani *et al.* (2008)).

The robustness of the inverse problem has been tested by adding white noise to the sampled self-potential data. The intensity of the white noise was 10% of the amplitude of the self-potential anomaly (this is typically the level of noise observed in the field data). The result of the inversion is shown on Figure 4.6a and 4.6b. We are able to retrieve the position of the source and its direction. This test demonstrates the robustness of our algorithm. The second test we perform concerns the resistivity distribution. If the resistivity distribution is unknown, one may be attempt to use a homogeneous resistivity distribution. For the present synthetic model, the result of the inversion is shown in Figures 4.6c and 4.6d. Clearly, the inverted result is grossly wrong, both in terms of amplitude and direction of the source current density vector. If resistivity tomograms are not available, it is certainly better to code the a prior geology, in terms of resistivity distribution than to use a homogeneous resistivity distribution.

We test now such type of algorithm of real data. Crespy *et al.* (2008) performed a controlled sandbox experiment that can be used to test our algorithm (Figure 4.7). In this experiment, ~0.5 ml of water was abruptly injected through a small vertical

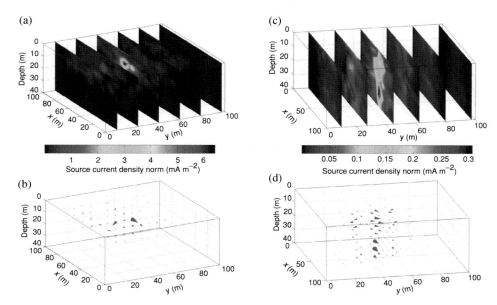

Figure 4.6. Inversion of the synthetic case for various cases. (a, b) Result of the inversion (amplitude and direction) with a white noise added to the sampled self-potential data (the amplitude of the white noise is equal to 10% of the self-potential anomaly). The result of the inversion shows that our approach is very robust to the presence of white noise (compare with Figure 4.3). (c, d) Inversion of the data without any prior knowledge of the resistivity distribution. Note that, in this case, both the amplitude and the direction of the inverted current density are grossly wrong. This shows how important the knowledge of the resistivity distribution is in order to invert properly self-potential sources even in the case of a layered system (see Jardani *et al.* (2008)). See also color plate section.

capillary with its outlet located at a depth of 15 cm (9 cm above the bottom of the tank) and located in the middle of the sandbox. The self-potential signals resulting from the pulse-injection of water were measured using 32 sintered Ag/AgCl electrodes located at the surface of the tank. These electrodes were connected to a voltmeter with a sensitivity of 0.1 μV, and an acquisition frequency of 1.024 kHz. The self-potential anomaly is negative, because of the source term $\Im = \nabla \hat{Q}_V \cdot \mathbf{U}$ in the Poisson equation at the outlet of the capillary ($\hat{Q}_V = 0$ in the capillary and $\hat{Q}_V > 0$ in the sand).

We applied our algorithm to the self-potential anomaly observed at the top surface of the tank when it reached its maximum amplitude (30 ms after the start of the injection). The result is shown in Figures 4.8 and 4.9. We use the first-order differential operator to regularize the inverse problem, as the source is assumed to be compact. In addition, we use not a prior model, but the depth-dependent normalization given by Eq. (4.42). The compact source is pointing upward and is

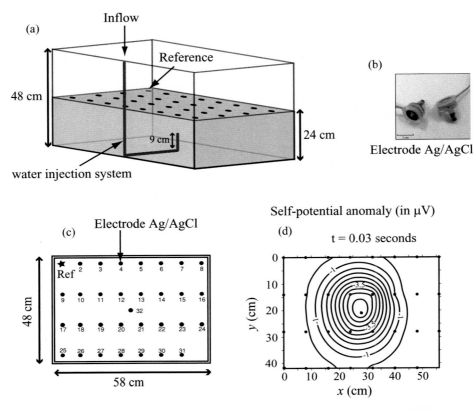

Figure 4.7. Sketch of the geometry of the sandbox experiment. A total of 32 non-polarizing electrodes are located at the top surface of the tank, which is partially filled with a well-calibrated sand and saturated by a solution of known composition and electrical conductivity. (a) Geometry of the tank. (b) Picture of the self-potential sintered Ag/AgCl electrodes developed by BioSemi. These electrodes are very sensitive thanks to a built in amplifier. (c) Sketch of the top surface of the tank showing the position of the electrodes in the vicinity of the top surface of the tank. The electrodes are located at a depth of 3 cm. "Ref" indicates the position of the reference electrode. Electrode #32 is located just above the inlet/outlet of the capillary. (d) Self-potential anomaly observed on the network of electrodes 30 ms after the pulse injection of water. Modified from Crespy *et al.* (2008).

located at source point S_p (29±2 cm, 24±2 cm, 10±1 cm). This was consistent with the actual position of the outlet of the capillary in the tank S_{true} (29 cm, 24 cm, 9 cm) and the physics of the primary flow problem. This illustrates the capability of our approach in locating the causative source of self-potential signals at depth when the support volume of the source is small with respect to the distance between the barycenter of the source and the observation stations; see Jardani *et al.* (2008).

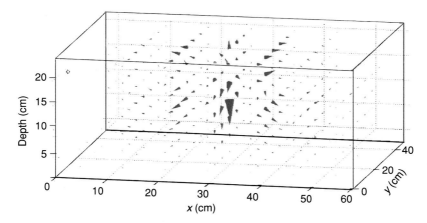

Figure 4.8. Determination of the direction and intensity of the streaming current density in the case shown in Figure 4.7. The largest arrow is located at 10 cm above the bottom of the tank (the end of the outlet is at 9 cm) and is directed downward.

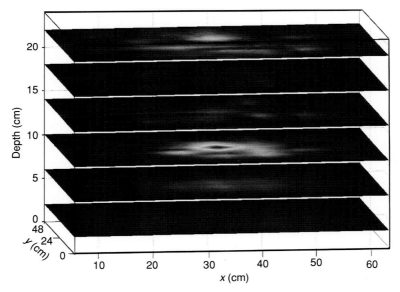

Figure 4.9. Distribution of the magnitude of the streaming (source) current density. The maximum of the distribution is equal to 8×10^{-5} mA m^{-2} and its location is consistent with the outlet of the capillary shown in Figure 4.7a.

4.3 Fully coupled inversion

As explained in Section 4.1, the forward modeling can be performed by computing the primary flow problem, then determining the streaming current density distribution, and finally solving the self-potential field. In this case, the inverse problem can be used to integrate the primary flow problem, which produces source distributions that are consistent with the physics of the primary flow problem. To emphasize such a solution of the inverse problem, we present the equations governing the physical processes of ground water flow and saline tracer transport in a heterogeneous unconfined aquifer. We also introduce the semi-coupled equations connecting the salt concentration to the electrical resistivity and to the source current density used to interpret self-potential data. We propose an inversion scheme, directly solving the primary flow problem, to invert for the distribution of the permeability of an heterogeneous aquifer.

4.3.1 Governing equations

We first present the transport equations corresponding to the primary flow problem, and their relationship to the resistivity and self-potential problems. In the steady-state condition, the governing ground water flow equation in a saturated and heterogeneous porous material is given by,

$$\nabla \cdot (K \nabla h) = 0, \tag{4.57}$$

subject to the boundary conditions

$$h = h_0 \text{ at } \Gamma_D, \tag{4.58}$$

$$-\hat{\boldsymbol{n}} \cdot K \nabla h = q_0 \text{ at } \Gamma_N, \tag{4.59}$$

where h denotes the hydraulic head (in m) and K is the hydraulic conductivity (m s^{-1}). We assume in this section that we are dealing with an isotropic aquifer. Equations (4.58) and (4.59) correspond to Dirichlet and Neumann boundary conditions, respectively. The hydraulic head h_0 denotes the head fixed at the upstream boundary Γ_D, q_0 is the hydraulic flux (m^2 s^{-1}) assumed at the Neumann boundary Γ_N, and $\hat{\boldsymbol{n}}$ is the outward unit vector normal to the boundary Γ_N.

The constitution of the transport equation of a salt tracer consists of the coupling of the Darcy's law and Fick's law for the flux of the salt $\mathbf{J}_d$ (in kg m^{-2} s^{-1}) (see Oltean and Buès (2002)):

$$\mathbf{U} = -K \nabla h, \tag{4.60}$$

$$\mathbf{J}_d = -\rho_f \phi \mathbf{D} \cdot \nabla c + \rho_f c \mathbf{U}, \tag{4.61}$$

where $\mathbf{U}$ denotes the Darcy velocity (m s^{-1}), $\mathbf{D}$ (in m^2 s^{-1}) denotes the hydrodynamic dispersion tensor, ϕ (unitless) denotes the connected porosity, c denotes the solute mass fraction (unitless), and ρ_f represents the solute bulk density (in kg m^{-3}).

The transport of the salt due to the injection of a salt tracer in the aquifer follows the advection–dispersion equation derived from a combination of the continuity equation (mass conservation equation for the salt) and a generalized Fick's law:

$$\frac{\partial(\rho_f \phi c)}{\partial t} + \nabla \cdot \mathbf{J}_d = 0. \tag{4.62}$$

Equation (4.62) can be rewritten as the following partial differential equation:

$$\phi \frac{\partial c}{\partial t} + \nabla \cdot (-\phi \mathbf{D} \cdot \nabla c) + \mathbf{U} \nabla c = 0, \tag{4.63}$$

which is subject to the following initial and boundary conditions:

$$c = 0, \forall \mathbf{x} \text{ at } t = 0, \tag{4.64}$$

$$c(\mathbf{x}, t) = c_0 \text{ at } \Gamma_1, \tag{4.65}$$

$$-\hat{\boldsymbol{n}} \cdot (\mathbf{D} \nabla c) = 0 \text{ at } \Gamma_2, \tag{4.66}$$

where Γ_1 and Γ_2 denote upstream and downstream boundaries associated with the position of the upstream well (injector) and the downstream (observation) well. In these equations, c_0 (unitless) is the solute mass fraction of the salt in the source term. In the Fickian model, the hydrodynamic dispersion tensor is given by

$$\mathbf{D} = \left[\frac{D_m}{\alpha} + \alpha_T v \right] \mathbf{I}_3 + \frac{\alpha_L - \alpha_T}{v} \mathbf{v} \otimes \mathbf{v}, \tag{4.67}$$

where D_m is the molecular (mutual) diffusion coefficient of the salt (in $m^2\ s^{-1}$) (for a NaCl solution, D_m is $1.60 \times 10^{-9}\ m^2\ s^{-1}$ at infinite dilution and is $1.44 \times 10^{-9}\ m^2\ s^{-1}$ at high salinities at $25\ ^{\circ}C$), $v = |\mathbf{v}|$, $\mathbf{a} \otimes \mathbf{b}$ represents the tensorial product between vectors $\mathbf{a}$ and $\mathbf{b}$, and α_L and α_T are the longitudinal (along $\mathbf{v}$) and transverse (normal to $\mathbf{v}$) dispersivities (in m), and α denotes the tortuosity of the pore space. This bulk pore space tortuosity can be obtained as the product of the electrical formation factor F entering into the electrical conductivity problem (see below) and the porosity ϕ (see Section 2.2).

The numerical solution of the hydrodynamic-transport problem is performed by using the package Comsol Multiphysics 4.3. In the first step, we determine the distribution of the Darcy velocity by solving Darcy's equation in steady-state conditions, and then we use the Darcy velocity to solve the advection–dispersion equation to determine the distribution of the salinity in space and time.

Electrical resistivity tomography consists of inverting apparent resistivity data obtained by injecting an electrical current with two (current) electrodes, and measuring the difference in the electrical potential at a set of (voltage) electrodes (at least two electrodes). For an isotropic and heterogeneous electrical 2D (two-dimensional) conductivity distribution can be defined with the potential spectrum $\tilde{V}(x, k_y, z)$ described as:

$$\nabla \cdot (\sigma \nabla \tilde{V}) - k_y^2 \sigma \tilde{V} = -\frac{I_s}{2} \left(\delta \left(\mathbf{x} - \mathbf{x}_s^+ \right) - \delta \left(\mathbf{x} - \mathbf{x}_s^- \right) \right), \tag{4.68}$$

with the following boundary conditions:

$$\tilde{V} = 0 \text{ at } \Gamma_d, \quad \|r\| \to \infty, \tag{4.69}$$

$$- \mathbf{n}.\sigma \nabla \tilde{V} = 0 \text{ at } \Gamma_N, \tag{4.70}$$

where $\tilde{V}$ denotes the electrical potential (in V) due to dipole source modeled in 3D, σ denotes the electrical conductivity (in S m^{-1}), $\mathbf{x}$ denotes the position vector of the source of current, I_s corresponds to the injected current (in A), $\mathbf{x}_s^+$ and $\mathbf{x}_s^-$ are the locations of the positive and negative current sources, respectively, and k_y denotes the real wave number. The boundary conditions are subdivided as a Neumann boundary condition Γ_N imposed at the insulating air–ground interface, and a Dirichlet boundary condition Γ_D at the other boundaries.

The electrical resistivity problem is solved numerically with the finite element approach for a set of the wave numbers k_y. The solution $\tilde{V}$ is transformed from the wave number domain to the spatial domain using an inverse cosine–Fourier transform,

$$V = \frac{2}{\pi} \int_0^\infty \tilde{V} \cos(k_y y) dk_y. \tag{4.71}$$

As shown earlier, the electrical conductivity of the soil is connected to the solute electrical conductivity according to

$$\sigma = \frac{\sigma_f}{F} + \sigma_s, \tag{4.72}$$

where $F = \phi^{-m}$ (unitless) denote the electrical formation factor, σ_f denote the pore water electrical conductivity (S m^{-1}), and the power-law cementation exponent m (>1, unitless), which ranges from 1.3 to 1.5 for clay-free unconsolidated sands and from 1.8 to 2.5 for consolidated sandstones, according to Revil *et al.* (1998).

The pore water conductivity is related to the concentration c by a simple linear equation:

$$\sigma_f = ac \left(\beta_{(+)} + \beta_{(-)} \right) e, \tag{4.73}$$

where c is salinity concentration of the pore water expressed in kg m^{-3}, e is the elementary charge of the electron ($e = 1.6 \times 10^{-19}$ C), $\beta_{(+)}(25 \,°C) = 5 \times 10^{-8}$ m^2 s^{-1} V^{-1} and $\beta_{(-)}(25 \,°C) = 7 \times 10^{-8}$ m^2 s^{-1} V^{-1} denote the mobility of the cations and anions in the pore water, respectively, and a is a coefficient to convert the unit of concentration from kg m^{-3} to molecules m^{-3}. From Eqs. (4.72) and (4.73), the macroscopic electrical conductivity measurements are related to the salinity by (see Chapter 3):

$$\sigma = ac\phi^m e \left(\beta_{(+)} + \beta_{(-)} \right) + \sigma_s. \tag{4.74}$$

Therefore electrical resistivity, through its dependence on salinity, provides information that is indirectly related to the hydraulic conductivity.

We now present the equation for the self-potential problem. The self-potential method consists of passively measuring the electrical potentials at the ground surface, or in boreholes, using a couple of unpolarizable electrodes and a multi-channel d.c.-Voltmeter, characterized by a high input impedance (>10 Mohm) and a sensitivity of 0.1 mV. The self-potential signal recorded during a salt tracer experiment is the sum of two distinct contributions for the source current density. The first component is associated with the flow of ground water, which drags the excess electrical charge contained in the pore water. This excess of charge is more precisely located in the electrical diffuse layer coating the surface of the minerals; see Jardani *et al.* (2009). This contribution is known as the streaming current density. The second contribution is related to the salinity gradient itself. It depends on the gradients of the activity of the charge carriers (ions) that are present in the pore water see Revil and Linde (2006), Maineult *et al.* (2006) and Martínez-Pagán *et al.* (2010). This contribution is known as the diffusion current density as discussed in Chapter 2 (Section 2.2).

In an isotropic heterogeneous porous material, the total current density **J** (in A m^{-2}) is the sum of a conductive current density (given by Ohm's law) plus the two components described above (see Chapter 2):

$$\mathbf{J} = -\sigma \nabla \psi + \hat{Q}_V \mathbf{U} - \frac{k_b T}{e} \sigma (2t_{(+)} - 1) \nabla \ln \sigma_f, \tag{4.75}$$

where ψ is the electrical self-potential field (in V) σ is the electrical conductivity of the porous material (in S m^{-1}), k_b is the Boltzmann constant (1.381×10^{-23} J K^{-1}), T is the absolute temperature (°C), U represents the Darcy velocity (in m s^{-1}), $t_{(+)} = 0.38$ (for NaCl) denotes the microscopic Hittorf number of the cations in the pore water (see Section 2.2), and $\overline{Q}_V$ (in C m^{-3}) denotes the excess charge density (due to the diffuse layer) of the pore water per unit pore volume. As discussed in Chapter 3, the charge density $\hat{Q}_V$ can be accurately predicted from the permeability (expressed in the m^2) according to $\log_{10} \hat{Q}_V = -9.2 - 0.82 \log_{10} k$ (Section 3.1).

Using the continuity equation for the current density (see Section 3.1), the self-potential problem is, therefore, defined by the following boundary value problem:

$$\nabla \cdot (\sigma \nabla \psi) = \nabla \cdot \mathbf{J}_S, \tag{4.76}$$

$$\mathbf{J}_S = \hat{Q}_V \mathbf{U} - \frac{k_b T}{e} \sigma (2t_{(+)} - 1) \nabla \ln \sigma_f, \tag{4.77}$$

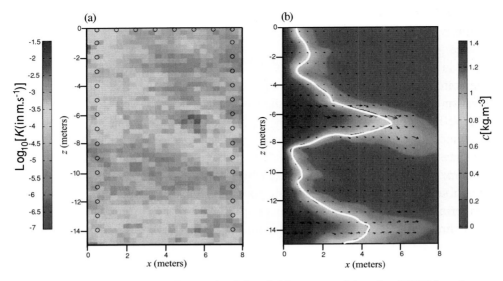

Figure 4.10. True hydraulic conductivity field generated by the SGSIM code (http://www.uofaweb.ualberta.ca/ccg//pdfs/2002%2053-sgsim_pw1.pdf). The salt movement is due to a constant hydraulic head gradient between the upgradient well (on the left) and the downgradient well (on the right). The left-hand panel shows the salt concentration distribution at $t = 20$ days. This figure shows the formation of preferential fingering associated with preferential flow pathways (the arrows denote the Darcy velocity). See also color plate section.

with the following boundary conditions:

$$\psi = 0 \text{ at } \Gamma_d, \quad \|r\| \to \infty, \tag{4.78}$$

$$-\hat{\boldsymbol{n}} \cdot \sigma \nabla \psi = 0 \text{ at } \Gamma_N, \tag{4.79}$$

the Neumann boundary condition Γ_N imposed at the air–ground interface and a Dirichlet boundary condition Γ_D at the other boundaries. The physical model described above illustrates that the self-potential field is related to the pathways of the salt plume migration within the aquifer and therefore to the hydraulic conductivity field.

4.3.2 Forward modeling in a heterogenous aquifer

The salt tracer test involves injecting an amount of the salt (NaCl) into an upgradient well and monitoring the perturbation of the electrical conductivity and self-potential associated with the migration of the salt tracer due to the natural hydraulic gradient plus dispersion and diffusion. The monitoring is performed until the salt tracer reaches a downstream well, where the salt concentration is sampled at eight depths. The geophysical data are measured with a network of electrodes located at the ground surface as well as in the two wells at eight different depths (see Figure 4.10).

For the synthetic model, we use a heterogeneous hydraulic conductivity field between the two wells with a distance of 8 m from each other. The two wells have the same depth, 15 m (Figure 4.10). The heterogeneous model for the permeability field is generated by the SGSIM software (see Deutsch and Journel (1992)) with an anisotropic spherical variogram with a range of 10, a sill of 0.8, an angle of 75° and an anisotropic ratio of 0.3. The porosity is assumed constant over the domain ($\phi = 0.35$). The values of the hydraulic properties were chosen to approximate a sandy alluvial environment; see Revil *et al.* (1998). The hydraulic conductivity is used to numerically solve the flow problem under steady-state conditions, and determine the distribution of the hydraulic heads, as well as the Darcy velocity distribution. The domain boundaries are placed at 50 m, far enough from the study-area located between the boreholes to minimize edge/boundary effects on the simulations. The hydraulic heads imposed at the boundaries on the left and right sides of the domain were selected in a manner in which the hydraulic gradient between the both boreholes is constant with a value of 0.02 m m^{-1}. We use an impervious boundary condition at the top and bottom boundaries of the system.

The solution of the ground water flow problem is used to evaluate the advection term (see Eq. (4.25)) and the streaming current density (see Eq. (4.39)). Assuming an injection of a saline (conservative) tracer (1 kg m^{-3}) in the upgradient well (on the left side of Figure 4.10), considered entirely screened, we simulate the advection–dispersion equation for the transport of the salt. We assume the values of longitudinal and transverse dispersivities are 0.5 m and 0.05 m, respectively, homogeneous, and taken in accordance with the scale of the investigated area. The transient transport equation is solved over a period of 100 days.

The salt transport in the aquifer has a direct impact on the distribution of the electrical conductivity of the pore water and, consequently, on the bulk electrical conductivity of the porous medium. A series of simulated electrical conductivity measurements has been carried out with 21 electrodes positioned on the surface and along the walls of the two boreholes (see Figure 4.10). The configuration technique used for the voltage measurements is the dipole–dipole array, which provides a virtual survey of the electrical conductivity perturbation during the salt plume movement. A series of the electrical conductivity snapshots at $t = [5, 10, 15, 20$ and 30 days] acquired to track the salt migration and to reconstruct hydraulic conductivity.

We also simulated the acquisition of self-potential measurements to numerically record the anomalies associated with the gradients of the salinity and hydraulic gradients. In real field conditions, the electrodes, used previously for electrical resistivity tomography in the downgradient well, can also be used to record the electrical potential field. The self-potential measurements are assumed to be collected once a day during a period of 30 days. In real field conditions, the self-potential distribution has to be recorded before the injection of the electrical current used to

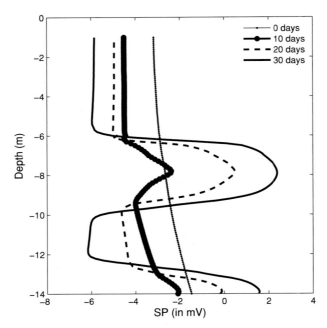

Figure 4.11. The self-potential profiles collected at four distinct times at the electrodes located along downgradient borehole. The curves illustrate that the self-potential anomalies present some important amplitudes (about 6 mV) when the plume comes closer to the electrodes. At the opposite, the change in the self-potential signals is only of 2 mV in areas characterized by low hydraulic conductivity values.

measure the apparent electrical resistivities. More often, the electrical resistivity equipment records the self-potential data as well, but this record is considered as noise by the practitioners of electrical resistivity tomography, and is only used to correct the apparent electrical resistivity data. The measured self-potential data are reported with respect to a unique reference location located remotely from the study area. The self-potential anomalies recorded in the downgradient well provide a very useful source of information in characterizing the transport parameters. Figure 4.11 shows that the self-potential signals at the electrodes located in the second well increase substantially when the plume approaches the viscinity of the electrodes. This is because of the decreasing distance between the electrical sources at the interface of the salt front and the measuring electrodes. In contrast, in regions of low hydraulic conductivity, where the salinity gradient is almost negligible, the self-potential is invariant.

4.3.3 Stochastic joint inversion

We employ a stochastic approach to predict the hydraulic conductivity field $\mathbf{m}$ from multiple sources of data $\mathbf{d}$ coming from saline concentration (see C in

Figures 4.16–4.20), electrical resistivity imaging (abbreaviated as ERI in Figures 4.16–4.20) and self-potential (abbreviated as SP in Figures 4.16–4.20). The pilot points are used as a parameterization technique to reduce the number of the model parameters by identifying a few key locations (called pilot points) in the aquifer where the values of the hydraulic conductivity can be assessed to reproduce the hydrogeophysical measurements. This is done by assuming certain spatial statistical characteristic of the hydraulic conductivity field. The method consists of hydraulic conductivity field, conditioned by the pilot point values generate from the SGSIM code (see Deutsch and Journel (1992)), perturbed at each iteration with the mean of the field. Then, the spatial distribution of the hydraulic conductivity field is written (from Kowalsky *et al.* (2005)) as

$$K(\mathbf{x}) = 10^{f_p(\mathbf{x})} \overline{K}, \tag{4.80}$$

where $\overline{K}$ is the mean value of the hydraulic conductivity (in m s^{-1}), $f_p(\mathbf{x})$ denotes a space random function with a mean of zero and a variance equivalent to the semi-variogram of the $\log_{10} K(\mathbf{x})$. Running conditional simulation of the values assigned to the pilot points yields the field magnitudes, which requires a known semi-variogram. The unknown parameters explored during the inverted reconstruction of the hydraulic conductivity are $\mathbf{m} = [\overline{K}, f_p(\mathbf{x})]$, the mean of the hydraulic conductivity $\overline{K}$ and pilot point values of $f_p(\mathbf{x})$. These parameters are perturbed at each iteration to honor both the hydrological and geophysical data (see Figure 4.12 and Jardani *et al.* (2012) for further details).

In a probabilistic framework, the inverse problem maximizes the conditional probability density of occurring $\mathbf{m}$ given the data vector $\mathbf{d}$. This corresponds to finding the best model $\mathbf{m}$, respecting the constraints posed by the prior model, and minimizing the difference between the observed and predicted data. We note that $P_0(\mathbf{m})$ is the prior probability density of parameters $\mathbf{m}$, and $P(\mathbf{d} \,|\, \mathbf{m})$ represents the probability corresponding to the data given the model $\mathbf{m}$, establishing a connection between the data predicted by the model $\mathbf{m}$ and measurable data $\mathbf{d}$. The a posteriori probability density $\pi(\mathbf{m} \,|\, \mathbf{d})$ of the model parameters $\mathbf{m}$, given the data $\mathbf{d}$ is obtained using Bayes' formula:

$$\pi(\mathbf{m} \,|\, \mathbf{d}) = \frac{P(\mathbf{d} \,|\, \mathbf{m}) P_0(\mathbf{m})}{P(\mathbf{d})}, \tag{4.81}$$

where $P(\mathbf{d})$ is a normalizing term known as evidence,

$$P(\mathbf{d} \,|\, M) = \int P_0(\mathbf{m} \,|\, M) P(\mathbf{d} \,|\, \mathbf{m}, M) \mathbf{dm}. \tag{4.82}$$

The posterior probability density $\pi(\mathbf{m} \,|\, \mathbf{d})$ of the model parameters $\mathbf{m}$ given the data $\mathbf{d}$ is written as

$$\pi(\mathbf{m} \,|\, \mathbf{d}) \propto P(\mathbf{d} \,|\, \mathbf{m}) P_0(\mathbf{m}). \tag{4.83}$$

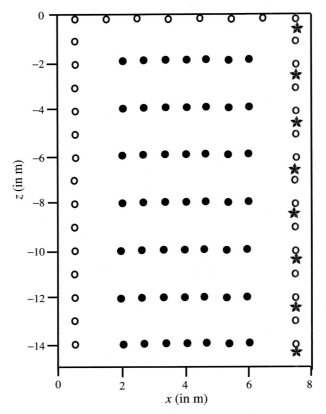

Figure 4.12. The filled circles indicate the spatial locations of the pilot points where we assigned the hydraulic conductivity values selected randomly during each iteration of the McMC sampler to build the field perturbation by using conditional simulation with a variogram assumed to be known. The white circles denote the locations of electrodes in both boreholes to monitor the variations in time of the apparent electrical resistivity and self-potential signals. The stars indicate the position of the monitoring of the salt concentrations (in situ port for sampling in the downstream well).

The presence of the multiple and independent data sets requires rewriting the likelihood function as a product of the partial likelihood of each data type:

$$P(\mathbf{d}_i \mid \mathbf{m}) = \prod_i^n P_i(\mathbf{d}_i \mid \mathbf{m}), \qquad (4.84)$$

$$\pi(\mathbf{m} \mid \mathbf{d}_i) \propto \left[\prod_i^n P_i(\mathbf{d}_i \mid \mathbf{m}) \right] P_0(\mathbf{m}), \qquad (4.85)$$

where i is the number of data sets. In this case, we have three types of data: salt concentration, C, resistivity imaging ERI, and self-potential SP. The likelihood function used to evaluate the quality of a model $\mathbf{m}$ to maximize a probability function is considered to be Gaussian:

$$P_i(\mathbf{d}_i \mid \mathbf{m}) = \frac{1}{[(2\pi)^N \det \mathbf{C}_{d_i}]^{1/2}} \exp\left[-\frac{1}{2}(g_i(\mathbf{m}) - \mathbf{d}_i)^T \mathbf{C}_{d_i}^{-1}(g_i(\mathbf{m}) - \mathbf{d}_i)\right],$$

(4.86)

where $g_i(\mathbf{m})$ is the forward modeling operator used to simulate a type of data referenced by i. The forward problem connects the generation of data $\mathbf{d}$ to a given hydraulic conductivity model $\mathbf{m}$, $\mathbf{d}$ is an N-vector corresponding to the observed data, and $\mathbf{C}_{d_i}$ denotes a $(N \times N)$ diagonal covariance matrix, accounting for the noise-to-signal ratio of the data in the inversion. In the salt tracer test, the forward problem comprises three partial forward problems. The first forward problem $g_1(\mathbf{m})$ generates the saline concentrations, the second $g_2(\mathbf{m})$ simulates the electrical resistivity imaging, and the third $g_3(\mathbf{m})$ models self-potential signals. The prior probability distribution, if available, is also taken as Gaussian:

$$P_0(\mathbf{m}) = \frac{1}{[(2\pi)^M \det \mathbf{C}_m]^{1/2}} \exp\left[-\frac{1}{2}(\mathbf{m} - \mathbf{m}_{prior})^T \mathbf{C}_m^{-1}(\mathbf{m} - \mathbf{m}_{prior})\right], \quad (4.87)$$

where $\mathbf{m}_{prior}$ denotes the prior value of the model parameters for each unit and $\mathbf{C}_m$ is the model diagonal covariance matrix incorporating the uncertainties related of the prior model of material properties.

To avoid unrealistic hydraulic conductivity values, the inequality constraints $\mathbf{b}_{min} \leq S(\mathbf{m}) \leq \mathbf{b}_{max}$ were included in the posterior probability. Therefore, we rewrite the constrained posterior probability distribution according to Michalak, (2008), as follows:

$$\pi(\mathbf{m} \mid \mathbf{d}) \propto P(\mathbf{d} \mid \mathbf{m}) P_0(\mathbf{m}) \left[\mathcal{H}(S(\mathbf{m}) - \mathbf{b}_{min})^T \mathcal{H}(\mathbf{b}_{max} - S(\mathbf{m}))\right], \quad (4.88)$$

where

$$\mathcal{H}(S(\mathbf{m}) - \mathbf{b}_{min}) = \begin{cases} 1 & S(\mathbf{m}) > \mathbf{b}_{min} \\ 1/2 & S(\mathbf{m}) = \mathbf{b}_{min} \\ 0 & S(\mathbf{m}) < \mathbf{b}_{min} \end{cases}, \quad (4.89)$$

where $\mathcal{H}$ denotes the Heaviside function, $K(x) = S(\mathbf{m})$ is a function generating the hydraulic conductivity field with the parameters $\mathbf{m}$ ($\bar{K}$ and $f_p(\mathbf{x})$), and $\mathbf{b}_{max}$ and $\mathbf{b}_{min}$ are the lower and upper limits of the hydraulic conductivity values. We point out that the posterior probability constrained (Eq. (4.88)) is going to be non-Gaussian (see Jardani *et al.* (2012) for further details).

In this study, the pilot points are placed on a regular grid, where their unknown values can be assessed with the mean of the hydraulic conductivity $\bar{K}$ through a

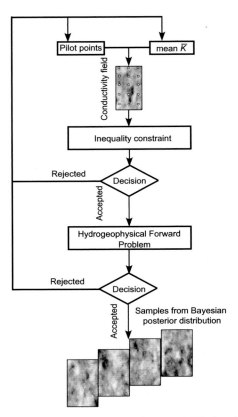

Figure 4.13. This sketch illustrates the use of the McMC algorithm used to reconstruct the hydraulic conductivity field from the inversion of the hydrogeophysical data (tracer concentration, apparent resistivity and self-potential data). The
pilot points technique is used as a parameterization method to perturb the mean
hydraulic conductivity tensor $\bar{K}$, which is unknown. Inequality constraints are
used to avoid the sampling of undesirable hydraulic conductivity values. The
routine provides a set of the realizations sampled from the posterior probability
distribution.

Markov chain Monte Carlo (McMC) algorithm, exploring the posterior probability
density $\pi(\mathbf{m} \,|\, \mathbf{d})$, which is expressed by Eq. (4.88). McMC algorithms are based on
the random generation routines of a set of models (in the present case the values of
the pilot point and the mean of the hydraulic conductivity $\bar{K}$). We then select the
best realizations of these models that are able to reproduce both the hydrogeological
and geophysical data (see Figure 4.13). We can infer the statistical properties of
the ensemble of the realizations such as the central tendency (median, mode and
mean) and the percentiles as dispersion parameters to estimate the uncertainty of
the solution of the inverse problem.

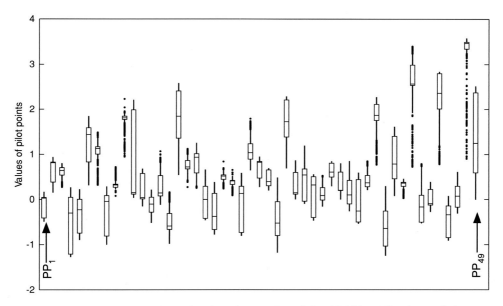

Figure 4.14. The box plots showing the results of the 20 000 realizations of the values of the each pilot point (#1 to #49) obtained from the McMC sampling to invert the concentration saline data in the second (downstream) well. The box plots represent inter-quartile ranges and the solid black line at the centre of each box denotes the median. The arms of each box extend to cover the central 95% of the sampled posterior distribution.

4.3.4 Results

To evaluate the effectiveness of the inversion with coupling hydrogeological and geophysical data, we compared several scenarios. We first examined the individual inversion of each type of data, and then we proceeded to combine two data sets. At the end, we combined all the hydrogeological and geophysical data sets together (time-lapse joint inversion) (see Jardani *et al.* (2013) for details).

Note that the inversion was performed with an inequality constraint on the hydraulic conductivity field with 10^{-7} and $10^{-1.5}$ m s^{-1} as the upper and lower limit respectively to facilitate a reasonable sampling of this field with a set of the 49 pilot points distributed on the regular grid (Figure 4.12). The results of the inversion are presented from the computation of the median and the both percentiles 25% and 75% of the simulations sampled (for example see Figure 4.14). The discussion of the inversion is based on a comparison between the real and inverted hydraulic conductivity, which is calculated from the median of the simulations sampled via posterior probability. We compared the spatial distribution of the real salted plume and the plume resulting from the hydraulic conductivity inverted at time

$t = 20$ days; we also added two saline fronts, which are determined from the hydraulic conductivity fields, the 75% and 25% quartiles of simulations sampled, respectively. In this inversion, the variogram parameters and porosity are assumed to be known.

At first, we evaluated the reconstruction of the hydraulic conductivity field from the salt concentration data collected at the downgradient borehole at eight depths (Figure 4.12) during the time lapse $t = 0$ to $t = 100$ days. The 20 000 simulations of the hydraulic conductivity fields were generated during the McMC process. The median field of the realizations sampled from a posteriori distributions varies between 10^{-7} and $10^{-1.8}$ m s^{-1} (the results of this inversion are named *inversion A1*; see Figure 4.15a). The median of the realizations sampled perfectly honors the saline concentration data (see Figure 4.16a). However, the spatial distribution of the salty plume at $t = 20$ days, reconstructed from this median field of the hydraulic conductivity and two quartiles, is far from the true postion of the saline plume (Figure 4.15b). We found that the self-potential signature predicted from this field remains consistent with the self-potential observed (Figure 4.16b). We note that the inversion of the concentration data did not delineate the main permeable areas; however, there is a lack of the similarity between the inverted hydraulic conductivity field and the true one (Figure 4.15c). Consequently, the downstream salt concentration measurements in a well do not carry enough information to reconstruct the hydraulic conductivity field.

Regarding the electrical resistivity method, we used a series of electrodes located in the two wells with a dipole–dipole array (Figure 4.12). The data was acquired during the following periods (in days) (5, 10, 15, 20 and 30) to track the variations of the electrical resistivity associated with the passage of the salt tracer. The results of this inversion is named *inversion A2* (Figure 4.15). The hydraulic conductivities estimated by the resistivity inversion are far from perfect, according to the difference between the true and inverted fields (Figure 4.15c). At the surface of the aquifer, where the hydraulic conductivity assumes high values, the reconstructed hydraulic conductivity field generates an erroneous distribution of the saline tracer (see Figure 4.16c).

The inversion of the self-potential measurements, recorded once per day during a period of 30 days, in the same locations as the electrodes used for ERI, yields a hydraulic conductivity field in the range (10^{-7} m s^{-1}, 10^{-2} m s^{-1}) (see inversion A2 in Figure 4.15). Visually, this reconstruction also reproduces the salinity data (Figure 4.15e). This reconstruction is better than the reconstructed field using the d.c. resistivity data; the hydraulic conductivity field inverted from the self-potential data yields an area of lower hydraulic conductivity in the vicinity of the upgradient well between the depths (6 m, −2 m) that does not exist in the true conductivity

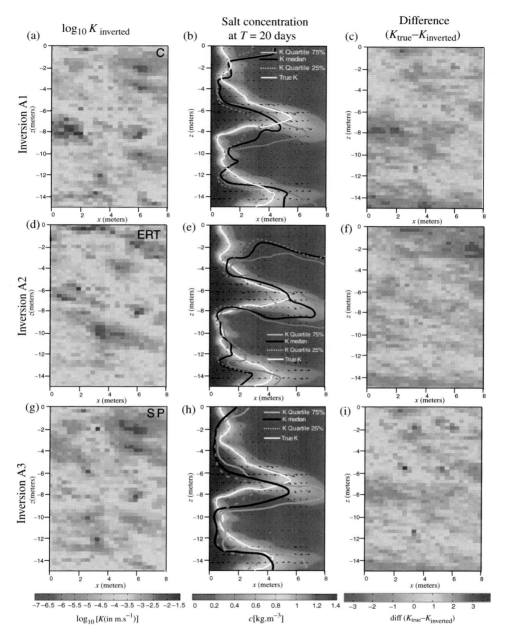

Figure 4.15. Uncoupled inversion of each data set to estimate the hydraulic conductivity field. The inversions A1, A2 and A3 denote the inversion of concentration data, resistivity and self-potential data, respectively. (a, d, g) Inverted hydraulic conductivity field. (b, e, h) True and inverted position of the salt plume at $t = 20$ days (the inverted position is from the median and the quartiles 75% and 25% of all realizations sampled during the inversions). (c, f, i) Difference (in log scale) between the real and inverted hydraulic conductivity field. See also color plate section.

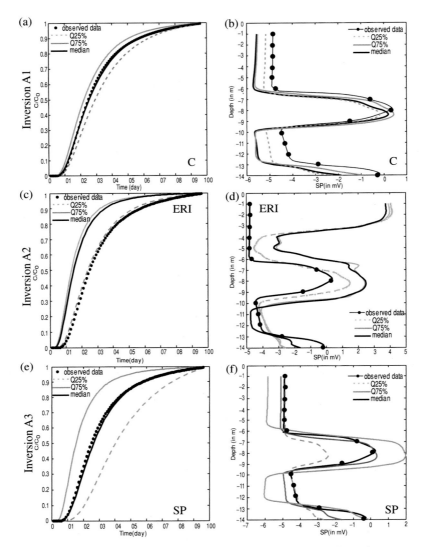

Figure 4.16. The inversions A1, A2 and A3 indicate the inversion of concentration data, resistivity and self-potential data respectively. For each inversion, we illustrate in the first and second columns the predicted data of the saline concentration at position P ($x = -7.5$, $z = -8$ m) and the self-potential data collected at the downgradient borehole and resulting from the inverted hydraulic conductivity field.

field (Figure 4.15f). The reconstructed hydraulic conductivity field is still far from the true field (see Figure 4.15f).

We present now the joint inversion results with the data sets taken two by two (Figure 4.17). We want to emphasize the result of the joint inversion of apparent

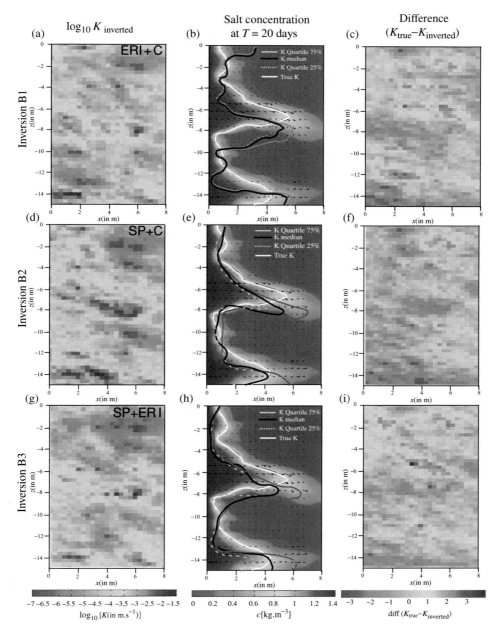

Figure 4.17. Results of the joint inversion of both data sets to estimate the hydraulic conductivity field. The inversions B1, B2, and B3 denote the inversion of the ERI + C, SP + C and ERI + SP data, respectively. (a, d, g) Hydraulic conductivity fields determined from the coupled inversion. (b, e, h) True position of the salt plume at $t = 20$ days and inverted position from the median and the quartiles 75% and 25% of all realizations sampled during the inversion. The third column for each inversion represents the difference (in log scale) between the true hydraulic conductivity field and the inverted hydraulic conductivity field. See also color plate section.

resistivity and concentration data (inversion B1) considerably improves the mapping of the hydraulic conductivity, compared to results obtained from the decoupled inversion data (see inversion B1 in Figure 4.17 and compare with the inversions A1 and A2 in Figure 4.15). In this joint inversion, we have eliminated the occurrence of the areas of high hydraulic conductivity near the surface and resulting from the inversion of the apparent resistivity data. However, the spatial distribution of the salt front at $t = 20$ days reveals that the estimated hydraulic conductivity does not reproduce perfectly the salt motion in the aquifer (see Figure 4.17b). Inversion B2 delineates the preferential flow path located in the center portion of the aquifer fairly well (see Figure 4.17d). This improvement is also observed in the joint inversion of the ERI and self-potential (SP) data (see inversion B3 in Figure 4.17g in which we can see that the spatial distribution of hydraulic conductivity and the motion of the salt plume have been improved by the joint inversion). The predicted salt concentration and self-potential data are also better fitted through the joint inversion by comparison with the uncoupled inversions (see Figure 4.18). *Inversion C* has improved the reconstruction of the hydraulic conductivity field, due to the complementary sensitivity of the methods to gradients in salinity and hydraulic head. The joint inversion of all the data sets together is shown in Figure 4.19, and a comparison with the data is shown in Figure 4.20. The joint inversion is clearly the best strategy to increase the resolution between the two wells, and to provide the most meaningful hydraulic conductivity distribution (see Jardani *et al.* (2013) for further details).

4.4 Conclusions

In this chapter we have discussed two distinct types of strategies to perform the self-potential inverse problem. One set of strategies is focused on inverting the self-potential alone, including d.c. resistivity tomography and eventually some constraints on the solution. Then, the solution of the inverse problem is interpreted in terms of relevant hydrogeological or hydrogeochemical parameters. This is a classical approach in geophysics that has been used for more than 40 years to address geophysical problems. Often, a better approach is to fully couple the inverse self-potential problem with the underlying physics of the primary flow response, for the occurrence of the self-potential anomalies. This approach is richer, and reduces the non-unicity of the inverse problem. It is a very efficient method for time-lapse problems, and for the joint inversion of various geophysical and non-geophysical data, in terms of a quest for information in a Bayesian framework as explained in the seminal papers of Tarantola and Valette. In the next four chapters (Chapters 5 to 8), we will present various applications in which self-potential data are inverted

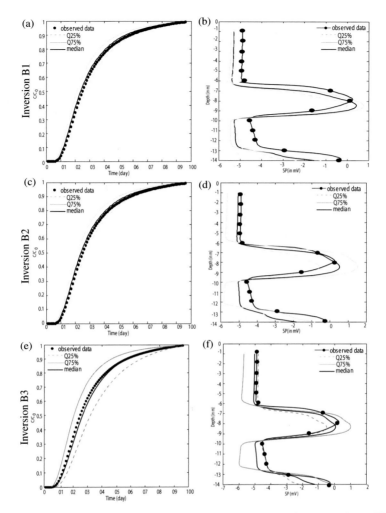

Figure 4.18. The inversion B1, B2 and B3 indicate the coupled inversions (ERI + C, SP + C and ERI +SP respectively). For each inversion, we represent in the first column the predicted data of the concentration sampled at position P ($x = -7.5$, $z = -8$ m) and in the second column the predicted data of SP data collected at the downgradient borehole at $t = 20$ days.

to address specific problems in hydrogeology, environmental engineering, and volcanology.

Exercises

Exercises 4.1. Self-potential distribution associated with a 1D flow experiment. In this exercise, we simulate the occurrence of self-potential generated by

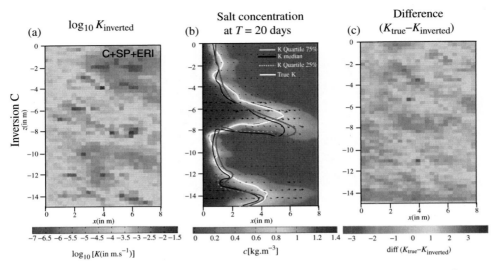

Figure 4.19. Result of the inversion for time-lapse joint inversion. (a) The hydraulic conductivity field estimated from the joint inversion of all data sets (concentration C, ERI and SP). (b) Spatial distribution of the saline plume at $t = 20$ days from the true conductivity field compared to the plume predicted from the inverted hydraulic conductivity field (with its confidence interval) using all the data (salinity, resistivity and self-potential). (c) Difference (in log scale) between the true and inverted hydraulic conductivity fields. See also color plate section.

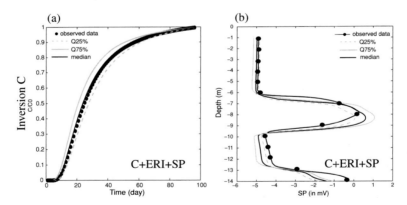

Figure 4.20. The predicted data obtained from the joint inversion of all data sets. (a) Predicted salinity recorded at position P ($x = 7.5$, $z = -8$) and inverted salinity. The predicted data is obtained from the joint inversion of the saline tracer test plus the geophysical information (electrical resistivity imaging and self-potential). (b) True versus predicted self-potential profiles at the downgradient borehole at $t = 20$ days.

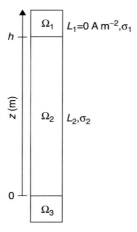

Figure 4.21. Geometry for a 1D (one-dimensional) flow through a column conducted to determine the streaming coupling coefficient C. The domains Ω_1 and Ω_3 correspond to the upstream and downstream reservoirs, respectively, which are filled with water. The domain Ω_2 corresponds to the porous material filled with the same solution. The vertical coordinate z is taken positive upward with the bottom of the porous material as datum.

the vertical flow of water by using an analog experiment conducted in a vertical tube. We divide the system into three sub-domains as shown in Figure 4.21. The second domain corresponds to the porous material. For solving the hydraulic problem, we impose a hydraulic head at $z = 0$ m, $h = 0$ m and a positive head h in the top of the second sub-domain. (a). Demonstrate that $h = a \times z$. (b) Show that self-potential can be written as: $\psi = -L_2 az/\sigma_2$ and $\Delta\psi/\Delta h = -L_2/\sigma_2 = C$, where C denotes the streaming potential coupling coefficient and L corresponds to the streaming current coefficient defined by $\boldsymbol{J}_S = -L\nabla h$. (c) Describe the position and the geometry of the electrokinetic sources.

Exercise 4.2. Inverse self-potential problem. The inverse self-potential problem consists of determining the distribution of the source current density vector. This requires the solution of a linear least-squares problem: $\Phi_T = \|\mathbf{Gm} - \psi^{obs}\|_2^2 + \lambda^2\|\mathbf{m}\|_2$, where the matrix $\mathbf{G}$ denotes the kernel of size $M \times N$ (N denotes the number of self-potential stations and M denotes the number of cells used to discretize the subsurface), ψ^{obs} denotes the self-potential data vector, and λ is the regularization parameter.

(1) Demonstrate that the cost function given above can be written in the following form:

$$\Phi_T = \|\tilde{\mathbf{G}}\mathbf{m} - \tilde{\psi}^{obs}\|_2^2. \qquad (4.90)$$

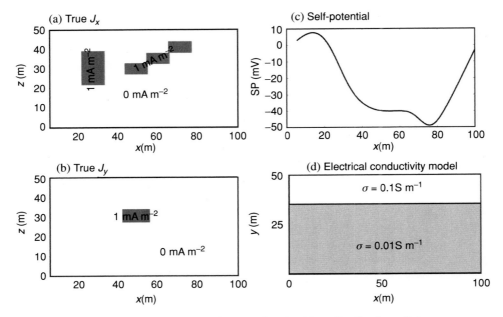

Figure 4.22. Self-potential anomaly associated with a distribution of the source current density in a conductive body. (a) Model of distribution of the horizontal (x-) component of the electrical density source that consists of a cube and a dipping slab anomalies. (b) Model of distribution of the vertical (y-)component of the electric density source where the anomaly taken a rectangle shape. (c) Self-potential anomaly recorded in the surface of the model. (d) The electrical conductivity distribution used in the forward problem and used to compute the kernel.

(2) Show that the minimum of the cost function Φ_T is of the form:

$$\hat{\mathbf{m}}(\lambda) = \left(\frac{\sigma_i^2}{\sigma_i^2 + \lambda^2} \right) \frac{u_i^T \psi^{obs}}{\sigma_i} v_i \tag{4.91}$$

$$G = \sum_{i=1}^{Q} u_i \sigma_i v_i^T \quad \text{where } Q = \min (N, M), \tag{4.92}$$

where $\mathbf{U} = (\mathbf{u}_1; : : :; \mathbf{u}_n)$ and $\mathbf{V} = (\mathbf{v}_1; : : :; \mathbf{v}_n)$ are matrices with orthonormal columns, $\mathbf{U}^T \mathbf{U} = \mathbf{V}^T \mathbf{V} = I_n$, where $\Sigma = \text{diag} (\sigma_1, \dots, \sigma_n)$ has non-negative diagonal elements.

Exercise 4.3. Self-potential tomography. We consider a model subdivided into several blocks (10×10 cells), in which a distribution of the electrical conductivity and the distribution of the electrical current density source are known (Figure 4.22).

The numerical self-potential signal corresponding to these distributions is contaminated by Gaussian noise, whose standard deviation is equal to 2% of the data magnitude, plus 1 mV. We dealt with this undetermined problem with 200 unknown parameters and 60 available measurements. Use the Tikhonov approach, with a depth-weighting matrix, enforced by a smoothing constraint. The regularization parameter must be estimated from the L-curve tool to reconstruct the distribution of the vector of the current density source.

5

Applications to geohazards

In this chapter, we describe four applications of the self-potential method to geo-hazard problems including (i) landslides and flank stability, (ii) the detection of sinkholes, (iii) the detection of cavities, and (iv) the study of leakages in dams and embankments using salt tracer tests and self-potential monitoring. In each case, we develop a specific approach to invert or interpret the self-potential field and we provide insights regarding the physical mechanisms at play.

5.1 Landslides and flank stability

In the past decade, an increasing number of works have been performed to better understand the physics of landslides and mudflows and to use geophysical methods to characterize and monitor their activity and predict the onset of sliding; see Schmutz *et al.* (2009). This is an important problem as landslides are responsible for thousands of deaths and injuries each year and enormous economic losses around the world (see Hutchinson *et al.* (2003)). The complex dynamics of the landslides require a broad range of observations at various scales and times involving various disciplines including geology, hydrogeology, soil science, geomorphology, climate sciences and, more recently, geophysics. Because of the crucial role of the ground water and the mositure content of the soil in the triggering of landslides, the self-potential method may appear as a promising non-intrusive method to monitor landslides; see Bogoslovsky and Ogilvy (1977), McCann and Forster (1990), Hack (2000), Mauritsch *et al.* (2000) and Lapenna *et al.* (2003, 2005). In this section, we show and discuss self-potential maps that have been aplied to the study of landslides in Southern Italy by the group of Vincenzo Lapenna.

Vincenzo Lapenna and co-workers observed that the self-potential method can be used for delineating the boundaries of landslides and the main features of the associated ground water flow. The Giarrossa landslide (Figure 5.1) is located close to the city of Potenza and is one of the greatest mass

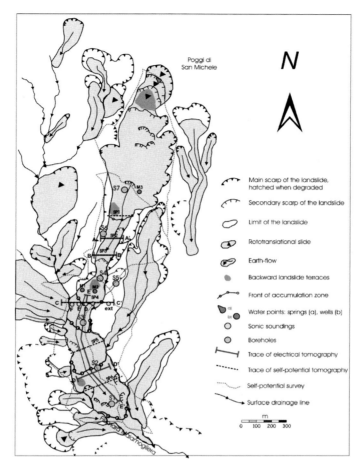

Figure 5.1. Geomorphologic map of the Giarrossa landslide showing location of geological and geophysical surveys; from Lapenna *et al.* (2005), reproduced with the authorization from the SEG and the authors. See also color plate section.

movements of the whole Lucanian Apennine. This landslide is about 2700 m long and about 60–420 m wide. It has a mean inclination of about 9°. The landslide involved the Argille Varicolori formations being composed of intensively tectonised and cracked mudstones and shales locally with blocks of lapideous rocks, marls, calcarenites and limestones. On December 1976, following heavy rains, the entire landslide moved and was responsible for serious damage to housing, communications, and roads; see Polemio and Sdao (1998).

Figure 5.1 summarizes the main geomorphological features of the landslide and shows the location of the geological and geophysical surveys performed by the group of Vincenzo Lapenna. The source area of this earth flow is broad and the

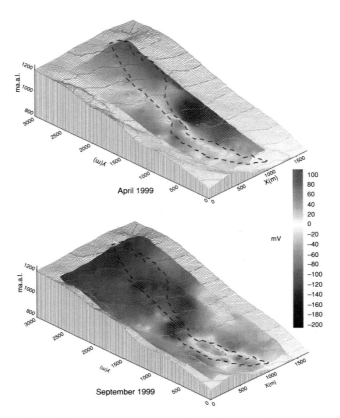

Figure 5.2. Maps of the self-potential anomalies relative to the areal survey carried out in April 1999 (above) and September 1999 (below). The self-potential profiles are in the range −200 to 100 mV and the dashed line represents contour of slide (reproduced with authorization from the SEG and the authors; from Lapenna *et al.* (2005)). See also color plate section.

main scarp is semicircular in plan and concave in the longitudinal direction. The thickness of the disturbed material is in the order of 35–40 m and the depth of the sliding surface varies between about 20 and 30 m, according to Lapenna *et al.* (2003). The accumulation zone is about 1050 m wide, with a mean inclination of about 7°. The thickness of material displaced varies from about 10 m in the lateral zones to 25–30 m in the central zones of the accumulation. Two self-potential maps have been carried out in different climatic conditions by the group of Lapenna; see Gallipoli *et al.* (2000) and Lapenna *et al.* (2003) for further details. Figure 5.2 shows two self-potential maps, which show broad self-potential anomalies, predominantly negative in the source zone and positive in the accumulation zone. These anomalies are consistent with the pattern of ground water flow in mountainous areas with the positive charges of the electrical diffuse layer (see Chapter 1) are dragged in the direction of the flow of the ground water; see, for example, Sharma (1997). In the map made in fall, the decrease and the migration downwards of the positive self-potential area may reflect the decrease of the piezometric surface inside the body of the landslide following the dry summer period (see Lapenna *et al.* (2005)). In the source area, a quite uniform negative self-potential anomaly can be observed. This anomaly indicates that the infiltration of the meteoric water is rather pervasive over a broad area. In the accumulation area, a clear separation is observed between the negative and positive self-potential. Moreover, strong self-potential horizontal changes from negative to positive self-potential values delimit the boundary of the accumulation zone of the landslide. According to Lapenna *et al.* (2005), such observations are consistent with water infiltration from the surrounding terrains into the disturbed material of the landslide characterized by a higher permeability because of the deformation and the formation of cracks and open conduits. This clear boundary in the self-potential map is therefore indicative of the dynamics of the ground water in the landslide.

Another interesting landslide investigated by the group of Vincenzo Lapenna corresponds to the Varco d'Izzo landslide. This landslide is located in the vicinity of the city of Potenza (Southern Italy). It has been recognized as one of the most dangerous landslides of the whole Lucanian Apennine Chain (see details in Cruden and Varnes (1996)). It is an active rototranslational landslide, i.e., complex mass movement occurring dominantly on a partly planar and partly curved detachment. The landslide is about 1400 m long, about 130–140 m wide and its extension is about 700 m. It is characterized by a mean inclination of about 10° (Figure 5.3). In the area of the landslide, clayey-marly formations outcrop. They are composed of intensively fractured shales and mudstones plus blocks of lapideous rocks including marls, calacarenitic rocks and limestones.

Figure 5.3 shows the main geomorphologic features of the landslide; see Perrone *et al.* (2004) and Lapenna *et al.* (2005). The source area of the landslide is a rototranslational slide showing an inclination mean of about 16°. It is about 300 m long, about 100–150 m wide and extends from 885 to 820 m above see level (a.s.l.). The flow track of the landslide is ~700 m long and, on average, its inclination is 12°. The thickness of the sliding material varies from 15 to 25 m. The zone of accumulation is the widest part of the landslide (about 500 m) and is 350 m long. It shows an inclination of 8°. The toe material has diverted the Basento riverbed. The thickness of the displaced material, as shown by the borehole data, varies from approximately 13 m in the upper zone to 32 m in the lower zone.

The self-potential map of the Varco d'Izzo landslide is shown in Figure 5.4. It shows a spatial pattern characterised by slightly negative self-potential values in the source area, a positive zone in the flow track area and a strong self-potential

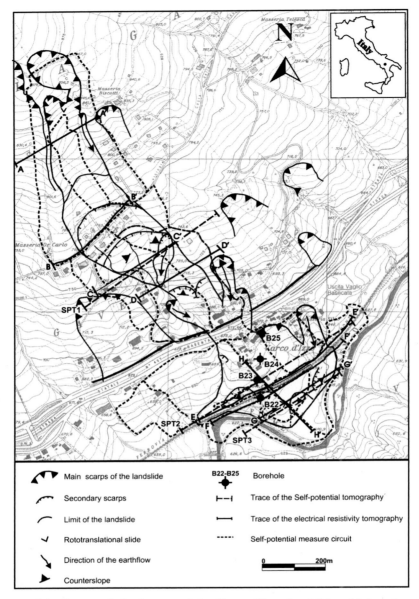

Main scarps of the landslide		**B22-B25**	Borehole
Secondary scarps		⊢–⊣	Trace of the Self-potential tomography
Limit of the landslide		⊢—⊣	Trace of the electrical resistivity tomography
Rototranslational slide		- - - - -	Self-potential measure circuit
Direction of the earthflow			
Counterslope		0 200m	

Figure 5.3. Geomorphologic map of the Varco d'Izzo landslide with location of the geological and geophysical surveys. SP mapping was carried out on the entire landslide body; seven electrical resistivity tomography profiles were performed transversal to the landslide and only one longitudinal; self-potential profiles were carried out transversal to the landslide body; from Lapenna *et al.* (2005), reproduced with authorization from the SEG and the authors.

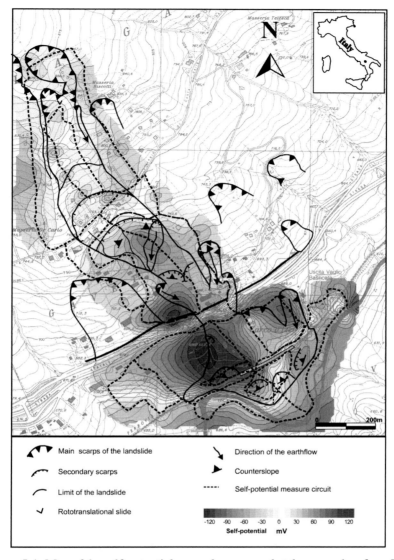

Figure 5.4. Map of the self-potential anomaly measured at the ground surface. Red zones are positive values, whereas blue zones are negative values; from Lapenna *et al.* (2005), reproduced with authorization from the SEG and the authors. See also color plate section.

negative zone in the accumulation area of the slide (Figure 5.4). The self-potential negative values in the source area are due to very pervasive water infiltration in the detachment zone. The self-potential positive anomaly in the flow track may be due to the containment walls and drainage structures built for the SS-407 Basentana road. These structures may indeed prevent ground water from flowing into the accumulation area. On the other hand, the self-potential negative sector in the accumulation zone, which seems to follow the slide outline perfectly, may be explained by a ground water flow towards the Basento River.

It will be interesting, in the future, to perform some ground water flow modeling of these landslides and to condition the ground water flow models to the resistivity and self-potential data on one side and the hydrogeological data on the other. This would, however, require some additional information regarding the electrical conductivity of the ground water and some piezometric data plus a petrophysical characterization of the material and scaling effects associated with the blocks. It would be also interesting to monitor the self-potential signals of these landslides together with performing time lapse resistivity monitoring. The recent development of new algorithms for time lapse complex resistivity tomography from Karaoulis *et al.* (2011) could be useful in this context.

5.2 Sinkhole detection

The detection of sinkholes in karstic areas is also a growing field of interest to hydrogeophysicists in many countries. These sinkholes are sometimes difficult to observe from the ground surface. For instance, in Normandy (France), a fraction of the sinkholes in the chalk substratum are masked by a loess and clay-with-flint sedimentary cover; see Laignel *et al.* (2004) (Figure 5.5). Locating sinkholes in the chalk substratum is important for several reasons. The sinkholes are responsible for the vulnerability of the chalk aquifer to agricultural waters, which contain pesticides and nitrates. In addition, the development of sinkholes over time is associated with the risk of collapse of the ground and represents therefore a strong geohazard in this region.

A fraction of the sinkholes existing in Normandy are visible from the ground surface. They appear as more or less pronounced circular depressions. The sinkholes are usually clusterized along fractures in the chalk bedrock; see Salvatia and Sasowskyb (2002), Some sinkholes, however, are not visible at the ground surface because they are not mature enough to create a depression. Indeed, sinkholes start as micro-caves formed by dissolution-enlargement of some initial fractures in the chalk, which correspond also to depressions of the clay-with-flint/loess interface. As the micro-caves enlarge, the overlying sediments are piped down into the chalk, creating a void within the clay-with-flint formation (Figure 5.6). At this stage, these sinkholes are called crypto-sinkholes and non-intrusive geophysical methods are needed to map their positions with the goal to quantify the risks associated with their location and their development over time.

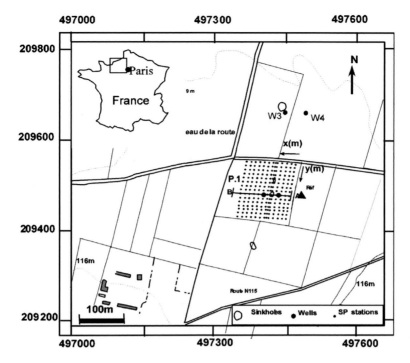

Figure 5.5. Location of the test site in Normandy (north-west of France). At this test site, the sinkholes are all aligned along a North–South trend; from Jardani *et al.* (2006a). The small dots correspond to the self-potential stations. "Ref" denotes the position of the reference self-potential station for the survey.

Several geophysical methods have been tested in the past to detect sinkholes. They include conductivity mapping with the EM-31 and EM-34 induction-based electromagnetic instruments (see Ahmed and Carpenter (2003)), ground-penetrating radar, and electrical resistivity tomography using multi-array electrodes (see van Schoor (2002) and Zhou *et al.* (2002)), and (micro)gravimetry (see Closson *et al.* (2003)). Surface geophysical data can also be combined with downhole measurements to locate sinkholes more accurately, according to Wedekind *et al.* (2005). Ground-penetrating radar can not be used if the overburden is conductive (e.g., in the presence of a clay-rich cover like in the present study). Also d.c.-electrical resistivity and gravimetry cannot be used to survey a large region due to the time required for the acquisition of the data. As shown below, the self-potential method appears therefore as a suitable reconnaissance method to survey broad areas. In addition, the self-potential method is the only geophysical method sensitive directly to ground water flow and therefore, because sinkholes

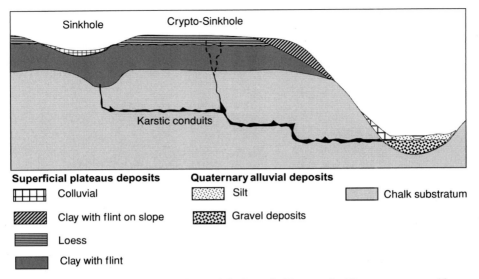

Figure 5.6. Sketch of the geology of the karst in Normandy. There are two aquifers, one perched aquifer in the loess and one deeper aquifer in the chalk. The sinkholes are visible from the ground surface. Crypto-sinkholes may exist that cannot be observed through visual inspection from the ground surface. They can be detected only through drilling or (non-intrusive) geophysical investigations.

are preferential flow conduits, they should be associated with clear self-potential (negative) anomalies.

Several works have focused on the use of the self-potential method to survey sinkholes and karst; see, for example, Erchul and Slifer (1987), Wanfang *et al.* (1999) and Wedekind *et al.* (2005). Jardani *et al.* (2006a) used numerical simulations to demonstrate that the self-potential method can be used qualitatively to detect the position of crypto-sinkholes and sinkholes because they form vertical preferential pathways for the flow of the ground water. This initial work was followed by several other works; see Jardani *et al.* (2006a, 2007b).

The test site discussed below is located in Normandy (France) in the Upper Cretaceous chalk karst of the Western Paris basin (Figure 5.5). In this region of extensive agricultural areas, sinkholes and crypto-sinkholes are frequently clustered in a thick chalk substratum. The cover of this substratum is composed of Pliocene clay-with-flint and loess materials resulting from the alteration of the chalk substratum (Laignel *et al.* (2004)). The thickness of this cover ranges between a few meters to ten meters outside sinkholes. It can reach ~ 15 m over some sinkholes. The permeability of the upper units is relatively low ($\sim 10^{-5}$ m s^{-1} for the loess cover to 10^{-10} m s^{-1} for the clay-with-flint cover; see Jardani *et al.* (2006a)). Owing to the contrast in permeability between the loess and the clay-with-flint covers, a shallow perched aquifer exists above the clay-with-flint formation in spring. Water flow is predominantly lateral in this upper flow system until a sinkhole is encountered. Sinkholes act as local vertical pathways through the clay, connecting the shallow upper flow system to the main aquifer located in the chalk formation at a depth of about 30 m. Five boreholes drilled at the test site demonstrate the existence of this shallow aquifer above the clay-with-flint cover (Figure 5.6).

As discussed in Chapter 1, the percolation of water through a porous material generates an electrical field of electrokinetic nature called the streaming potential because of the drag of the excess of charge of the electrical diffuse layer. As described in Chapter 4, several methods can be used to connect these self-potential signals to the geometry of the ground water flow pathways; see, for example, Fournier (1989), Aubert and Atangana (1996), Revil *et al.* (2003) and Leroy and Revil (2004).

In Jardani *et al.* (2006a), non-polarizing electrodes consisting of a bare copper cylinder immersed in a supersaturated copper sulfate solution in a porous ceramic cup were used for the self-potential surveys described below. The self-potentials between the reference station and the scanning electrode were measured with a calibrated Metrix MX-20 voltmeter with a sensitivity of 0.1 mV and an internal impedance of 100 MOhm. The distance between two measurement stations was 10 m and 5 m in the outer and central parts of the investigated areas. A total of 225 measurements were performed over a surface area of 15 400 m^2. The standard deviation of the self-potential measurements at this specific site was 1 mV. This means a very high signal-to-noise ratio and a high reproducibility of the measurements (see Figure 1.7). Using finite element modeling, Jardani *et al.* (2006a) showed that the sinkholes are associated with negative self-potential anomalies of a few tens of millivolts associated with the downward infiltration of meteoric waters (Figure 5.7).

5.2.1 Spring data

The self-potential map realized in spring 2005 shows negative anomalies (Figure 5.7a). Spring is the rainy season in Normandy. In spring 2005, the cumulative rain amounted to 400 mm. As modeled by Jardani *et al.* (2006a, 2007b), the percolation of water into the sinkholes in spring was responsible for negative self-potential anomalies of about -30 mV (see the self-potential anomalies A1 and A2 in Figure 5.7a). The position of these self-potential anomalies provides a clear footprint signature of the location of the sinkholes. The density of self-potential stations used in this study is insufficient to allow the shape of the sinkholes to be precisely determined. Wider anomalies (B1, B2 and B3) correspond to topographic

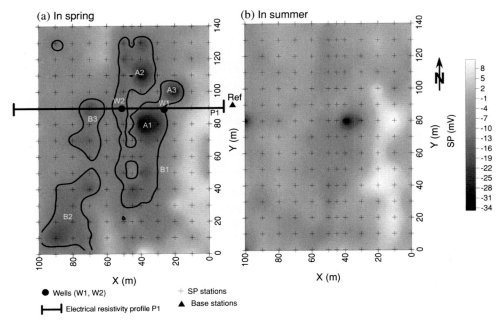

Figure 5.7. Self-potential map of the test site shown in Figure 5.5. The stars correspond to the stations where the self-potential was measured ("Ref" indicates
the position of the self-potential reference station). The self-potential data were
obtained (a) in spring during the rainy season and (b) in summer during the dry
season. W1 and W2 denote two boreholes. AB is a multi-electrode resistivity
profile termed P1. The self-potential anomalies A1 and A2 are associated with
the presence of two sinkholes (visible from the ground surface). They are located
on the depression of the clay-with-flint/loess interface. Anomalies B1, B2 and
B3 denote crypto-sinkholes (see explanation in Figure 5.6) that is located outside
this depression (see Jardani *et al.* (2006a)).

lows of the clay-with-flint/loess interface. All these anomalies are clustered along
a North–South trend at the test site (Figure 5.7). This trend is also the trend along
which the visible sinkholes are clustered (Figures 5.7 and 5.8). The anomaly termed
A3 is potentially a crypto-sinkhole. The areas near the anomalies termed B2 and B3
(amplitude of about −20 mV) are topographic lows of the clay–loess interface; see
Jardani *et al.* (2006a, 2007b). They were interpreted as potential crypto-sinkholes.

Based on the assumption proposed first by Jackson and Kauahkana (1987)
(negative self-potential anomalies are proportional to the thickness of the vadose
zone, see also Chapter 6), Aubert and Atangana (1996) proposed to determine
the bottom of the vadose zone from the following formula (see Aubert and Atangana (1996) and Aubert *et al.* (2000)): $H(x, y) = h(x, y) - e_0 - \psi(x, y)/c$, where
$H(x, y)$ is the altitude (with respect to a datum, e.g., the sea level) of the bottom of

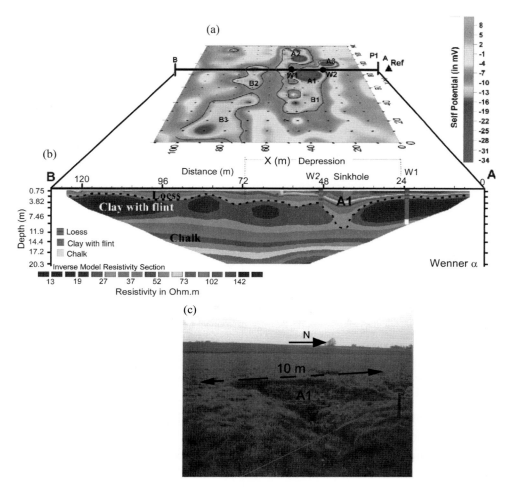

Figure 5.8. Comparison between the self-potential map and the d.c.-electrical resistivity imaging. (a) Self-potential map. Electrical resistivity tomography along profile P1 (electrode spacing of 3 m, Wenner-α array). The electrical resistivity tomography is shown at the fourth iteration (RMS error: 1.39%). (b) Electrical resistivity tomography along profile P2 (electrode spacing of 4 m, dipole-dipole array). (c) Picture of Sinkhole A1, which has a diameter of 10 meters. The depth of the depression is about 2 m. Modified from Jardani *et al.* (2007b). See also color plate section.

the vadose zone below the measurement station $P(x, y)$, $h(x, y)$ is the altitude of the ground surface at station $P(x, y)$, e_0 is the thickness of the vadose zone below the reference self-potential station (where by definition the self-potential signals is equal to zero, see Chapter 1), ψ is the measured self-potential signal at station $P(x, y)$, and c is an apparent voltage coupling coefficient usually expressed in mV

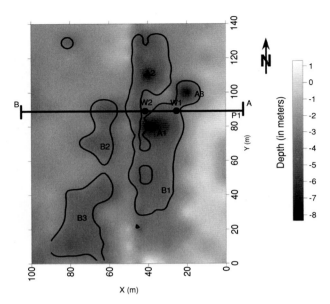

Figure 5.9. Determination of the depth (expressed in meters) of the SPS surface using the self-potential data measured in spring. This interface corresponds to the interface between the clay-with-flint cover and the loess formation (see Figures 5.6 and 5.8). This interface is also the top of the aquitard (the clay with the clay-with-flint formation) and the bottom of the perched aquifer (located in the bottom part of the loess formation) (from Jardani *et al.* (2007b)).

m^{-1}. As explained above, the vadose zone corresponds to the loess layer. It follows that the bottom of the vadose zone corresponds also to the interface between the clay-with-flint cover and the loess formation. We know the position of this interface from two boreholes. It is also very well marked on the d.c. resistivity tomograms (see Figure 5.8). Jardani *et al.* (2006a) showed that by taking $c = -5.7$ mV m^{-1} and $e_0 = 2$ m, it was possible to reconstruct the position of the bottom of the vadose zone (called the SPS surface in the terminology of Aubert and co-workers, see Chapter 6). The result is reported on Figure 5.9. This map was consistent with the well data in terms of the position of the interface between the clay-with-flint and loess along which the perched aquifer is located.

5.2.2 Summer data

The summer 2005 was a very dry period in France. In summer, the self-potential map evidences much fewer anomalies than in spring (compare Figures 5.7a and 5.7b). However, the summer self-potential map shows the position of the main sinkholes at the test site (e.g., sinkhole A1). The wells show no free water above the clay-with-flint formation (no perched aquifer). Consequently, there is still a residual percolation of water through this sinkhole but this infiltration comes from the slow flow of water through the clay-with-flint formation and not from the percolation of water from the perched aquifer lying just above the clay-with-flint/loess interface. The smaller permeability of the clay-with-flint layer is compensated by the higher surface of discharge offered for such an infiltration. In addition, the monitoring of the self-potential signals in summer shows positive signals during the day (not shown here) that are likely due to evaporation of water in the vadose zone.

As for landslides, it would be interesting to setup a permanent array of electrodes to monitor the fluctuations of the self-potential signals over time and use this information to invert the hydrogeological characteristics of the infiltration in these sinkholes and their evolution over long time series.

5.3 Detection of cavities

Locating cavities in the ground is an important step in drawing geohazard maps especially in populated areas. In High Normandy (North-West of France), anthropic cavities have been dug in the chalk from the fifteenth century until the 1950s to extract chalk blocks used in agriculture to improve the agricultural productivity of acidic soils. Nowadays, these cavities have been abandoned and their shafts plugged. It is estimated that more than 100 000 marl-pits could exist in Normandy. Most of the time, the position of these marl-pits is unknown. One can therefore easily imagine the cost induced by their collapse and the need for efficient and cost-effective methods to locate them not only in this region but also in other regions of the world where similar problems exist (see Jardani *et al.* (2006b)).

Various geophysical methods have been used to detect caves. Microgravimetry, seismic and electrical resistivity surveys are expensive while georadar does not perform well if the sediment cover is electrically conductive (e.g., in the presence of clay-bearing sediments) as in the case reported below. Self-potential tomography is shown below to be an efficient method to locate caves. Previous works (see, for example, Lange (1999, 2000)) have shown that water-saturated caves are responsible for positive self-potential signals at the ground surface while air-filled caves are responsible for negative self-potential anomalies. Some anomalies show sometimes an M- or W-shape signal centered on the position of the cave in either case. While Quarto and Schiavone (1996) and Lange (1999, 2000) suggested that these self-potential anomalies are due to the streaming potential resulting from the percolation of the ground water, Green (2000) suggested they result of steady-state telluric currents and the contrast of electrical resistivity between the cave and the

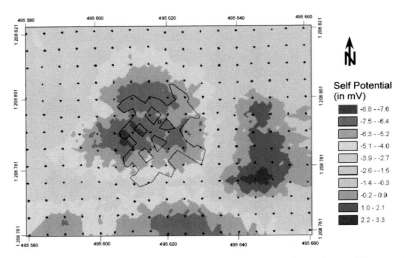

Figure 5.10. Self-potential map above a cave in the investigated area. The spacing between the measurements is 5 m and the *x* and *y* coordinates denote Cartesian coordinates (in meters). The mine shaft corresponds to the small open circle ($x = 618$ m, $y = 795$ m) located near the center of the negative self-potential anomaly. The black thin lines indicate the shape of the cave including the presence of five pillars (see Jardani *et al.* (2006b)). See also color plate section.

surrounding medium. In the following, we show how both the percolation of water and the distribution of the electrical resistivity are responsible for the self-potential anomalies observed at the ground surface of the Earth.

The test site we used to test the usefulness of the self-potential method is located in Normandy, in the north-west of France. A high-resolution self-potential map (Figure 5.10) was realized in the Spring of 2005 using non-polarizing Cu/CuSO$_4$ electrodes and a high internal impedance (100 MOhm) Metrix MX20 voltmeter (sensitivity of 0.1 mV). The self-potential map of Figure 5.10 shows a negative self-potential anomaly centered around the shaft of the cavity (note the position of the cavity was unknown when the field survey was performed and the shaft was plugged and not visible from the ground surface). The density of the measurements has to be high enough to avoid aliasing of the data. The repeatability of the measurements was 1 mV and the determination of the standard deviation of the measurement was equal to 0.8 mV because of the excellent contact between the electrodes and the ground (see Jardani *et al.* (2006b)).

As shown in Chapter 4, the potential at an observation point P depends on a primary source term corresponding to the divergence of the source current density and a secondary source term associated with the gradient in resistivity. In the case of a cavity, there is a divergence of the source current density and there is also is a strong contrast in the resistivity distribution between the cavity and the surrounding medium. Revil *et al.* (2001) and Iuliano *et al.* (2002) proposed a Source Element Occurrence Probability (SEOP), which works as follows.

The electrical field $\mathbf{E}(\mathbf{r})$ due a single dipole with a moment $\mathbf{d}$ is written as, $\mathbf{E}(\mathbf{r}) = \mathbf{d}\nabla\mathbf{G}$ where $\mathbf{G}$ is the Green function of the dipolar source. We note E, the norm of the electrical field vector $\mathbf{E}(\mathbf{r})$. The power associated with the electrical field distribution is

$$\wp(E) = \int_S E^2(\mathbf{r})dS, \tag{5.1}$$

$$\wp(E) = \sum_v d_v \int_S \mathbf{E}(\mathbf{r})\frac{\partial\mathbf{G}}{\partial v_P}dS, \tag{5.2}$$

where $v = x, y, z$ and $v_p = x_p, y_p, z_p$. The projection of S onto the (x, y) horizontal plane is adapted to a rectangle with sides of total length $2X$ and $2Y$ along the x- and y-axis, respectively. The cross-correlation product is defined by

$$\eta_v(r_p) = C_v^p \int_{-X}^{X} \int_{-Y}^{Y} \mathbf{E}(\mathbf{r})\frac{\partial\mathbf{G}(\mathbf{r}_p - \mathbf{r})}{\partial v_p}dxdy, \tag{5.3}$$

$$C_v^p = \left[\int_{-X}^{X} \int_{-Y}^{Y} E^2(\mathbf{r})dxdy \int_{-X}^{X} \int_{-Y}^{Y} \frac{\partial\mathbf{G}(\mathbf{r}_p - \mathbf{r})}{\partial v_p}dxdy \right]^{-1/2}, \tag{5.4}$$

where C_v^p denotes the normalization constant. The semblance function is therefore the normalized scalar product between the form-anomaly factor indicated by the electrical potential measurements at the ground surface and the form-factor associated with the dipolar source located in the source volume. These cross-correlation densities have the following property $-1 \le \eta_v(r_p) \le 1$. The norm of the cross-correlation vector $\boldsymbol{\eta}(\eta_x, \eta_y, \eta_z)$ is given by $\eta(r_p) = (\eta_x(r_p)^2 + \eta_y(r_p)^2 + \eta_z(r_p)^2)^{1/2}$. This function can be contoured to provide a tomographic image of the source current density. Note that this algorithm does not distinguish between self-potential sources associated with primary signals (the existence of a source of current in the ground) or secondary sources associated with the existence of resistivity contrasts.

In the field case reported below, the cavity is filled with air and therefore very resistive; see Jardani *et al.* (2006b). This implies a strong contrast of resistivity between with cavity and the host rock (chalk). The application of the self-potential tomography algorithm to the synthetic cases reported above indicates that the maximum of the SEOP is associated with the roof of the cavity at a depth of 9 m (Figures 5.11 and 5.12). Once the shaft was localized and the map of the cavity

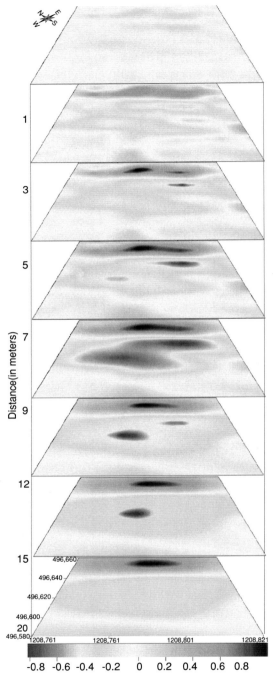

Figure 5.11. 3D-self-potential tomography using a dipolar cross-correlation approach (from Jardani *et al.* (2006b)). The normalized dipole charge density is comprised between (-1) for negative charge accumulations and ($+1$) for positive charge accumulations. The negative charge accumulations are consistent with the position of the roof of the cave located about 9 m below the ground surface (see Figure 5.12).

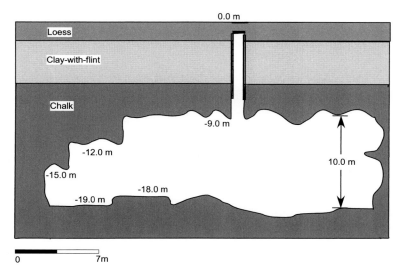

Figure 5.12. Real shape of the cave. Note that the roof of the cavity is roughly located at a depth of 9 –10 m below the ground surface (see Jardani *et al.* (2006b)).

established, we found that 9 m was effectively the depth of the roof of the air-filled cavity (see Figure 5.12). This shows that this very simple cross-correlation approach is able to locate the roof of cavities in the ground with a minimum of prior knowledge.

5.4 Leakages in dams and embankments

Self-potential mapping has been applied for decades to qualitatively detect preferential fluid flow pathways in embankments and earth dams (see Ogilvy *et al.* (1969), Gex (1980), Sill and Killpack (1982), Merkler *et al.* (1989), Wilt and Corwin (1989), Wilt and Butler (1990), Al-Saigh *et al.* (1994) and Sheffer and Howie (2001, 2003)). Typical steady-state self-potential anomalies associated with earth dams amount to several tens of mV (32 mV in the earth dam investigated by Panthulu *et al.* (2001), 80 mV in the embankment dam analyzed by Bolève *et al.* (2009), 80 mV in Rozycki (2009)) and sometimes hundreds of millivolts (300 mV in the earth dam investigated by Rozycki *et al.* (2006), 400 mV for the case study reported by Asfahani *et al.* (2010), over 200 mV at Dana Lake, see Moore *et al.* (2011), and 170 mV at the Hidden dam, see Minsley *et al.* (2011)).

We discuss a method based on the injection of salt upstream of a heterogeneous porous material and the time-lapse monitoring of the self-potential response at the ground surface to detect and image preferential fluid flow pathways between an upstream reservoir and a downstream reservoir. We call this test "SMART"

(Self-potential Monitored sAlt tRacer Test). In this section, we review the theory behind the SMART test to characterize the change in the self-potential signals associated with the advection–dispersion of a saline tracer in a porous material.

5.4.1 Theory for salt tracer tests

The equations of transport for the simulation of the migration of a salt tracer in a water-saturated porous material were given in Section 4.1 and will not be repeated here. The same applies for the source current density, which is in this case associated with two contributions: the pore water flow and the gradient of concentration (examples of applications can be also found in Revil and Jardani (2010a) and Martínez-Pagán *et al.* (2010)):

$$\mathbf{J}_S = \hat{Q}_V \mathbf{U} - k_b T \sum_{i=1}^{N} \frac{t_i \sigma}{q_i} \nabla \ln\{i\}. \tag{5.5}$$

The first term of the right-hand side of Eq. (5.5) corresponds to the streaming current density and the second term corresponds to the diffusion current density. The variable T represents the absolute temperature (in K), k_b denotes the Boltzmann constant (1.381×10^{-23} J K^{-1}), $\hat{Q}_V$ (in C m^{-3}) is the effective charge of the electrical diffuse layer per unit pore volume that can be dragged by the flow of the pore water, q_i (in C) is the charge of species i dissolved in water, and t_i (dimensionless) is the microscopic Hittorf number of the ionic species i in the pore water. This microscopic Hittorf number represents the fraction of electrical current carried by species i in the aqueous phase. The term $\{i\}$ represents the activity (concentration times the activity coefficient, unitless) of the ionic species i. In the case reported below, the complete dissociation of NaCl(s) provides two ionic species Na$^+$ and Cl$^-$ to the pore water.

The charge per unit pore volume $\hat{Q}_V$ is salinity dependent and, to be consistent with the Helmholtz–Smoluchowski equation at thermodynamic equilibrium between the pore water and the mineral surface, this salinity dependence has to be the same as the salinity dependence of the zeta potential, the inner potential of the electrical double layer. Revil *et al.* (1999a,b) showed that the zeta potential (hence the charge density $\hat{Q}_V$) changes by one order of magnitude over six orders of magnitude in salinity change for silica sands. Therefore the salinity dependence of $\hat{Q}_V$ can be neglected at first approximation. There is also another reason to keep this parameter constant in the following analysis: the change in the zeta potential or in $\hat{Q}_V$ with the salinity is controlled by the sorption of cations in the Stern layer (the inner part of the electrical double layer). Usually the salt tracer experiments reported below takes only few minutes while the kinetics of sorption of sodium on silica takes a few tens of minutes to several hours.

Revil (1999) showed that, in a diffusion problem, one might replace the gradient of the logarithm of the activity of the salt by the gradient of the logarithm of the electrical conductivity of the salt. Using this approach, we can rewrite the total source current as (see Chapters 2 and 3),

$$\mathbf{J}_S = \hat{Q}_V \mathbf{U} - \frac{k_b T}{e} \sigma (2t_{(+)} - 1) \nabla \ln \sigma_f, \tag{5.6}$$

where e is the elementary charge of the electron, $t_{(+)}$ is the microscopic Hittorf number of the cation (see values in Revil (1999), $t_{(+)} = 0.38$ for a sodium chloride solution), and σ_f is the conductivity of the pore water (in S m^{-1}), which is proportional to the salinity at a given temperature. Eq. (5.6) has been successfully used in a number of recent studies, such as those by Revil and Jardani (2010a), Martínez-Pagán *et al.* (2010) and Woodruff *et al.* (2010).

In a clay-free sand at low Dukhin numbers (see Chapter 3; see also Crespy *et al.* (2008) and Bolève *et al.* (2007a)), the conductivity of the sand σ is linearly related to the conductivity of the pore water σ_f (see Archie (1942) and Clavier *et al.* (1984)) by

$$\sigma = \frac{\sigma_f}{F}, \tag{5.7}$$

The formation factor is related to the connected porosity ϕ by Archie's law: $F = \phi^{-m}$ (see Archie (1942)), where m is called the cementation exponent (typically 1.3 for well-sorted clean sand as used in the following experiment, e.g., Hallenburg (1998), p. 127). From Eqs. (5.6) and (5.7), the total source current density can be rewritten (see Ikard *et al.* (2012)) as

$$\mathbf{J}_S = \hat{Q}_V \mathbf{U} - \frac{k_b T}{Fe} \left(2t_{(+)} - 1 \right) \nabla \sigma_f. \tag{5.8}$$

Combining Eqs. (5.6) and (5.7), the self-potential field ψ is the solution of the following Poisson equation described in Chapter 1,

$$\nabla \cdot (\sigma \nabla \psi) = \nabla \cdot \mathbf{J}_S, \tag{5.9}$$

where the source term (the right-hand side of Eq. (5.9)) can be directly related to the Darcy velocity $\mathbf{U}$ and to the gradient of the conductivity of the pore water. Both the Darcy velocity and the salinity are obtained by solving the primary flow problem as discussed in Chapter 4, including the choice of appropriate boundary conditions.

In the case of the laboratory experiment presented below, the boundary conditions include insulating boundary conditions for the electrical potential at the top surface of a tank and on the side boundaries and impervious boundary conditions except between the reservoirs and the sandbox containing the sand. An important point is that the self-potential field is never measured in an absolute sense. The measured electrical potentials at a given set of non-polarizing electrodes are measured with respect to a reference electrode for which the electrical potential is considered to be, by definition, equal to zero. As explained in Chapter 3, the position of the reference electrode needs to be accounted for when comparing numerical modeling with the experimental results.

5.4.2 Laboratory experiment

The goal of this experiment was to determine if, by measuring the fluctuations of the electrical field at the top surface of a sandbox, we could visualize non-intrusively a preferential flow pathway that is illuminated by the advective transport of salt dissolved in water. The tank consists of two reservoirs, upstream reservoir 1 for injection and downstream reservoir 2 for pumping, and the sand medium in between (Figure 5.13). The two reservoirs are separated by a distance of 0.99 m, as shown in Figure 5.13. The sandbox comprises a central channel of coarse sand (sand A), bounded by two flanking banks of fine sand (sand B) (Figure 5.13d). The sands were placed dry in layers of 20 mm, and tamped.

The material properties of the two sands are described in Table 5.1. The properties of the tap water used for the experiment are reported in Table 5.2. There is no cross-water flow at the boundaries between the channel and the bank sand. The sand is separated from the upstream and downstream reservoirs by a permeable membrane made of plastic with a square cell size of 100 μm (Figure 5.13). The flow of water in the sandbox is controlled by pumping (outflow) and injection (inflow) rates to produce a constant hydraulic head gradient across the tank, allowing for measurement of the steady-state self-potential distribution. During steady-state conditions, the difference in head between the two reservoirs is 22.3 cm over a distance $L = 99$ cm, so the hydraulic head gradient is 0.225. In each sand, the permeability is assumed to be isotropic and therefore the permeability is defined as a scalar denoted by k. Using the measured hydraulic conductivity $K = 1.52 \times 10^{-2}$ m s^{-1} for the coarse sand (Table 5.1), it follows that the mean Darcy velocity is given by $u = 3.4 \times 10^{-3}$ m s^{-1}. As $u = v \phi$, the mean velocity of the pore water in the coarse sand channel is given by $v = 8.3 \times 10^{-3}$ m s^{-1}. Therefore, the computed residence time is approximately $\tau = L/v = 119$ s in the coarse sand channel. A similar calculation yields $u = 2.7 \times 10^{-4}$ m s^{-1}, $v = 6.6 \times 10^{-4}$ m s^{-1}, and a corresponding a residence time of 25 min in the fine sand. We also introduced red food dye to the upstream reservoir to independently assess the residence time. The observed residence time of the dye in the permeable channel was 167 s versus

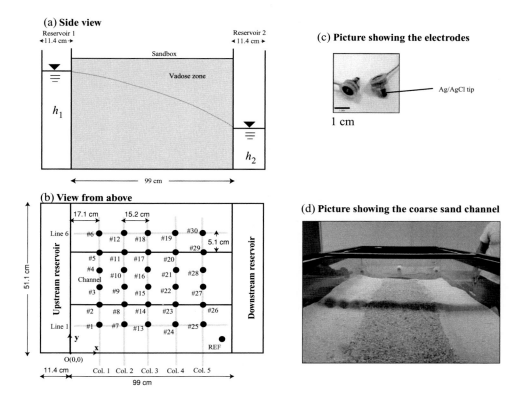

Figure 5.13. Sketch of the experimental setup showing the position of the permeable channel and the positions of the non-polarizable electrodes (small filled circles, the true size of the electrodes being much smaller than the size of the filled circles) located at the top surface of the sand. The hydraulic gradient is defined by the difference between the heads h_1 and h_2 in the two reservoirs located 99 cm apart. The width of the tank is 51.1 cm. REF denotes the position of the reference electrode where the potential is assigned to zero all along the experiment (reference electrode). (a) Side view of the tank. (b) Top view of the tank (not to scale). (c) Picture showing the size of the Biosemi Ag/AgCl electrodes with the amplifiers built in. (d) Picture of the tank showing the coarse sand channel located in between two fine-grained banks.

16 min in the fine sand, implying a reduced mean velocity $v = 5.9 \times 10^{-3}$ m s^{-1} in the high-permeability coarse sand channel.

The electrical potential measurements were recorded over a period of 19 min in a partially electrically shielded laboratory, mitigating the influence of external sources of electrical noise and eliminating the effects of radio frequency interferences. Voltages were recorded at a sample rate of 512 Hz with the BioSemi EEG system by using 30 sintered Ag/AgCl electrodes with integrated amplifiers. The positions of the electrodes are shown in Figure 5.13b. They are located at the top

Table 5.1. *Properties of the two sands used in the experiment.*

Properties	Coarse sand (Channel, #08)	Fine sand (Banks, #30)
Mean grain diameter d_{50} (m) (1)	1.51×10^{-3}	5.00×10^{-4}
Porosity ϕ (–) (1)	0.398	0.410
Formation factor F (–)	3.63	3.48
Hydraulic conductivity K (m s^{-1})	1.52×10^{-2}	1.2×10^{-3}
Permeability k (m^2)	1.98×10^{-9}	2.47×10^{-10}
Charge density $\hat{Q}_V$ (C m^{-3}) (1)	8.10×10^{-3}	4.45×10^{-2}
Conductivity σ (S m^{-1}) (2)	3.90×10^{-3}	4.00×10^{-3}

(1) Using $\log_{10} \hat{Q}_V = -9.23 - 0.82 \log_{10} k$ (see Revil and Jardani (2010a,b), Chapter 3, Figure 3.11).
(2) Using $\sigma = \sigma_f / F$ with $\sigma_f = (4.9 \pm 0.2) \times 10^{-2}$ S m^{-1} at 25 °C.

Table 5.2. *Composition of the tap water with the assumption that hardness is due to calcium. This yields a TDS of 245 ppm ($\sim 5 \times 10^{-2}$ S m^{-1} at 25 °C). Measurement made in April–May 2009; pH = 8.4.*

Component	Concentration (mmol l^{-1})
Ca^{2+}	0.95
K^+	0.09
Na^+	1.44
Cl^-	1.30
SO_4^{2-}	0.82
HCO_3^-	0.75

surface of the sand and are therefore not in contact with the pore water or the salt flowing in the saturated portion of the sand. No clays were added to improve the contact resistance between the electrodes and the sand. Thus, the self-potential data are used as a non-intrusive method here in contrast to the experiments performed by Maineult et al. (2006) for which the electrodes are immersed inside the water-saturated sand. The electrodes used in this study have sintered Ag–AgCl tips (Figure 5.13c), providing low noise, low offset voltages, and stable d.c. performances. Specifications of the BioSemi EEG system can be found in Crespy et al. (2008) and Haas and Revil (2009) for laboratory applications (see also http://www.biosemi.com/). All electrical potentials are measured with respect to the reference electrode denoted "REF" (see Figure 5.13). In addition, a background self-potential distribution was recorded over a 100 s time window prior to salt injection. These data are used to establish a background distribution (which include the unknown electrode-to-electrode static potential differences), which will be removed in the time-lapse mapping of the anomaly associated with the transport of the salt tracer. Therefore, all self-potential anomalies during salt transport are measured with respect to the mean values and trends of the background self-potential distribution. The experiment was repeated three times and the self-potential distribution were found to be reproducible inside 1 mV. We injected the salt (NaCl) instantaneously in the upstream reservoir (16.5 g NaCl mixed with a very small amount of water). The volume of the upstream reservoir was 20.8 l (Figure 5.13).

The self-potential snapshots shown in Figure 5.14a were obtained according to the following steps: (1) the raw BioSemi data collected in 30 channels at 512 Hz were loaded and then (2) converted from units of microvolts to volts. (3) We removed the gain factor of 31.25 from the data. (4) We decimated the data by a factor of 20. First, an eighth-order Chebyshev type 1 low-pass filter was applied to the data with a cutoff frequency of 0.8 $(f_s/2)/20 = 10.24$ Hz, where f_s is the original sampling frequency of the signals (512 Hz). The input data were filtered in forward and reverse directions to eliminate all phase distortions. Second, the smoothed signal was resampled at a lower rate equal to $f_s/20 = 512$ Hz/20 = 25.6 Hz. (5) A linear function is fitted to the first 120 s of data in each channel (prior to the salt injection) and removed from the entire data string for each channel to remove the background distribution and trends. This linear trend is usually associated with a slow drift of some of the electrodes because of their aging. (6) Time-lapse surface self-potential maps were created for data collected after salt injection at $t = 120$ s (note in Figure 5.14, $t = 0$ s corresponds to the time of salt injection). (7) We contoured the data to create surface maps with an isotropic kriging approach based on a uniform variogram. These maps are shown in Figure 5.14a. Figure 5.14b shows the time-lapse change in the electrical potential at one electrode above the coarse sand channel. The displacement of the self-potential anomaly agrees fairly well with the velocity of the flow of the dye through the channel between the upstream and the downstream reservoir $(0.0059 \text{ m s}^{-1})$.

We first need to prove that the traveling self-potential anomaly is indeed caused by the passage of the saline plume. If we first neglect the concentration gradient, the total current density is given by

$$\mathbf{J} = -\frac{\sigma_f}{F}\nabla\psi - \hat{Q}_V K \nabla h, \tag{5.10}$$

where h the hydraulic head. Therefore, the so-called streaming potential coupling coefficient (which represents a change in electrical potential with

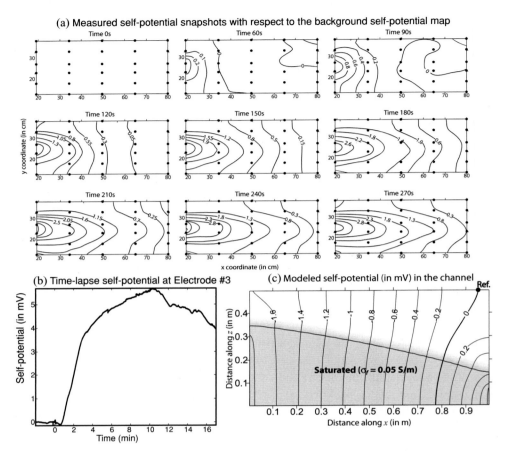

Figure 5.14. Observed self-potential anomalies after the injection of salt upstream. (a) Kriged measured self-potential anomaly contours (expressed in mV) at different elapsed times ($t = 0$ corresponds to the infiltration of the salt in the upstream reservoir). The background potential measured prior to salt injection has been removed. This explains that, at $t = 0$ s, there is no self-potential anomaly. Constant flow conditions are maintained for the duration of the experiment. (b) Self-potential (mV) versus time at electrode #3. (c) Numerical modeling with Comsol Multiphysics of the electrical equipotentials (expressed in mV) associated with the flow of the water in the tank just prior to injection of the salt. The electrical equipotentials are perpendicular to the water table (materialized by the plain line) and the electrical field is higher downstream because of the increase of the velocity downstream. Note the equipotentials are also normal to the upper and lower boundaries because of the insulating boundary conditions (modified from Ikard *et al.* (2012)).

respect to a change in the hydraulic head) is defined (see Chapter 3, Section 3.1) as

$$C = \left. \frac{\Delta\psi}{\Delta h} \right|_{j=0} = -\frac{\hat{Q}_V K F}{\sigma_f}. \tag{5.11}$$

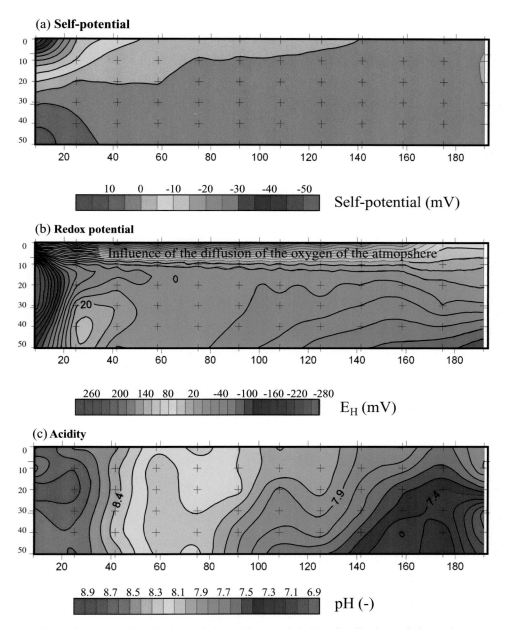

Figure 3.13. (a) Distribution of the self-potential (b), distribution of the redox potential, and (c) distribution of the pH at the end of the experiment (distributions are shown 6 weeks after the introduction of the iron bar). Note that the self-potential signals exhibit a clear dipolar distribution with a negative pole located near the surface of the tank and a positive pole located at depth.

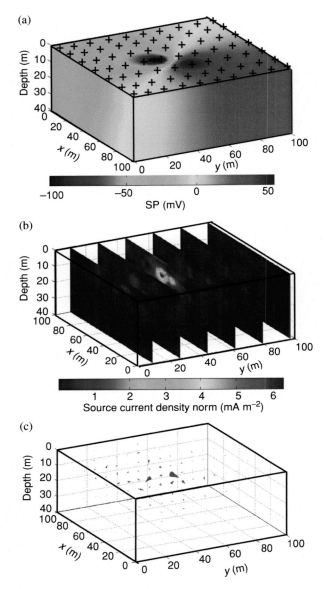

Figure 4.3. Result of the 3D-inversion of the source current density. (a) Distribution of the self-potential over the medium. The crosses correspond to the self-potential stations. (b) Result of the inversion in terms of distribution of the magnitude of the current density (the maximum of the recovered distribution is 6.3 mA m^{-2}). (c) Result of the inversion in terms of direction of the source current density (see Jardani *et al.* (2008)). For full caption, see text (p. 126).

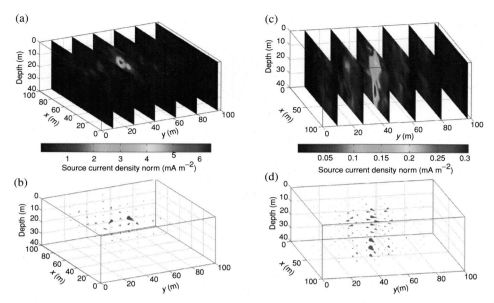

Figure 4.6. Inversion of the synthetic case. (a, b) Result of the inversion (amplitude and direction) with a white noise added to the sampled self-potential data (the amplitude of the white noise is equal to 10% of the self-potential anomaly). Our approach is very robust to the presence of white noise (compare with Figure 4.3). (c, d) Inversion of the data without any prior knowledge of the resistivity distribution. Note that, in this case, both the amplitude and the direction of the inverted current density are grossly wrong. (See Jardani *et al.*, 2008.) For full caption, see text (p. 129).

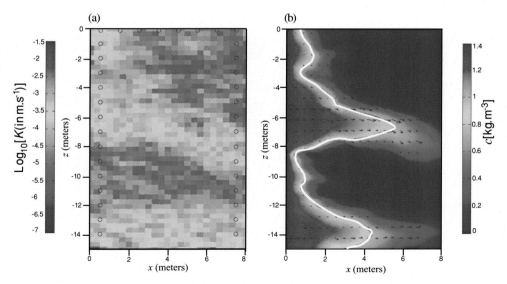

Figure 4.10. True hydraulic conductivity field generated by the SGSIM code (http://www.uofaweb.ualberta.ca/ccg//pdfs/2002%2053-sgsim pw1.pdf). The salt movement is due to a constant hydraulic head gradient between the upgradient well (on the left) and the downgradient well (on the right). The left-hand panel shows the salt concentration distribution at $t = 20$ days. For full caption, see text (p. 136).

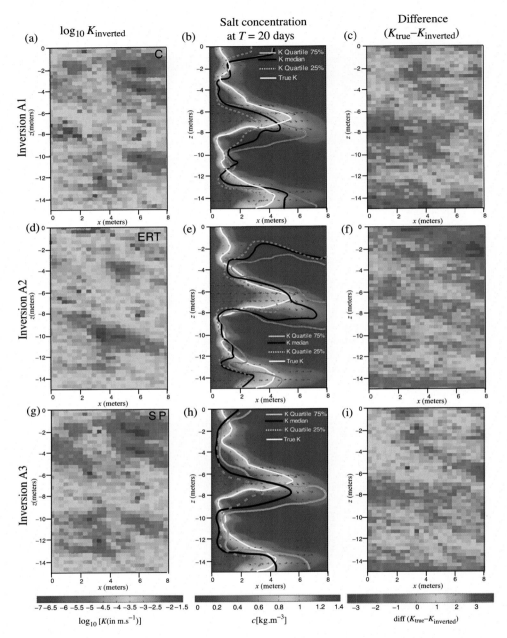

Figure 4.15. Uncoupled inversion of each data set to estimate the hydraulic conductivity field. The inversions A1, A2, and A3 denote the inversion of concentration data, resistivity, and self-potential data, respectively. (a, d, g) Inverted hydraulic conductivity field. (b, e, h) True and inverted position of the salt plume at $t = 20$ days (the inverted position is from the median and the quartiles 75% and 25% of all realizations sampled during the inversions). (c, f, i) Difference (in log scale) between the real and inverted hydraulic conductivity field.

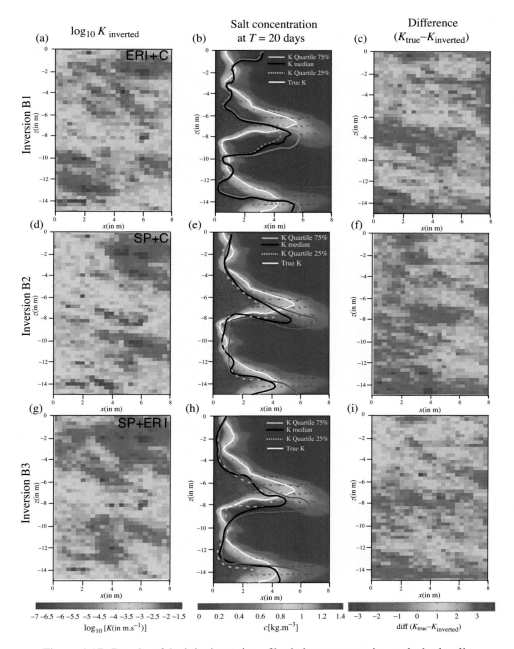

Figure 4.17. Results of the joint inversion of both data sets to estimate the hydraulic conductivity field. The inversions B1, B2, and B3 denote the inversion of the ERI + C, SP + C and ERI + SP data, respectively. (a, d, g) Hydraulic conductivity fields determined from the coupled inversion. (b, e, h) True position of the salt plume at $t = 20$ days and inverted position from the median and the quartiles 75% and 25% of all realizations sampled during the inversion. The third column for each inversion represents the difference (in log scale) between the true hydraulic conductivity field and the inverted hydraulic conductivity field.

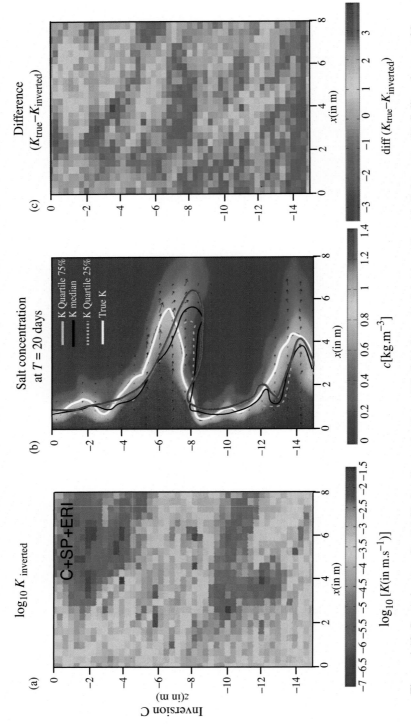

Figure 4.19. Inversion results. (a) Hydraulic conductivity (K)-field from the joint inversion of all data sets. (b) Saline plume at 20 days from the true K-field compared with the prediction from the inverted K-field. (c) Difference between the true and inverted K-fields. For full caption, see text (p. 150).

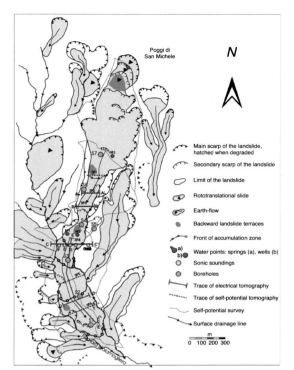

Figure 5.1. (Left) Geomorphologic map of the Giarrossa landslide; from Lapenna *et al.* (2005), reproduced with the authorization from the SEG and the authors. For full caption, see text (p. 155).

Main scarp of the landslide, hatched when degraded

Secondary scarp of the landslide

Limit of the landslide

Rototranslational slide

Earth-flow

Backward landslide terraces

Front of accumulation zone

Water points: springs (a), wells (b)

Sonic soundings

Boreholes

Trace of electrical tomography

Trace of self-potential tomography

Self-potential survey

Surface drainage line

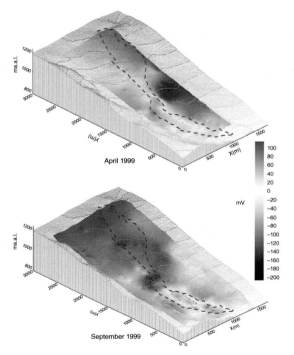

Figure 5.2. (Left) Self-potential (SP) anomalies in April 1999 (above) and September 1999 (below). The dashed line represents contour of the slide (reproduced with the authorization from the SEG and the authors; from Lapenna *et al.* (2005)). For full caption, see text (p. 156).

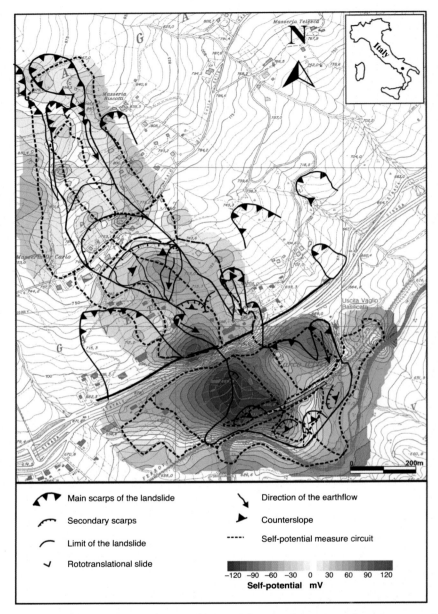

Main scarps of the landslide

Secondary scarps

Limit of the landslide

Rototranslational slide

Direction of the earthflow

Counterslope

Self-potential measure circuit

-120 -90 -60 -30 0 30 60 90 120
Self-potential mV

Figure 5.4. (Below) Map of the self-potential anomaly measured at the ground surface; from Lapenna *et al.* (2005), reproduced with authorization from the SEG and the authors. For full caption, see text (p. 159).

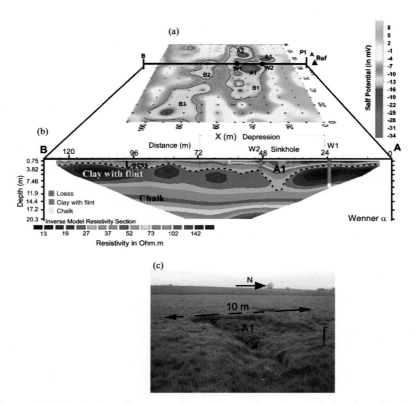

Figure 5.8. Comparison between the self-potential map and the d.c.-electrical resistivity imaging. Electrical resistivity tomography along profile P1. The electrical resistivity tomography is shown at the fourth iteration (RMS error: 1.39%). (c) Picture of Sinkhole A1, which has a diameter of 10 meters. The depth of the depression is about 2 m. Modified from Jardani *et al.* (2007b). For full caption, see text (p. 165).

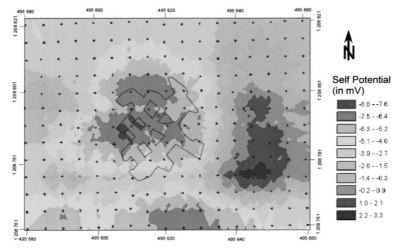

Figure 5.10. Self-potential map. The spacing between the measurements is 5 m. The mine shaft corresponds to the small open circle ($x = 618$ m, $y = 795$ m) located near the center of the negative self-potential anomaly. The black thin lines indicate the shape of the cave, including the presence of five pillars (see Jardani *et al.*, 2006b). For full caption, see text (p. 168).

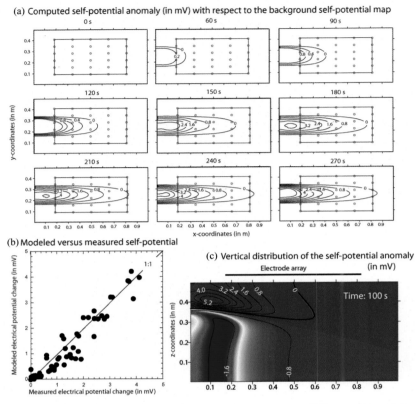

(a) Computed self-potential anomaly (in mV) with respect to the background self-potential map

Figure 5.15. Computed self-potential anomaly contours at different elapsed times ($t = 0$ corresponds to the injection of the salt in the upstream reservoir). (a) Result of the model. (b) Prediction of the model versus measured data. (c) Vertical distribution of the SP field. For full caption, see text (p. 183).

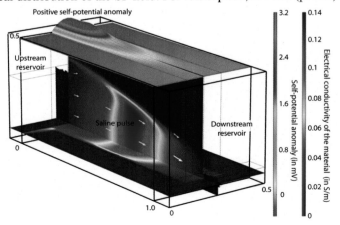

Figure 5.16. Snapshot of the 3D distribution of the resistivity and surface self-potential anomaly during the saline pulse experiment. For full caption, see text (p. 184).

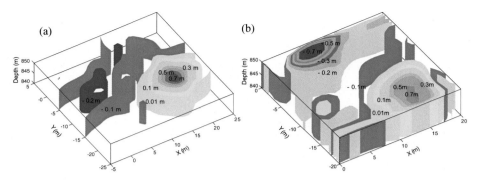

Figure 6.10. Posterior hydraulic head distributions (a) Dipole tests C4–C1. (b) Corrected posterior distribution of the steady-state hydraulic heads for dipole test C4–C1. We have not used the self-potential data in the vicinity of the pumping well; see Jardani *et al.* (2009).

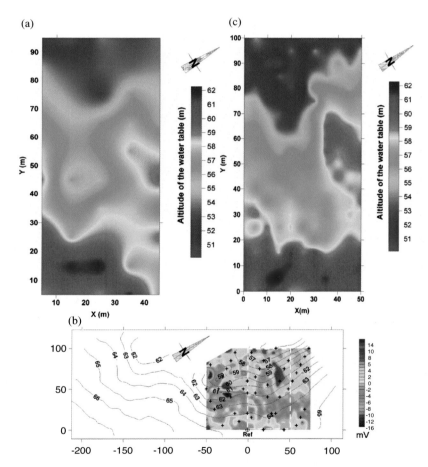

Figure 6.18. (a) Altitude of the water table from vertical electrical resistivity soundings. (b) Self-potential map of the investigated area and position of the measurement stations. Data from Nicolas Florsch. (c) Altitude of the water table using the Parker algorithm ($P = 0.45$ mV m^{-1} and $Z_0 = 4$ m).

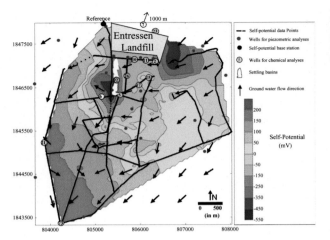

Figure 6.28. (Left) Self-potential map (mV) over the Entressen Landfill. For full caption, see text (p. 237).

Figure 6.31. (Right) Streaming potential component of the self-potential signals (mV). For full caption, see text (p. 240).

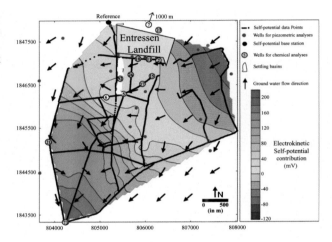

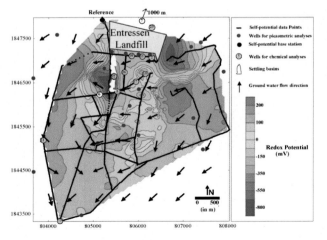

Figure 6.33. (Left) Map of the redox potential (mV). The small black circles represent the location of the self-potential station. For full caption, see text (p. 242).

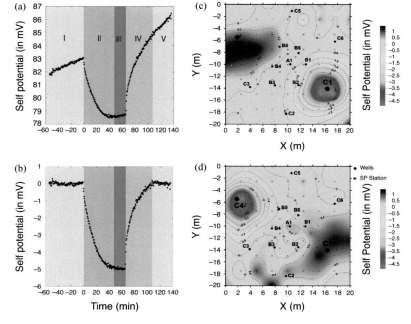

Figure 6.7. Self-potential (SP) data during a dipole pumping/injection test. (a) Raw SP time series. (b) Detrended SP time series. (c, d) SP maps obtained under steady-state conditions for the C4 pumping –C1 injection test (c) and for the reciprocal test. For full caption, see text (p. 206).

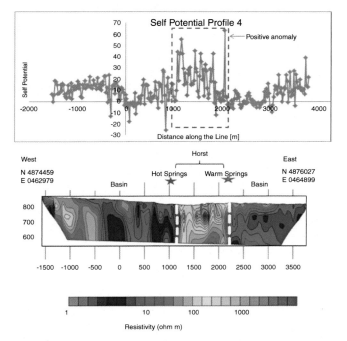

Figure 7.18. Resistivity tomogram for the Neal Hot Springs area and associated SP anomalies (in mV). For full caption, see text (p. 279).

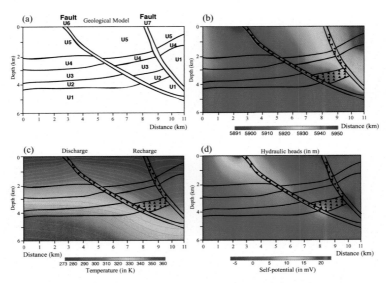

Figure 7.1. Synthetic model and forward numerical simulations. (a) The synthetic model comprises seven hydraulic units (including two faults U6 and U7). (b) Distribution of the hydraulic heads. The arrows represent the direction of the Darcy velocity. (c) Distribution of the temperature in steady-state condition of ground water flow. (d) Distribution of the self-potential voltages associated with the ground water flow pattern. (See Jardani and Revil (2009b)). For full caption, see text (p. 250).

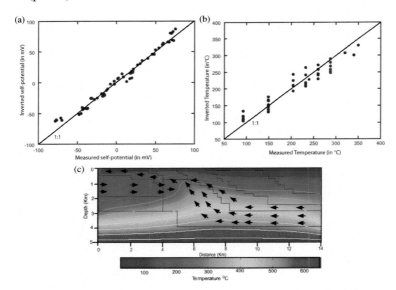

Figure 7.12. Results of the inversion. (a) Best fit of the self-potential data from Figure 7.10a. (b) Best fit of the temperature data from Figure 7.11. (c) Flow pattern associated with the highest posterior probability densities of the permeabilities values of each geological unit. Note the upflow of the hot water in the central part of the system and the horizontal flow in the Western part of the system. The arrows represent the direction of the Darcy velocity with a threshold value of 5×10^{-9} m s^{-1}. For full caption, see text (p. 267).

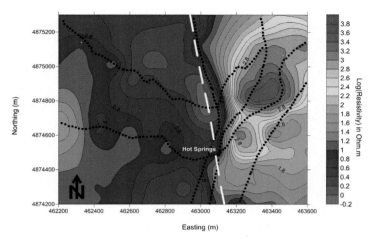

Figure 7.19. Result of the resistivity inversion at a depth of 50 m. The East side of this figure shows high resistivities in the range 100–5000 ohm m corresponding to the horst. The West side is more conductive and corresponds to the sedimentary infilling of the graben. The filled black circles correspond to the position of the electrodes. Fault A bounds the resistive horst. For full caption, see text (p. 280).

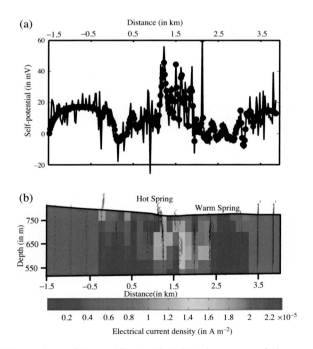

Figure 7.20. Inversion of the self-potential data in terms of the source current density distribution, which in turn can be related to the ground water. (a) Fit of the self-potential data (RMS data misfit 8%). The line corresponds to the (noisy) data, while the black filled circles correspond to the reconstructed self-potential profile. (b) Tomogram of the source current density. For full caption, see text (p. 281).

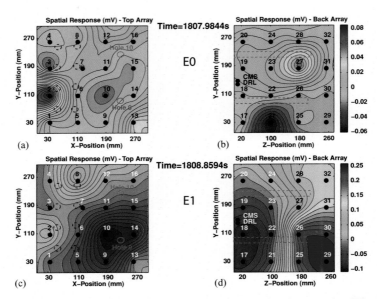

Figure 8.16. Self-potential spatial voltage distributions for the snapshot E0 and the event E1. The black dots denote the position of the electrodes. (a) and (b) Spatial electrical potential distribution for snapshot E0 showing the spatial variations associated with the background noise. Note the very small color bar voltage scale (+0.08 mV to −0.06 mV). (c) and (d) Voltage distribution for event E1. For full caption, see text (p. 331).

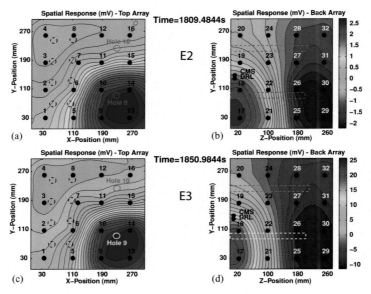

Figure 8.17. Self-potential spatial voltage distributions of events E2 and E3. (a, b) Event E2 voltage distribution showing the peak voltage associated with the second hydraulic pulse during constant flow injection. (c, d) Event E3 spatial voltage distribution showing the peak voltage associated with the highest magnitude pulse during constant flow injection. For full caption, see text (p. 332).

This sensitivity coefficient is inversely proportional to the electrical conductivity of the pore water and, as explained above, $\hat{Q}_V$ increases linearly with the logarithm of the salinity but this dependence is neglected. An order of magnitude of the expected self-potential anomaly can be obtained by measuring the streaming potential coupling coefficient and using the difference of hydraulic head between the two reservoirs. From Eq. (5.11) and the parameters reported in Table 5.1, we obtain a value of the streaming potential coupling coefficient of -9.5 mV m^{-1}. The estimate is consistent with the results (streaming potential coupling coefficient versus salinity) reported by Revil *et al.* (2003) for a variety of porous materials. We have also measured directly the streaming potential coupling coefficient using the approach described by Bolève *et al.* (2007a) using an NaCl solution at 0.05 S m^{-1} (25 °C). The result is -12 ± 2 mV m^{-1}, again a value consistent with the previous estimate.

Using a difference of head between the two reservoirs of 22.3 cm, a first-order magnitude of the potential distribution is given by the product of the streaming potential coupling coefficient and the head difference. This yields a value of 2.1 mV. We cannot, however, use our pre-injection test potential distribution because these data are very spatially noisy. The reason is that the difference of potential between electrodes is unknown, as mentioned above. That said, as these differences of electrical potentials are constant (or linearly dependent) during the course of the experiment, it is easy to remove them, as explained below, in order to enhance the change in the electrical potential distribution associated with the migration of the salt.

From Eq. (5.11), an increase in salinity implies a decrease of the magnitude of the (negative) streaming potential coupling coefficient (as it becomes less negative). Therefore, this change implies that, during the passage of the salt plume, the intensity of the self-potential signals decreases. Because in Figure 5.14, each channel is referenced with respect to its potential prior to the introduction of the salt, we expect the traveling self-potential anomaly observed above the channel to be positive as observed. As we introduce the saline water, the amplitude of the self-potential anomaly decreases. Because we subtract the zero-referenced pre-injection anomaly (which was negative), we obtain a positive anomaly. In other words, it is like we are progressively "erasing" the pre-injection anomaly along the channel as the salt tracer progresses through the tank.

The observed change in self-potential distribution associated with the advection–dispersion of the salt is on the order of 4 mV (Figure 5.14), which is higher than the magnitude of the anomaly given above (2.1 mV). The 2.1 mV is the maximum possible change based on the streaming potential component. Note that, in addition to this component, there is also the contribution from the diffusion potential associated with the salinity gradient, and we will show below that this second contribution generates also a positive self-potential anomaly at the top surface of the tank of the same magnitude as the electrokinetic component (see Revil and Jardani (2010a)).

5.4.3 Numerical modeling

In order to provide more insight into the origin of the measured self-potential signals, we performed a numerical simulation of the sandbox experiment. We use the finite element package Comsol Multiphysics 4.2 to solve the PDEs resulting from the combination of the Richards, advection–dispersion and Poisson equations as explained below. The reason is that we need to estimate the water content distribution in the vadose zone in order to obtain a realistic electrical resistivity distribution in the unsaturated sand. Therefore, we need to account for the effect of the capillary fringe, which cannot be neglected at the scale of the sandbox experiment.

For unsaturated conditions, we solve the Richards equation with the van Genuchten–Mualem model (see van Genuchten (1980) and Mualem (1986)) for the capillary pressure and the relative permeability of the water phase. This approach offers a simple first-order model to describe unsaturated flow. The governing equation for the flow of the water phase (see Richards (1931)) is

$$[C_e + S_e S]\frac{\partial \Psi}{\partial t} + \nabla \cdot [-K\nabla(\Psi + z)] = 0, \tag{5.12}$$

where z is the elevation above a datum, Ψ is the capillary head (in m), C_e denotes the specific moisture capacity (in m^{-1}) defined by $C_e = \partial \theta / \partial \Psi$ where θ is the water content (dimensionless), S_e is the effective saturation, that is related to the saturation of the water phase by $S_e = (S_w - S_w^r)/(1 - S_w^r)$ ($\theta = S_w \phi$ where ϕ represents the connected porosity of the material), S is the storage coefficient (m^{-1}) and t is time. The hydraulic conductivity is related to the relative permeability k_r and K_s, the hydraulic conductivity at saturation, by $K = k_r K_s$. With the van Genuchten–Mualem model, the porous material is saturated when the fluid pressure reaches the atmospheric pressure ($\Psi = 0$) at the water table). The effective saturation, the specific moisture capacity, the relative permeability, and the water content are defined by

$$S_e = \begin{cases} 1/[1 + |\alpha\Psi|^n]^m, & \Psi < 0 \\ 1, & \Psi \geq 0 \end{cases} \tag{5.13}$$

$$C_e = \begin{cases} \dfrac{\alpha m}{1 - m}(\phi - \theta_r)S_e^{\frac{1}{m}}\left(1 - S_e^{\frac{1}{m}}\right)^m, & \Psi < 0 \\ 0, & \Psi \geq 0 \end{cases} \tag{5.14}$$

Table 5.3. *Properties of the two sands used in the numerical modeling. The dispersivities are modeled with* $\alpha_L = \alpha_T = \alpha_d$.

Properties	Coarse sand	Fine sand
Median grain diameter d_{50} (m)	1.51×10^{-3}	350×10^{-6}
Porosity ϕ (–)	0.398	0.410
Formation factor F (–)	3.63	3.48
Hydraulic conductivity K (m s^{-1})	1.52×10^{-2}	1.20×10^{-3}
Permeability k (m^2)	1.98×10^{-9}	2.47×10^{-10}
Charge density $\hat{Q}_V$ (in C m^{-3})	8.6×10^{-3}	4.8×10^{-2}
Peclet number Pe (–)	1.8	1.0
Salt diffusion coefficient D (in m^2 s^{-1})	1.5×10^{-9}	1.5×10^{-9}
Dispersivity α_d (m)	8.25×10^{-4}	4.5×10^{-4}
Irreducible water content θ_r (–)	0.023	0.032
van Genuchten parameter α (m^{-1})	12.5	5.0
van Genuchten exponent n (–)	9.03	6.57
van Genuchten exponent L (–)	1.0	1.0

$$k_r = \begin{cases} S_e^l \left[1 - \left(1 - S_e^{\frac{1}{m}} \right)^m \right]^2, & \Psi < 0 \\ 1, & \Psi \geq 0 \end{cases} \tag{5.15}$$

$$\theta = \begin{cases} \theta_r + S_e(\phi - \theta_r), & \Psi < 0 \\ \phi, & \Psi \geq 0 \end{cases}. \tag{5.16}$$

The variable θ_r represents the residual water content ($\theta_r = S_w^r \phi$), and α, n, $m = 1 - 1/n$, and l are parameters that characterize the porous material (see van Genuchten (1980) and Mualem (1986)). The exponents m and n in these equations are NOT the two Archie's exponents.

The values of the material properties used for the simulation are reported in Table 5.3. In each sand compartment, the permeability is assumed to be homogeneous and isotropic, and therefore is defined with a constant permeability. This permeability is related to the mean grain diameter and the formation factor F (see Revil and Cathles (1999)) by

$$k = \frac{d_{50}^2}{24 F^3}, \tag{5.17}$$

where d_{50} is the median of the grain size distribution and the hydraulic conductivity at saturation is given by $K_S = k \rho_f g / \eta_f$. Note that Eq. (5.17) has been generalized recently by Revil and Florsch (2010) for an arbitrary grain size distribution. The electrical conductivity is determined by:

$$\sigma = \frac{1}{F} S_w^2 \sigma_f. \tag{5.18}$$

Note that if S_w goes to zero on the top part of the tank, the resistivity would tend toward infinity. To avoid such a problem in the numerical computations, we keep the water content to the irreducible water content in Eq. (5.18) (Table 5.3). We obtained a irreducible water saturation of 5.8% for the coarse sand and 7.8% for the fine sand. In turn, this yields a resistivity of 22 kOhm m in the coarse sand and 11 kOhm m in the fine sand above the capillary fringe. Note that the capillary fringe is much thicker above the fine sand than above the coarse sand owing to the difference in the size of the pores.

All calculations were performed in 3D for the flow and in 3D + time when solving the advection–dispersion equation and the resulting self-potential distribution. The Richards equation is solved for stationary flow conditions and, therefore, it simplifies to a Laplace-type partial differential equation,

$$\nabla \cdot [K \nabla(\Psi + z)] = 0. \tag{5.19}$$

Regarding the way the model addresses salt injection, we assume complete and instantaneous mixing in the upper reservoir and use a boundary condition corresponding to an exponentially decaying salinity on the boundary of the upstream reservoir. Although it is only an approximation of the true change in the upstream salinity, it represents the time-variant nature of the dilution of the injection water throughout the experiment. The salt plume moves through the sandbox according to the advection–dispersion equation. We use insulating boundary conditions at the different boundaries where this condition applies (sides, top and bottom boundaries, and end-sides of the two reservoirs) as well as impervious boundary conditions at these boundaries except where the flux of water is injected and retrieved.

The numerical simulation of the self-potential distribution prior to salt injection is shown in Figure 5.19c. This distribution corresponds to the computed streaming potential distribution associated with the flow of the pore water. We see that the equipotentials are normal to the water table. As with the real data, we remove this contribution to focus on the change in self-potential signals over time (what we call the time-lapse self-potential anomaly below).

After the salt injection, the time-lapse distribution of the self-potential anomaly at the top surface of the tank is shown in Figure 5.15a. The background self-potential signals (i.e., the steady-state electrical response measured prior to the salt injection, see Figure 5.19c) are removed. The self-potential signals are mainly confined to

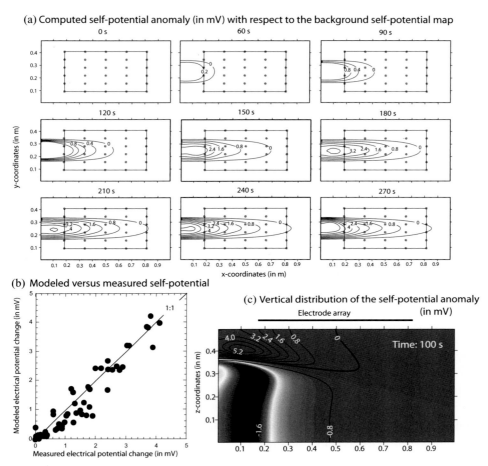

Figure 5.15. Computed self-potential anomaly contours at different elapsed times ($t = 0$ corresponds to the injection of the salt in the upstream reservoir). (a) Result of the model. The outer rectangle corresponds to the dimension of the tank while the inner rectangle corresponds to the area covered by the electrodes. Constant flow conditions are maintained during the numerical experiment and insulating boundary conditions are applied at the top, sides, and bottom of the tank. (b) Comparison between the prediction of the model and the measured data for the two lines of electrodes just above the channel ($R = 0.95$). (c) Vertical distribution of the self-potential field. The colors correspond to the electrical conductivity of the sand. This conductivity is modified by the presence of the salt tracer in the pore space of the sand (modified from Ikard *et al.* (2012)). See also color plate section.

the high-permeability coarse sand channel due to the strong resistivity contrast in the vadose zone between the preferential flow channel and the flanking low permeability domains. This phenomena is also evident in the sandbox experiment, but the equipotentials are less confined to the permeable channel, indicating that the

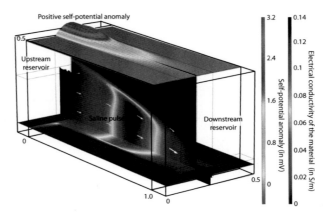

Figure 5.16. Snapshot (120 s after the salt injection, see Figure 5.15) of the 3D distribution of the resistivity and map of the self-potential anomaly at the top surface of the tank during the saline pulse experiment. The arrows correspond to the direction of the Darcy velocity (their lengths are proportional to the intensity of the flow density vector **U**). See also color plate section.

resistivity distribution in the sandbox experiment may be different in the numerical simulation and the sandbox experiment for the reasons mentioned above. In Figure 5.15b, we plot the measured self-potential data (for all the snapshots shown in Figure 5.14) versus the modeled self-potential data shown in Figure 5.15a. The numerical model captures well the self-potential fluctuations above the coarse sand channel both in polarity and amplitude.

In Figure 5.15c, we show a cross-section of the electrical potential distribution in the middle of the channel in the flow direction. Figure 5.16 shows how the self-potential anomaly lags behind the in-situ saline pulse traveling in the permeable channel. The electrical distribution is rather complex. If we disregard the diffusion potential contribution, the current density associated with the flow is salinity independent. This means that the distorsion of the field lines is coming from the change in resistivity in the aquifer. That said, the second contribution to the current density produced a positive self-potential distribution that adds to the streaming potential contribution.

We investigate now the importance of the diffusion potential contribution. As the self-potential anomaly is positive and higher than the pre-injection test electrokinetic anomaly (see Figures 5.15c and 5.16), there is clearly more than a simple erasure of the electrokinetic anomaly by the advection–dispersion of the salt in the tank. These results imply that there is another contribution that generates a positive self-potential at the top surface of the tank (see Martínez-Pagán *et al.*

(2010) and Revil and Jardani (2010a)). This additional anomaly is due to the diffusion potential associated with the salt salinity gradient.

One limitation of the sandbox experiment is that we have to put a reference electrode somewhere in the sandbox where it is susceptible to spurious effects. Indeed, the absolute potential at the reference electrode changes over time, but we assign a zero value to this location for all sampled times. Therefore this requires an additional processing step to correct the data. The situation is different for field conditions, because the reference electrode can be placed far enough from the investigated area to avoid such spurious effects. In addition, the proper boundary conditions to be applied in the field are different from the boundary conditions applied in the laboratory. In our laboratory experiment, the boundary condition for the electrical current density at the side walls of the tank does not influence the electrical potential distribution because the changes in electrical field is mainly constrained by the position of the permeable channel. This result is explained by the high resistivity existing above the channel by comparison with the resistivity of the vadose zone above the fine sand.

5.4.4 Application to earth dams and embankments

The field test reported by Bolève *et al.* (2011) concerns a salt tracer injection performed between two reservoirs separated by an embankment (see Figure 5.17a). The embankment between the two reservoirs is leaking, probably because of a high permeability channel possibly associated with internal erosion (see Wan and Fell (2008) for a description of the various processes of internal erosion). A resistivity profile is also shown in Figure 5.18.

A self-potential map was performed in the upstream basins using a reference electrode on the bank of the basin, and having the scanning electrode in contact with the floor of the basin. The measurements were performed along profiles with a spacing of 2.5 m and using Petiau electrodes. The self-potential map reveals a negative self-potential anomaly in the upstream basin just above the area where two leakages can be observed downstream (areas A1 and A2). The static self-potential anomaly was modeled by Bolève *et al.* (2011) and the salt tracer test was investigated by Ikard *et al.* (2012). The self-potential map reveals a pre-injection self-potential anomaly of -55 mV in the leaking area.

In Figure 5.17a, we show the position of the salt tracer injection. The resulting self-potential monitoring (using a set of 32 Pb/PbCl$_2$ non-polarizing Petiau electrodes) is shown in Figure 5.18a. In Figure 5.18b, we show the results of the monitoring of the conductivity of the water at the leaking areas A1 and A2 downstream. The self-potential signals are recorded with a sampling frequency of 200 Hz. As for the laboratory test described above, the data are referenced with respect the

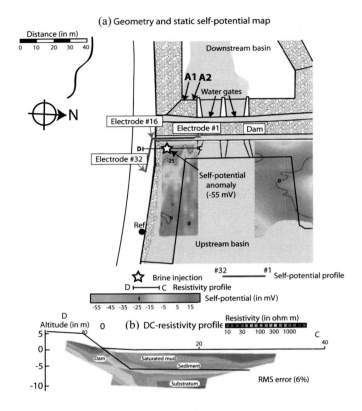

(a) Geometry and static self-potential map

Figure 5.17. Sketch of the test site. (a) Setting showing the position of the upstream and downstream basins, the outflow of the leaking areas (A1 and A2), the position of the monitoring self-potential profile (#1 to #32), the position of the brine injection (white star), and the position of the reference or base station (Ref) for the self-potential map. Note that the pre-injection self-potential anomaly is on the order of −55 mV. (b) The d.c.-resistivity tomography (profile d.c., data inverted with RES2DINV using a Gauss–Newton approach; see Loke and Barker (1996)).

self-potential values prior the salt injection to blank the self-potential distribution existing prior the salt tracer injection.

During the transport of the salt tracer, Bolève *et al.* (2011) observed a positive anomaly with an amplitude of 50 mV associated with the passage of the salt tracer as discussed below. The salt tracer transport progressively erases the pre-injection anomaly along the permeable pathways as the salt tracer progresses through the dam. The distance between the self-potential profile and the two leakage areas A1 and A2 is ~30 m. This implies a pore water velocity of 4.5×10^{-2} m s^{-1} inside the permeable pathways, the hydraulic gradient is on the order of 0.3 (see Figure 5.17). Using Darcy's law, this implies an apparent hydraulic

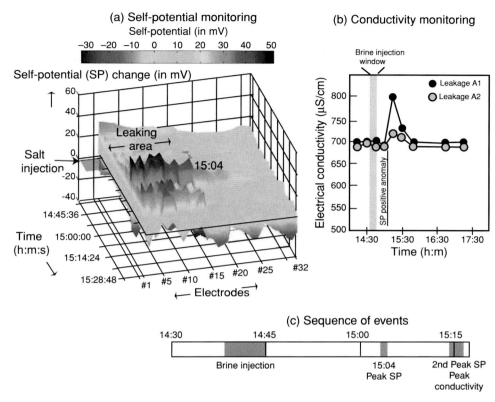

Figure 5.18. Results of the monitoring experiment reported in Bolève *et al.* (2009). (a) Time-lapse self-potential signals (referenced to the pre-injection values). Note that the self-potential positive anomaly is on the order of 50 mV (followed by a negative anomaly of −20 mV), which is consistent with the value of the pre-injection anomaly (−55 mV) shown in Figure 5.17. Note the higher electrical noise level in the leaking area. (b) Monitoring of the conductivity of the water in the downstream reservoir at the position of leaking areas A1 and A2, respectively.

conductivity of 0.15 m s^{-1} in the permeable channel (an equivalent apparent permeability of 1.5×10^{-8} m^2). Because this apparent permeability is rather high, it is important to check if there is an effect of the Reynolds number in this case. The Reynolds number is defined by Re $= \rho_f U \Lambda / \eta_f$, where U denotes the amplitude of the Darcy velocity and Λ denotes a mean pore size called the dynamic pore radius; see Johnson and Sen (1988). We use the following values for the parameters entering the expression of the Reynolds number: a Darcy velocity of 2.7×10^{-2} m s^{-1} (porosity times mean velocity given above according to the Dupuit equation) and Λ given by a $\Lambda = \sqrt{8kF}$ (see Johnson and Sen (1988)) (taking $F = 2$, from Archie's law with $m = 1.5$, and an apparent permeability $k_a = 1.5 \times 10^{-8}$ m^2, yields $\Lambda = 4.9 \times 10^{-4}$ m). This yields a Reynolds number of 13, which is not negligible

Table 5.4. *Material properties for the synthetic case*
study ($\alpha_L = 0.1$ *m and* $\alpha_T = 10^{-3}$ *m).*

Material properties	Resistivity (Ohm m)	Permeability (in m^2)	Excess charge density (C m^{-3})
Water	15	–	0
Rock	2000	0	0
Dam	2000	0	0
Leak	150	2×10^{-8}	1.2×10^{-3}
Seal	500	0	0

as much higher than unity. According to Bolève *et al.* (2007a, b), the apparent and true permeabilities are related to each other by $k_a = k/(1 + \text{Re})$. This yields a true permeability $k = 2.3 \times 10^{-7}$ m^2.

In order to test our approach and to see the effectiveness of inverse modeling, we develop a 2D synthetic model inspired from the test study reported in Section 5.1. Figure 5.19a shows a simple 2D model that is a numerical analog of this case study. Our goal is to show that, with this 2D simulation, we can use the surface self-potential data (contaminated with noise) and we can invert the permeability of the preferential fluid flow pathways using a stochastic approach. The material properties for this synthetic case study are reported in Table 5.4 together with the dispersivities. The excess charge density $\hat{Q}_V$ can be obtained directly from the permeability as shown in Chapter 3 (see Figure 3.11). This empirical relationship has been tested for both saturated (see Revil and Jardani (2010a, b)) and unsaturated materials (see the recent work of Jougnot *et al.* (2012)), but there is most likely a small influence of salinity, pH and soil type in this relationship, which effects are not considered in this simple relationship between the effective charge density and the permeability. For the conduit, this yields a charge density of 1.2×10^{-3} C m^{-3} for the material in the conduit (reported in Table 5.4).

We use Comsol Multiphysics to perform a finite element modeling of the transport of a salt tracer injecting the salt tracer directly at the entrance of the pipe as a boundary condition (see Figure 5.19c). The (single) recording electrode is placed in the middle of the crest of the dam. The self-potential synthetic data are referenced to a point located upstream as shown in Figure 5.19b (the position of this point is actually arbitrary as the potential can be shifted to zero using the potential value prior the salt injection). Insulating boundary conditions at applied at the boundaries of the domain shown in Figure 5.19a. The self-potential field created by the flow before the injection of the salt is shown in Figure 5.19b. The hydraulic gradient in the conduit is on the order of 0.17 and the average velocity in the conduit is on the order of 0.017 m s^{-1}. The pre-injection self-potential anomaly at the inlet of

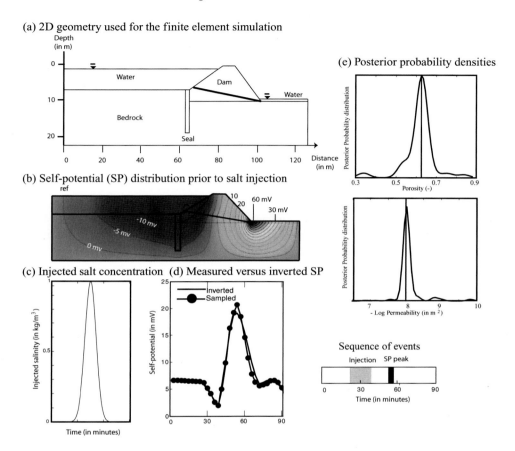

Figure 5.19. Numerical modeling of the synthetic case. (a) Sketch of the 2D geometry used for the simulation (thickness in the strike direction of 1 m). The material properties used in the numerical simulation are reported in Table 5.4. The recording electrode is at the crest of the dam and the reference electrode is assumed at infinity. The reference for the electrical potential is taken arbitrarily at $x = 0$. (b) Simulation of the self-potential signals before salt injection. (c) Salt concentration injected upstream the pipe. (d) Resulting self-potential changes at the crest of the dam. (e) Posterior probability distributions of the two model parameters. Porosity of the preferential fluid flow pathway. Permeability of the preferential fluid flow pathway. The vertical bars correspond to the true values of the model parameters (modified from Ikard *et al.* (2012)).

the preferential flowpath is approximately -25 mV, which can be compared to the measured -55 mV anomaly at the inlet.

Figure 5.19d shows the resulting self-potential variation over time for an electrode located at the crest of the dam while Figure 5.19c shows the salt injected over time upstream (at the entrance of the flowpath). During the passage of the salt tracer below the crest of the dam, the scanning electrode (located in the middle of the crest) records a positive self-potential anomaly with an amplitude change of 15 mV with respect to the pre-injection value (Figure 5.19d). This positive self-potential can only be explained by having a strong diffusion current density that dominates locally the self-potential response. Indeed, as shown in Martínez-Pagán *et al.* (2010), the diffusion self-potential signal associated with the diffusion of a NaCl salt is responsible for a positive anomaly. Note that the self-potential anomaly that would be generated by subtracting the modeled pre-injection self-potential value from the modeled self-potential time series is bipolar, which is consistent with the actual field case (see Figure 5.18). The complex behavior shown in Figure 5.19d indicates that the streaming and diffusion current densities generate a complex behavior and show that depending on the conditions of the experiment, both contributions to the source current density have to be accounted for. Therefore unified models like the one produced recently in Chapter 2 are very important in that respect.

We test the adaptive Metropolis (AMA) algorithm (see Chapter 4) to invert the posterior probability distribution of a set of model parameters controlling the occurrence of the self-potential anomalies. We assume that the shape of the self-potential anomaly measured on a set of electrodes located along the crest can be used to locate the preferential flow pathway using cross-correlation (see Rozycki *et al.* (2006) for the development of such a method and case studies). We assume also that the pre-injection resistivity distribution is known through resistivity tomography. The AMA algorithm has been introduced by Haario *et al.* (2001) and recently used by Jardani *et al.* (2010) and Woodruff *et al.* (2010) for some geophysical applications.

In a general case, the parameters to invert would include the geometry of the flow pathway and its material properties (porosity and permeability). That said, in the present case, the position of the pathway is pretty straightforward as the inlet is characterized upstream by self-potential mapping and the outlet is observable downstream. Therefore, we limit the model parameters to the permeability k and the porosity ϕ of the conduit. We use $\log k$ and $\mathrm{logit}(\phi) = \log(\phi/(1 - \phi))$ as model parameters to impose the constraints that the permeability is positive and that the porosity is a concentration of voids comprised between 0 and 1. We are solving the steady-state flow equation and the advection–dispersion equation for the salt tracer. We use 1000 realizations and the convergence of the chain was reached very quickly after 106 iterations (as determined from the standard deviations on the realizations of the model parameters). Once the McMC has converged, we use the statistics to build the posterior probability density functions on the two model parameters (Figure 5.19e). There is a very good agreement between the peaks of the posterior probability distributions and the true value of the model parameters. The peak values of the PDF of the two model parameters are used to compute the inverted self-potential versus time curve, which compares well with the true curve in Figure 5.19e.

For highly permeable channels, the previous example illustrates that time-lapse self-potential is potentially more appropriate than a time-lapse d.c.-resistivity tomography. The reason is that d.c. resistivity is an active method, which takes time (often longer than the transit time of the tracer through the area of interest) to obtain 4D snapshots of the subsurface. This is clearly the case for the previous example as the residence time of the salt over a vertical cross-section was less than 20 min. This renders time-lapse resistivity not very efficient as d.c. resistivity data acquisition is more time consuming than the process of advective salt transport in this case. On the contrary, the self-potential method is made in real-time because it is a passive method. A background resistivity tomogram can be taken prior to the salt injection and used with the self-potential data to locate the source of the electrical disturbance during the transport of the salt tracer (see Rozycki *et al.* (2006)).

5.5 Conclusion

In this chapter, we have investigated four applications of the self-potential method to geohazards including landslides, the detection of cavities and crypto-sinkholes, and the detection of leakages in levees and dams. For all these problem, the self-potential method comes as a cheap reconnaissance tool that brings information that is complementary to d.c. resistivity and induced polarization surveys. One of the main topics of research in the future will be the automatized monitoring of hydrogeological conditions using a combination of passive and active geophysical methods together with in-situ sensors measurements of the piezometric level and infiltration in the vadose zone.

6

Application to water resources

Four applications of the self-potential method to hydrogeological problems are described in this chapter. The first application concerns the use of the self-potential method for providing information related to the reconstruction of the piezoemtric head surface during pumping tests (Section 6.1). The second application concerns the possibility of using the self-potential method to reconstruct the thickness of the vadose zone assuming that the self-potential anomalies are mainly due to the percolation distance of the residual water in the vadose zone (Section 6.2). The third application concerns the use of the self-potential method to reconstruct the water table at the scale of a catchment using a probabilistic (Bayesian) analysis of the self-potential data (Section 6.3). Finally, we will see how the self-potential signals can be used to delineate contaminated ground water downstream of a landfill (Section 6.4).

6.1 Pumping tests

6.1.1 Theory: forward and inverse modeling

As explained in Chapters 1 and 2, the self-potential signals associated with the flow of ground water are due to the drag of excess electrical charge contained in the pore water by this flow. In an isotropic heterogeneous media, the total current density $\mathbf{J}$ (in A m^{-2}) is given by the sum of a conductive current (Ohm's law) plus a source current of electrokinetic nature (see Chapter 2),

$$\mathbf{J} = -\sigma \nabla \psi - L \nabla h, \tag{6.1}$$

where h is the hydraulic head (in m), ψ is the electrical (self-) potential (in V), σ is the d.c.-electrical conductivity of the porous material (in S m^{-1}), and L is the streaming current coupling coefficient (in A m^{-2}). The term $\mathbf{J}_S = -L\nabla h$ is termed the streaming current density. The streaming potential coupling coefficient is defined by

$$C = \left(\frac{\partial \psi}{\partial h}\right)_{J=0} = -\frac{L}{\sigma}. \tag{6.2}$$

This coefficient is usually the coupling coefficient that is measured in the laboratory as discussed in Chapter 3. It is expressed in mV per meter of hydraulic head (see Chapter 3 for further details).

The continuity equation for the electrical charge is $\nabla \cdot \mathbf{J} = 0$ ($\mathbf{J}$ is conservative). The combination of this continuity equation and the constitutive equation, Eq. (6.1), yields the following Poisson's equation;

$$\nabla \cdot (\sigma \nabla \psi) = \Im, \tag{6.3}$$

where $\Im = -\nabla \cdot (L \nabla h)$ is the volumetric source term (in A m^{-3}). In order to gain a better understanding of the various contributions of the source term of Eq. (6.3), we expand this source term as $\Im = -L \nabla^2 h - \nabla L \cdot \nabla h$. The Laplacian of the hydraulic head is given, in steady-state conditions, by

$$\nabla^2 h = -\frac{Q}{K} \delta(\mathbf{r} - \mathbf{r}_S) - \nabla \ln K \cdot \nabla h, \tag{6.4}$$

where $\mathbf{r}_S$ is the position of the source where water is infiltrated or removed from the ground, $\delta(\mathbf{r} - \mathbf{r}_S)$ is the Kronecker delta ($\delta(\mathbf{r} - \mathbf{r}_S) = 1$ when $\mathbf{r} = \mathbf{r}_S$ and $\delta(\mathbf{r} - \mathbf{r}_S) = 0$ for $\mathbf{r} \neq \mathbf{r}_S$). Therefore, the source term in the electrostatic Poisson equation can be written as

$$\Im = L \frac{Q}{K} \delta(\mathbf{r} - \mathbf{r}_S) - \nabla L \cdot \nabla h + L \nabla \ln K \cdot \nabla h. \tag{6.5}$$

There are two distinct source terms in Eq. (6.5). The first one corresponds to the primary source term associated with the pumping/injection rate, Q. The second source term is associated with the heterogeneity of the streaming current coupling coefficient of the ground in the presence of a hydraulic gradient. This heterogeneity can be assessed independently with an electrical resistivity tomogram.

If we consider N sources and sinks of water supply or removal inside the ground, the distribution of the electrical potential is given by solving the following Poisson's equation (see Jardani *et al.* (2009)):

$$\nabla^2 \psi = \sum_{i=1}^{N} L \frac{Q_i}{K} \delta(\mathbf{r} - \mathbf{r}_i^S) - \frac{1}{\sigma} \nabla L \cdot \nabla h - \nabla \ln \sigma \cdot \nabla \psi + L \nabla \ln K \cdot \nabla h, \tag{6.6}$$

which shows that spatial variation of electrical conductivity can be also considered as an additional secondary self-potential source (see discussion in Chapter 4).

The influence of the Reynolds number upon the streaming potential has been studied by different researchers (see, for example, Watanabe and Katagishi (2006) and Kuwano *et al.* (2006)). Bolève *et al.* (2007a) have studied recently the effect of the Reynolds number upon the magnitude of the streaming potential. For a capillary of radius R, U being the strength of the seepage velocity, the Reynolds number is defined (see Batchelor (1972)) by $Re = \rho_f U R / \eta_f$. At Reynolds numbers between 1 and few hundred (the critical value corresponding to turbulence), the flow is said to be in the inertial laminar flow regime. In this regime, the apparent streaming current coupling coefficient and the apparent hydraulic conductivity are given by Bolève *et al.* (2007a) as

$$L = L_0/(1 + Re), \tag{6.7}$$

$$K = K_0/(1 + Re), \tag{6.8}$$

where L_0 and K_0 are the value of the streaming current coupling coefficient and hydraulic conductivity in the viscous laminar flow regime characterized by $Re \ll 1$. In particular,

$$K_0 = \frac{d_0^2 \rho_f g}{24 F^3 \eta_f}, \tag{6.9}$$

from Revil and Cathles (1999), where d_0 is the mean grain diameter, F is the electrical formation factor, ρ_f is the mass density of the pore fluid, g is the acceleration of the gravity, and η_f is the dynamic shear viscosity of the pore fluid) and

$$L_0 = \hat{Q}_V K_0, \tag{6.10}$$

from Bolève *et al.* (2007a, b) and where $\hat{Q}_V$ represents the excess of electrical charge of the diffuse layer per unit pore volume; see Revil and Linde (2006). According to Bolève *et al.* (2007a), the Reynolds number can be determined from the following expressions:

$$Re = \frac{1}{2}(\sqrt{1 + c} - 1), \tag{6.11}$$

$$c = \frac{\beta \rho_f^2 g}{\eta_f^2} \frac{d_0^3}{F(F-1)^3} \nabla h, \tag{6.12}$$

where $\beta \approx 2.25 \times 10^{-3}$ is a numerical constant and c is a function of both the texture and head gradient. Here, we see that the Reynolds number depends on ∇h so in this regime, the current density and the Darcy velocity are non-linear with respect to the pressure heads. As a result, the apparent hydraulic conductivity and the streaming current coupling coefficient also depend on the gradient of the hydraulic head.

We now come back to the discussion of the forward problem. The normal component of the total current density vanishes at the surface of the Earth because air is insulating. In self-potential monitoring, it is also important to remember that all the electrodes are connected to a reference electrode where the potential is taken to be equal to zero over time. The forward model was solved with Comsol Multiphysics 3.4 coupled with Matlab®. In forward numerical modeling, it is therefore important to remove the potential obtained at the location of the reference electrode in order to compare the measured self-potential distribution and the distribution resulting from numerical modeling.

Whatever the method used to discretize the forward model, the model is linear between the head values and the self-potential signals as long as the flow occurs in the laminar flow regime. Equation (6.3) can be written as (see Chapter 4)

$$\bar{\psi} = \overline{\overline{K}} \, \bar{h}, \tag{6.13}$$

where $\bar{h}$ is a column vector of hydraulic heads, $\bar{\psi}$ is a column vector of the resulting self-potential at the ground surface, and $\overline{\overline{K}}$ is the kernel matrix that depends on the distributions of the electrical conductivity and streaming current coupling coefficient and the distance between each elementary source and the self-potential stations. In Section 6.3, we will assume that the material properties (especially the streaming current coupling coefficient and the hydraulic conductivity) are independent of the values of the hydraulic heads themselves. However, according to Eqs. (6.9) and (6.10), this is true only in the viscous laminar flow regime. In the following, we use a Bayesian approach to estimate the hydraulic heads from self-potential data incorporating some information about the distribution of electrical resistivity into the kernel. Geostatistical constraints will also be used to account for spatial structure or heterogeneity in the model.

First, we briefly present the basis for the Bayesian approach. If we consider that two states A and B exist, the probability that state B will follow, or overlie, state A, $P(A\,|\,B)$ (where $P(A\,|\,B)$ is called a conditional probability), is the probability that both states occur, $P(A,\,B)$ (joint probability), divided by the probability that state B occurs: $P(A\,|\,B) = P(A,\,B)/P(B)$. Similarly, we can write $P(B\,|\,A) = P(B,\,A)/P(A)$. It follows that the two conditional probabilities $P(A\,|\,B)$ and $P(B\,|\,A)$ are related to each other by Bayes' theorem,

$$P(B\,|\,A) = \frac{P(A\,|\,B)P(B)}{P(A)}. \tag{6.14}$$

We note now $\bar{m}$ a vector column of M unknown model parameters (in our case $\bar{m} = \bar{h}$, where $\bar{h}$ is the vector of hydraulic heads). We have a vector $\bar{d}$ of observations (in our case, $\bar{d}$ represents a vector column of N self-potential observations). Bayes' theorem, written in terms of probability distribution between the model parameters and the data, is

$$P(\bar{h} \mid \bar{d}) = \frac{P(\bar{d} \mid \bar{h})P(\bar{h})}{P(\bar{d})}. \tag{6.15}$$

The posterior probability density on the space of models, $\sigma(\bar{h})$, is the product of two terms (see Tarantola and Valette (1982))

$$\sigma(\bar{h}) = L(\bar{h})\rho(\bar{h}), \tag{6.16}$$

where the likelihood function $L(\bar{h})$ represents the probability density associated with the fit of the data while the second term, $\rho(\bar{h})$, represents the a priori probability density. This probability density incorporates information about the subsurface that is independent of the observed data from which the inferences are being made. The goal of the Bayesian approach is to construct the probability density $\sigma(\hat{h})$ where $\hat{h}$ is a posterior estimate of the hydraulic head distribution.

Equation (6.16) is rather general. We consider below that all uncertainties in the problem can be described by stationary Gaussian distributions. This yields

$$L(\bar{h}) = \sqrt{\frac{(2\pi)^{-N}}{\det \overline{\overline{C}}_d}} \exp\left[-\frac{1}{2}(\overline{\overline{K}}\bar{h} - \bar{d}_{obs})^T \overline{\overline{C}}_d^{-1}(\overline{\overline{K}}\bar{h} - \bar{d}_{obs})\right], \tag{6.17}$$

$$\rho(\bar{h}) = \sqrt{\frac{(2\pi)^{-M}}{\det \overline{\overline{C}}_m}} \exp\left[-\frac{1}{2}(\bar{h} - \bar{h}_{prior})^T \overline{\overline{C}}_m^{-1}(\bar{h} - \bar{h}_{prior})\right], \tag{6.18}$$

where $\overline{\overline{K}}$ is the kernel (see Chapter 4 and Eq. (6.13) above), $\overline{\overline{C}}_d$ is the data covariance matrix, $\bar{d}_{obs}$ is the observed self-potential data, which are measured at the ground surface, $\bar{h}_{prior}$ is the a priori value of the hydraulic heads, $\overline{\overline{C}}_m$ is the model covariance matrix, and N and M are the number of observations and the number of model parameters, respectively.

In the case where all the uncertainties are Gaussian, the posterior probability density is the normalized product of Eqs. (6.17) and (6.18). Because the forward operator $\overline{\overline{K}}$ is linear (in the inertial laminar flow regime), then this distribution is itself a Gaussian, given by

$$\sigma(\bar{h}) \propto \exp\left[-\frac{1}{2}(\overline{\overline{K}}\bar{h} - \bar{d}_{obs})^T \overline{\overline{C}}_d^{-1}(\overline{\overline{K}}\bar{h} - \bar{d}_{obs}) - \frac{1}{2}(\bar{h} - \bar{h}_{prior})^T \overline{\overline{C}}_m^{-1}(\bar{h} - \bar{h}_{prior})\right]. \tag{6.19}$$

The matrices $\overline{\overline{C}}_d$ and $\overline{\overline{C}}_m$ incorporate the uncertainties in the data (modeling and observational errors) and uncertainties related to the hydraulic heads, respectively. Linde *et al.* (2007a, b) showed from field data that a self-potential measurement can be considered as a random process with a probability density that is Gaussian and they explained how $\overline{\overline{C}}_d$ can be obtained from mapping the self-potential data at the scale of a catchment. For self-potential monitoring, $\overline{\overline{C}}_d$ can be obtained by looking at the noise level for all the data taken before the start of a dipole pumping test for example.

We maximize the a posteriori probability density on the model parameters, a procedure called Maximum A Posteriori (MAP) estimation in the literature (Gouveia and Scales, 1998). Maximizing the likelihood corresponds with minimizing the associated negative log-likelihood. Consequently, we have to minimize the following cost function, $\Psi = -\ln(\sigma(\bar{h}))$, given by:

$$2G = (\overline{\overline{K}}\,\bar{h} - \bar{d}_{obs})^T \overline{\overline{C}}_d^{-1}(\overline{\overline{K}}\,\bar{h} - \bar{d}_{obs}) + (\bar{h} - \bar{h}_{prior})^T \overline{\overline{C}}_m^{-1}(\bar{h} - \bar{h}_{prior}). \quad (6.20)$$

The maximization of the previous cost function is obtained by $\partial G/\partial \bar{h} = 0$ (see Tarantola and Valette (1982)), which yields

$$\hat{\bar{h}} = \bar{h}_{prior} + \overline{\overline{C}}_m \overline{\overline{K}}^T \left[(\overline{\overline{K}}\,\overline{\overline{C}}_m \overline{\overline{K}}^T + \overline{\overline{C}}_d)^{-1}(\bar{d}_{obs} - \overline{\overline{K}}\bar{h}_{prior}) \right], \quad (6.21)$$

$$\bar{\sigma}(\bar{h}) \propto \exp\left[(\bar{h} - \hat{\bar{h}})^T \hat{\overline{\overline{C}}}_m^{-1}(\bar{h} - \hat{\bar{h}}) \right], \quad (6.22)$$

where the prior model of hydraulic heads $\bar{h}_{prior}$ is given by the linear model of Fournier (1989). The posterior covariance of the hydraulic heads is given by

$$\hat{\overline{\overline{C}}}_m = \left(\hat{\overline{\overline{K}}}^T \overline{\overline{C}}_d^{-1} \hat{\overline{\overline{K}}} + \overline{\overline{C}}_m^{-1} \right)^{-1}, \quad (6.23)$$

which is a resolution measure that quantifies the uncertainties about the hydraulic heads and where $\hat{\overline{\overline{K}}}$ denotes the Fréchet derivatives of $\overline{\overline{K}}$ evaluated at $\hat{\bar{h}}$. In the present case, $\hat{\overline{\overline{K}}} = \overline{\overline{K}}$ because the problem is linear. However, this would not be the case if the effect of the Reynolds number had been taken into account.

Inversion of self-potential data is an underdetermined problem, and an infinite set of hydraulic head distributions can reproduce the same self-potential distribution measured at the ground surface. To reduce the non-uniqueness of the solution, constraints have to be used. The method of inversion described above can incorporate constraints such as the adjustment of the data balanced by covariance of the uncertainty on the data, the adjustment of a model balanced by the uncertainty on the model parameters, the weighting of the model parameters according to their depth, and constraints provided by in-situ measurements of the hydraulic heads in a set of piezometers.

The adjustment of a model balanced by the covariance of the parameters allows us to reconstruct the hydraulic head distribution. These covariances can be given as an a priori probability distribution from geology of the site or from its topography. The covariance matrix can be determined from a variogram. Many variables in nature are not randomly distributed and follow a certain degree of resemblance when they are close to each other. We note M the variable considered. The value of the semi-variogram between two distant localizations of d is calculated by, for example, Deutch and Journel (1992) as

$$\gamma(d) = \frac{1}{2N(d)} \sum_{i=1}^{N(d)} [M(x_i) - M(x_i + d)]^2, \tag{6.24}$$

where $N(d)$ is the number of pairs of observations of separation distance d. The assumption of stationarity of the second order of the semi-variogram (Matheron, 1965) implies that the semi-variogram is invariant by translation (i.e., the semi-variogram will not depend on the particular locations of the observation points but rather depends only on the distance which separates them). Once calculated, the semi-variogram reveals three characteristics of the spatial correlation structure: (1) the range, (2) the sill and (3) the nugget effect. The range is the distance to which two observations are no longer correlated; the value of the semi-variogram for this distance corresponds to the variance of the variable. The sill C is the value of the semi-variogram once the range has been reached; this value corresponds to the variance σ^2 of the variable. The nugget effect represents very small scale variations at separation distances less than those available in the data set of measurements.

There is a simple relationship between the model of $C(d)$ covariance and the model of the semi-variogram:

$$\gamma(d) = \sigma^2 - C(d), \tag{6.25}$$

where $C(d)$ is the covariogram (or autocorrelation function). The self-potential signals are generated by the divergence of the streaming current density, which is in turn related to the flow of the ground water. We assume that this source is located at the water table. If capillary fringes can be neglected (this will be the case for the field data reported below), the distribution of the electrical resistivity tomography can also be used to determine the domain of strong probability of the presence of the source term for the self-potential signals. This solution makes it possible to calculate the kernels quickly and the solution will be relatively well localized. Among the many other possible scenarios to reduce the non-uniqueness of the solution, we can calculate the covariogram based on surface topographic variations within an area of study that exhibits no strong lithological discontinuities that could create secondary self-potential sources or that have significant relationships to the hydraulic heads.

For pumping tests, the semi-variogram of the piezometric data can be determined from the semi-variogram of the self-potential measurement. Indeed, we will show

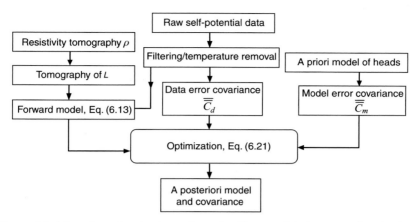

Figure 6.1. Flowchart for the inversion of self-potential data associated with a pumping, injection, or infiltration test. The a priori model of hydraulic heads is given by the linear relationship between the heads and the self-potential data; see Linde *et al.* (2007a, b) and Straface *et al.* (2007).

that the self-potential data follow a quasi-linear trend with respect to the piezometric levels. A flowchart summarizing the different steps used to perform the inversion of the self-potential data is given in Figure 6.1.

6.1.2 Synthetic tests

Flow in the viscous laminar flow regime

The first model corresponds to a synthetic case in which the flow of ground water occurs entirely in the viscous laminar flow regime. The geometry of the system is shown in Figure 6.2a. The medium is characterized by two layers and a rectangular heterogeneity placed in the vicinity of the pumping well. The hydraulic problem is solved in 3D with Comsol Multiphysics 3.4 assuming that the water table is a free boundary and by imposing the heads on the sides of the system (see Jardani *et al.* (2009)).

The numerical finite-element simulation of steady-state pumping is shown in Figure 6.2b. The distribution of the resulting self-potential anomaly that is determined at the ground surface is shown in Figure 6.2c. Interestingly, the self-potential signals recorded at the ground surface are linearly correlated to the hydraulic heads at depth (Figure 6.2d). This is in support of the basic models used by Rizzo *et al.* (2004) and Straface *et al.* (2007). Therefore, we use the variogram of the self-potential map to determine the semi-variogram of the a priori hydraulic heads (Figure 6.3). This semi-variogram is fitted with a Gaussian model. Then we used the methodology developed in Section 6.3 (see Figure 6.1) to invert the a posteriori

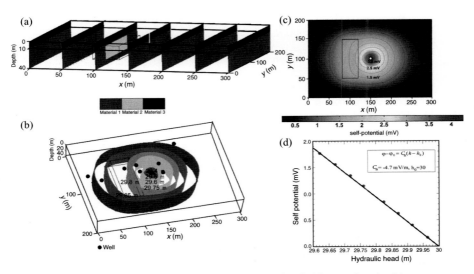

Figure 6.2. Synthetic modeling of the self-potential field associated with a pumping test. (a) Synthetic model used to invert self-potential data associated with a
pumping test under steady-state conditions. The base of the aquifer occurs at a
constant depth of 40 m from the flat ground surface. The heterogeneity is a rectangular block with dimensions: $\delta h = 20$ m, $\delta x = 40$ m, $\delta y = 100$ m. The initial depth
to the top of the aquifer is 10 m. Pumping is performed from a partially penetrating
well (diameter of 1 m, thickness of the pumping interval is 8 m (black zone in
well)) located at $x = 150$, $y = 100$. We impose a constant hydraulic head of 30 m
above the base datum at all the side boundaries of the aquifer. (b) Distribution of
the piezometric head (as meters above the datum) under steady-state conditions
(the pumping rate is 10^{-5} s^{-1}). The filled circles correspond to the positions of a
set of arbitrarily located observation wells. (c) Distribution of self-potential at the
ground surface under steady-state conditions. Note the non-symmetric distribution
of the electrical equipotentials associated with the presence of the heterogeneity.
(d) Correlation between the self-potential data at the ground surface and the
hydraulic heads (see Jardani *et al.* (2009)).

distribution of the hydraulic heads; the result is shown in Figure 6.4. Figure 6.4b
is a plot of the true hydraulic heads of the synthetic model versus the inverted
hydraulic heads which shows the high degree of correlation; in this regard, the
scatter is the result of numerical noise; see Jardani *et al.* (2009).

Flow in the inertial laminar flow regime

A second synthetic case is developed to study the self-potential response around
a pumping well in a case where Re > 1 in the vicinity of the pumping well. In
this case, the flow of ground water in the vicinity of the pumping well occurs in
the inertial laminar flow regime. The geometry is the same as that in the previous
case. We used material properties that favor a high Reynolds number close to

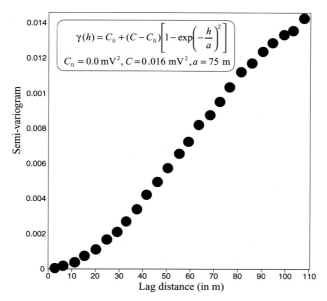

Figure 6.3. Semi-variogram of the a priori hydraulic head model determined from the linear relationship $\psi - \psi_0 = C_a (h - h_0)$. The semi-variogram is fitted with a Gaussian model (Jardani *et al.* (2009)).

the pumping well. In Figure 6.4b we plot the distribution of the self-potential at the ground surface when the effect of the Reynolds number is not taken into account. In Figure 6.5, we plot (a) the distribution of the self-potential at the ground surface when the effect of the Reynolds number is taken into account and also (b) the distribution of the Reynolds number. The result is striking. When the effect of the Reynolds number is taken into account, the self-potential anomaly is no longer centered on the pumping well. This observation is in agreement with the field results of Jardani *et al.* (2009).

6.1.3 Field application

Jardani *et al.* (2009) performed a series of coordinated hydraulic tomography, self-potential, and electrical resistivity tomography field experiments at the Boise Hydrogeophysical Research Site (BHRS). This site is located on a gravel bar adjacent to the Boise River, 15 km from downtown Boise, Idaho. The BHRS has been established to develop combined hydrogeological and geophysical methods to determine the distributions of hydraulic properties (e.g., transmissivity, storativity and specific yield) of naturally heterogeneous aquifers in the shallow subsurface. Eighteen fully penetrating wells have been cored and completed at the BHRS

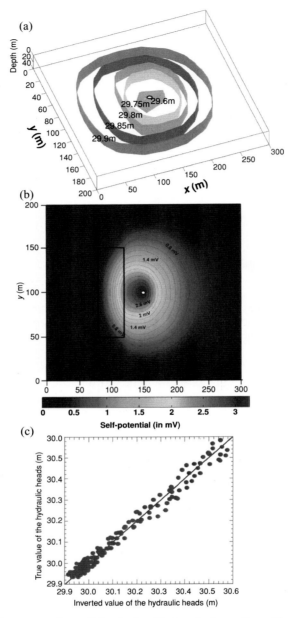

Figure 6.4. Reconstruction of the hydraulic heads from the self-potential data. (a) Reconstruction of the posterior distribution of the hydraulic heads. (b) Self-potential field obtained by neglecting the effect of the Reynolds number. In this case, the self-potential anomaly is centered on the pumping well (the white circle). The pumping rate is 10^{-5} s^{-1}. (c) True value of the hydraulic head versus inverted values under steady-state conditions of pumping ($R^2 = 0.97$) (see Jardani *et al.* (2009)).

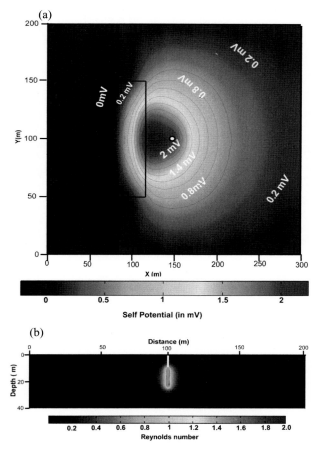

Figure 6.5. Influence of the Reynolds number. (a) Self-potential field computed by accounting for the distribution of the Reynolds number. In this case, the self-potential anomaly is not anymore centered on the pumping well (white dot) and its magnitude is smaller than in previous case (compare with Figure 6.4). (b) Distribution of the Reynolds number determined from the hydraulic head gradient. The pumping rate is 10^{-5} s^{-1} (see Jardani *et al.* (2009)).

to provide access for detailed characterization and testing (see Barrash *et al.* (1999)).

The aquifer at the BHRS consists of Pleistocene to Holocene coarse fluvial deposits that are unconsolidated and unaltered, and that are underlain by a red clay formation (Barrash and Reboulet, 2004). The thickness of the fluvial deposits is in the range of 18–20 m, and saturated thickness of the aquifer is generally about 16–17 m. The aquifer at the BHRS is known to be heterogeneous with multi-scale, hierarchical sedimentary organization (Barrash and Clemo, 2002) whereby the

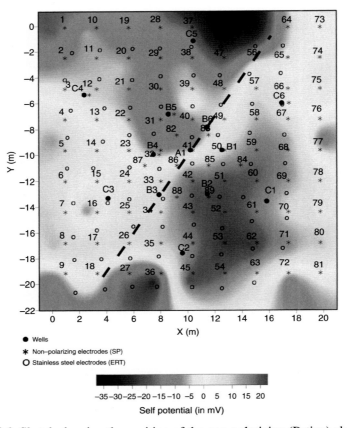

Figure 6.6. Sketch showing the position of the non-polarizing (Petiau) electrodes (self-potential stations) used for the self-potential monitoring (represented by stars), the stainless steel electrodes for the resistivity tomography (open circles), and the position of wells (the filled circles) in the central area of the BHRS. The self-potential map corresponds to the self-potential distribution priori to dipole pumping/injection tests. Relatively lower self-potential corresponds with portions of the aquifer that include sand channel deposits characterized by a relatively higher transmissivity or thickness-averaged hydraulic conductivity.

aquifer as a whole includes layers, and lenses within layers, that can be recognized with a variety of geological, hydrological and geophysical methods.

   Ten dipole tests were performed in which water was pumped from one well and injected into another. For the present investigation, we use data from two dipole tests with wells C1 and C4 as pumping and injection wells (alternatively). The self-potential signals were measured at the ground surface in a set of 88 electrodes with respect to a reference electrode placed 90 m from the region of investigation (see Figure 6.6, seven additional electrodes were placed in boreholes). The dashed line on Figure 6.6 separates two regions of high and low transmissivities determined from previous studies, such as that of Jardani *et al.* (2009).

Each electrode (stainless and non-polarizing) was placed inside a hole 10 cm deep filled with a moistened bentonite and gypsum mixture to ensure good contact between the electrode and the ground, and stones were placed above the electrodes. Measurements of the self-potential signals were carried out with a Keithley 2701 multichannel voltmeter and we used non-polarizing Pb/PbCl$_2$ (Petiau electrode) electrodes; see Perrier *et al.* (1997) and Petiau (2000). The voltmeter was connected to a laptop computer where the data were recorded. All the electrodes were scanned during a period of 30 s.

Here we note that in-field temperature changes caused drift in the self-potential measurements. To account for this drift, an additional reference electrode was located 50 m from the middle of the investigated region, and temperature of the packing medium around the reference electrode was measured periodically. According to SDEC (the company manufacturing the Petiau electrodes), the temperature dependence of the Petiau electrodes is 0.210 mV °C^{-1}. During the day, we measured variations of temperature ranging from a few degrees Celsius to 15 °C. A difference in temperature of 10 °C is responsible for a drift of the self-potential measurement of 2 mV, and therefore this effect cannot be neglected.

Figure 6.7a shows the time variation of the measured self-potentials (raw data) on one electrode during the dipole test. This potential measurement is divided into five time intervals denoted I to V. The data can be considered to be the sum of the time variation due to the pumping/injection test and its recovery plus a drift. The drift itself is believed to be generated by the variation of the temperature over time and can be adjusted by a polynomial of the third order with the coefficients of this polynomial fitted in time intervals I, III and V. This trend is then removed from the raw data, which gives the data shown in Figure 6.7b. These data are used to build self-potential maps. Rizzo *et al.* (2004) used a filtering operation to improve the signal-to-noise ratio of these data. However, we found that the self-potential data collected during the dipole tests at the BHRS in 2007 are of an exceptionally high quality (the standard deviation is typically of 0.1 mV) and no filtering is applied to them (see Jardani *et al.* (2009)).

Under steady-state conditions, the self-potential data exhibit a linear relationship with the position of the water table. A plot of the self-potential data as a function of the hydraulics heads (measured in the wells) is shown in Figure 6.8 and will be interpreted in Section 6.4. In addition to the self-potential data, electrical resistivity measurements were also collected during the steady-state conditions of the dipole tests. These measurements were taken with stainless steel electrodes (their positions are shown in Figure 6.6) and an IRIS-PRO System.

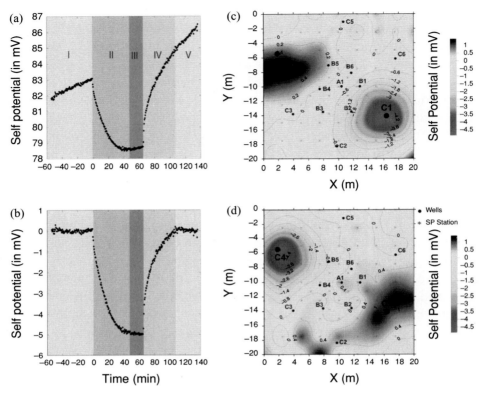

Figure 6.7. Example of self-potential data during a dipole pumping/injection test. (a) Raw self-potential time series. (b) Detrended self-potential time series. We can recognize five phases in the pumping test. Phase I corresponds to the data obtained prior the start of the pumping test. Phase II is the transient phase during pumping. Phase III denotes the steady-state phase. Phase IV denotes the rapidly changing portion of the recovery phase. Finally, Phase V corresponds to the slowly changing to steady-state portion(s) of the recovery phase. (c, d) Self-potential maps obtained under steady-state conditions (Phase III for the pumping/injection tests C4 pumping–C1 injection (c) and for the reciprocal test C1 pumping–C4 injection (d). Note that for the pumping well in both tests, the self-potential anomaly is not centered on the pumping well itself like in the synthetic case (Figure 6.5). However, the self-potential anomaly associated with the injection well is centered on the injection well for both tests; see Jardani *et al.* (2009). See also color plate section.

Permeability, electrical resistivity, and streaming potential coupling coefficient were measured on 16 core samples from wells at the BHRS. Only the fine fraction of the material (below 5 mm) was used. The methodology used to measure these properties is the same as that reported in Chapter 3. The average value of the streaming potential coupling coefficient is -13.2 mV m^{-1} for a brine conductivity

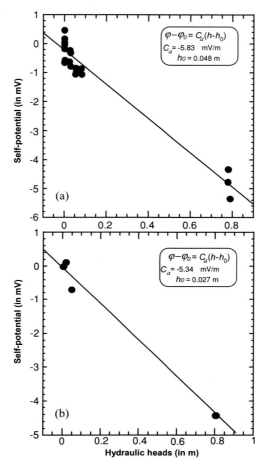

Figure 6.8. Self-potential data from electrodes collocated with transducers in wells versus the value of the piezometric levels in those wells during the steady-state portions of dipole tests. Data shown at three different times (50, 53 and 56 min after the beginning of the pumping/infiltration). (a) Pumping/infiltration test C4–C1. (b) Pumping/infiltration test C4–C1.

of 0.0221 ± 0.0005 S m^{-1} (at 25 °C), which is the conductivity of the ground water measured in one of the wells at the BHRS during the field experiments in June 2007. Because $C = -L/\sigma$, we can use the electrical resistivity tomography and the value $C = -13.2$ mV m^{-1} to map the distribution of L in 3D.

First, we can plot the self-potential data, under steady-state conditions, as a function of the hydraulic head for both the dipole hydraulic test C4–C1 (pumping from C4 and injecting into C1) and its reciprocal C1–C4 (pumping from C1 and injecting into C4; see Jardani *et al.* (2009) for further details). According to Rizzo

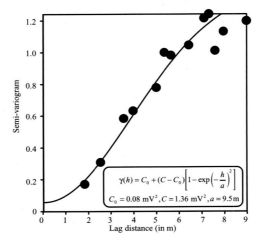

Figure 6.9. Semi-variogram of the a priori hydraulic head model determined from the linear relationship $\psi - \psi_0 = C_a (h - h_0)$. The semi-variogram is fitted with a Gaussian model (angle 55°).

et al. (2004) and Suski *et al.* (2006), we should observe a linear relationship between these two variables that can be written as $\psi - \psi_0 = C_a(h - h_0)$, where C_a is an apparent coupling coefficient (in V m^{-1}), ψ_0 is the reference electrical potential (0 mV), and h_0 and is the reference hydraulic head. This is indeed the case with $C_a =$ -5.8 mV m^{-1} for the dipole test C4–C1 and $C_a = -5.3$ mV m^{-1} for the dipole test C1–C4 (Figure 6.8). According to Suski *et al.* (2006), C_a and C are approximately related by $C_a = C/2$. Because the average value of C is -13.4 mV m^{-1}, this yields $C_a = -6.7$ mV m^{-1}, which can be considered as a reasonable estimate of the value of the apparent coupling coefficient given above. Then, we use this value of C_a to reconstruct the semi-variogram of the a priori hydraulic head model determined from the linear relationship $\psi - \psi_0 = C_a(h - h_0)$ (Figure 6.9). The semi-variogram shown in Figure 6.9 is fitted with a Gaussian model.

The optimization of the self-potential data using the flowchart shown in Figure 6.1 yields the posterior distribution of the hydraulic heads shown in Figure 6.10a for dipole test C4–C1. However, this estimate is wrong in the vicinity of the pumping well because we did not account for the effect of the Reynolds number. To obtain a better estimate, we took the posterior distribution of the hydraulic heads reconstructed from the self-potential measurements only in the area close to the injection well while only the piezometric data are used in the vicinity of the injection well. The result is shown in Figure 6.10b. This distribution can be used to determine the transmissivity distribution in the aquifer in the vicinity of the pumping and injection wells.

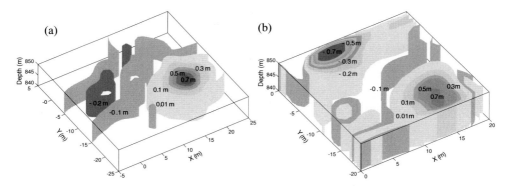

Figure 6.10. Posterior hydraulic head distributions (a) Dipole tests C4–C1. (b) Corrected posterior distribution of the steady-state hydraulic heads for dipole test C4–C1. We have not used the self-potential data in the vicinity of the pumping well; see Jardani *et al.* (2009). See also color plate section.

6.2 Flow in the vadose zone

6.2.1 Position of the problem

In unsaturated conditions, the streaming current density is given by Linde *et al.* (2007a) as

$$\mathbf{J}_S = \left(\frac{\hat{Q}_V}{S_w}\right) \mathbf{U}, \qquad (6.26)$$

where $\hat{Q}_V$ is the excess of electrical charge contained in the pore water at saturation, S_w is the saturation of the water phase ($s_w = 1$ at saturation), and $\mathbf{U}$ is the seepage velocity. We assume a vertical percolation of water through the vadose zone. The seepage velocity is determined using the van Genuchten formulation.

At a recording electrode P located at position P on the ground surface, the electrical potential is

$$\psi(P) = \int_{\Omega} \mathbf{K}(P, M) \cdot \mathbf{J}_S(M) dV, \qquad (6.27)$$

where $\mathbf{K}(P, M)$ is the kernel that summarizes the geometry of the system, the influence of the electrical resistivity distribution and establish the link between the source distribution in the source volume Ω (Figure 6.11), and the measured self-potential signals. The kernel can be determined by solving the forward problem successively with vertical unit dipoles at various locations and accounting for the electrical resistivity distribution of the system. There are several ways to invert self-potential data but some of them can be based on the linear formulation corresponding to Eq. (6.27).

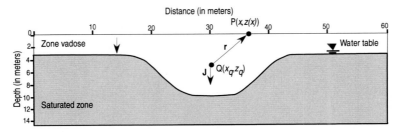

Figure 6.11. Sketch of the ground divided into two compartments, the vadose zone and the unconfined aquifer. The self-potential signal measured at point P located at the ground surface (or possibly in a borehole) is the convolution of all the dipolar electrokinetic sources associated with flow in the vadose zone driven by the gravity field.

6.2.2 *Numerical modeling*

To gain a better understanding of the electrical potentials resulting from the percolation of the ground water, and especially the relationship between the thickness of the vadose zone and the intensity of the self-potential signals, we first solve the hydraulic problem for the total pressure head h over a closed box by imposing a flux at the top boundary of the box (Figures 6.11 and 6.12).

The side boundaries of this box are characterized by impermeable boundary conditions. For the side boundaries of the box, one could wonder if this is a big limitation as this boundary condition imposes a vertical seepage in agreement with the assumption made above. Therefore the impermeable boundary condition adopted here for the side boundaries is not a serious limitation. Then the continuity equation is solved with an imposed distribution of the effective charge density $\hat{Q}_V / S_w$, electrical resistivity $\rho(x, z)$, and hydraulic conductivity $K(x, z)$. We also use the following insulating boundary condition $\hat{\mathbf{n}} \cdot \nabla \psi = 0$ at the ground surface and on the other boundaries; see Suski *et al.* (2006). In the case of the model of Figure 6.11, the Darcy flow equation is solved by fixing the flux of the water at the ground surface (set equal to 4×10^{-6} m^2 s^{-1} in the simulations reported below).

We consider that the electrical resistivity and the permeability of the vadose zone are homogeneous. In this case the polarization vector associated with the percolation of the ground water in the zone vadose can be regarded as constant with a downward orientation. This is basically the assumption used by Zhang and Aubert (2003), but while they did not presented a physical picture of their model, we shall see here how this assumption translates into an assumption of the material properties implied in the coupled hydroelectric problem.

Aubert and Atangana (1996) proposed the following empirical formula for the vadose zone; it is called the SPS model,

$$\psi(\mathrm{P}) - \psi_0 = c(e - e_0), \tag{6.28}$$

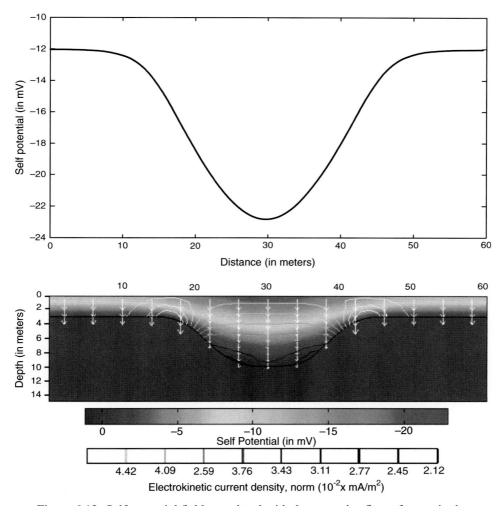

Figure 6.12. Self-potential field associated with the pervasive flow of water in the vadose zone. The self-potential field (in mV) is computed using a finite-element numerical model. The contour lines show the external current density norm J (in mA m^{-2}) while the arrows indicate the vertical direction of the ground water flow (Darcy velocity).

where $\psi(P)$ is the self-potential at station P, ψ_0 is the self-potential at the reference station, c is an apparent streaming potential coupling coefficient, $e = z(x, y) - h(x, y)$ represents the thickness of the vadose zone below the observation station $P(x, y, z(x, y))$, $z(x, y)$ describes the ground surface topography, $h(x, y)$ describes the water table (taken from an arbitrary datum, usually the sea level), and $e_0 = (z_0 - h_0)$ is the thickness of the vadose zone below the reference station.

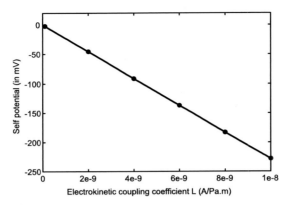

Figure 6.13. Correlation between the self-potential signals and the thickness of the vadose zone $e(x, y)$ ($R = 0.98$) for the simulation shown in Figure 6.12. The apparent voltage coupling coefficient c (-0.79 mV m^{-1}) is determined from the slope of this linear trend.

Numerical simulations show that the self-potential recorded at the surface of our box is proportional to the thickness of the domain characterized by vertical downward percolation of water (see Figures 6.12 and 6.13). The vertical percolation of the ground water produces a negative self-potential anomaly if the self-potential reference is taken at the outcrop of the water table. Therefore our model provides a clear and simple physical explanation of the SPS model.

We perform here a sensitivity analysis of the three material properties entering the coupled hydroelectric problem. They are the effective charge density $\hat{Q}_V/s_w$, the electrical conductivity σ, and the hydraulic conductivity K of the vadose zone. The current coupling coefficient controls the intensity of the conversion between the hydraulic heads and the self-potential signals, the hydraulic conductivity controls the seepage velocity, while the electrical conductivity controls the distribution of the equipotentials in the conductive body.

Influence of the hydraulic conductivity

Our sensitivity analysis (see Figure 6.14a) shows that the magnitude of the self-potential anomalies is inversely proportional to the hydraulic conductivity. This is easily explained by the fact that in Darcy's law, the hydraulic head gradient varies inversely with the hydraulic conductivity when the flux is imposed. The proportionality between the current density and the hydraulic gradient explains the relationship between the self-potential anomalies and the hydraulic conductivity under the assumption stated in Section 6.2. This finding is interesting because it implies that the self-potential method can be used to distinguish between permeable and impermeable grounds.

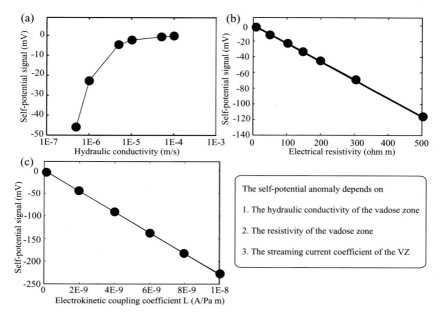

Figure 6.14. Sensitivity to the factors controlling the self-potential signals. (a) Influence of the hydraulic conductivity of the vadose zone upon the intensity of the self-potential signals computed at the ground surface. (b) Influence of the electrical resistivity of the vadose zone upon the self-potential signals computed at the ground surface of the Earth. (c) Influence of the streaming current coupling coefficient over the self-potential signals.

Influence of the streaming potential coupling coefficient

This coefficient determines the intensity of the current density associated with the pressure head gradient. The parameters influencing this coefficient were analyzed by Revil *et al.* (1999a, b) and this coefficient can be easily measured independently in the field or in the laboratory using core samples. Figure 6.14b shows that the magnitude of the self-potential signals are proportional to the magnitude of the current coupling coefficient.

Influence of the electrical resistivity

The electrical resistivity is assumed to be homogeneous in the vadose zone and therefore secondary sources associated with contrasts of electrical resistivity in the vadose zone are equal to zero. Figure 6.14c shows that the self-potential signals recorded at the ground surface depend on the electrical resistivity of the vadose zone. The magnitude of the self-potential anomalies increases with the electrical resistivity, which can be obtained independently in the field using electrical

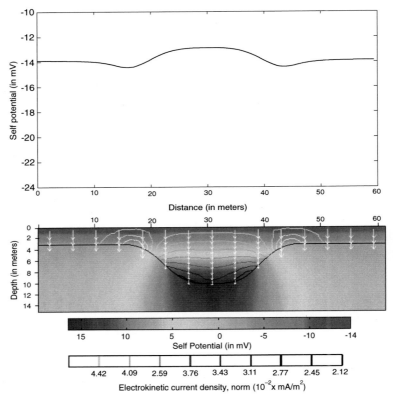

Figure 6.15. Distribution of the self-potential signals when the saturated zone is more resistive than the vadose zone (resistivity ratio: 1/100, same scale as Figure 6.12). In this case, the magnitude of the self-potential signals are very small (2 mV) and can not be used anymore to determine the thickness of the vadose zone.

impedance tomography. Therefore, our sensitivity analysis shows that the magnitude (but not the shape) of the self-potential anomalies depends on the three material properties entering the two coupled constitutive equations between the current density and the seepage velocity.

We now study how the contrasts in the petrophysical properties between the vadose zone and the aquifer influence the self-potential signals. Under the assumptions stated above, a contrast of permeability or a contrast of the streaming potential coupling coefficient between the vadose zone and the aquifer are *not* responsible for a change of the magnitude or the polarity of the self-potential signals. By contrast, a resistivity contrast between the vadose zone and the aquifer controls both the shape and the magnitude of the self-potential signals measured at the ground surface (Figures 6.15 and 6.16). If the vadose zone is more conductive than the aquifer (which is generally not the case), the self-potential anomalies vanish at the ground

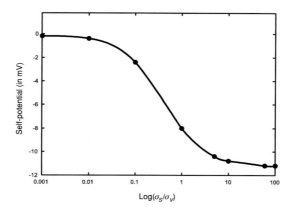

Figure 6.16. Self-potential (in mV) (relative to a reference station located at the ground surface) versus the logarithm of the ratio between the resistivity contrast between the saturated zone (subscript S) and the vadose zone (subscript V). When the saturated portion of the ground is much more resistivive than the vadose zone, the self-potential signals are linearly related to the thickness of the vadose zone. If the vadose zone is more conductive than the aquifer (which is generally not the case), the magnitude of the self-potential signals decreases and they are no longer related to the thickness of the vadose zone.

surface. In other words, a conductive vadose zone masks the effect associated with the thickness of the vadose zone. Therefore, we can determine the thickness of the vadose zone only if the vadose zone is more resistive by comparison with the resistivity of the aquifer and, hopefully, this is generally the case.

6.2.3 Inverse modeling

Parker (1973) showed that a series of Fourier transforms can be used to calculate the gravity anomaly caused by an uneven interface between two uniform layers of materials. Gomez-Ortiz and Agarwal (2005) observed that the forward modeling equation can be rearranged to obtain the depth to this interface iteratively from a given gravity profile. The iterative inversion is, however, non-unique owing to the inherent ambiguities in potential field inversion. The datum level to the interface and density are the two free parameters that control the topography of the inverted interface. The inversion becomes unstable at high frequencies, requiring a low-pass filter. In this section, we discuss the usefulness of the Parker algorithm to determine the interface between the vadose zone and the aquifer with self-potential data as input data.

To be consistent with the assumptions stated above regarding the homogeneity of the vadose zone, the polarization vector **P** is taken as constant in the vadose zone.

In this case, Eq. (6.27) can be rewritten, according to Zhang and Aubert (2003) as

$$\psi(P) = \mathbf{P} \cdot \int_{\Omega} \nabla \frac{1}{x} dV, \tag{6.29}$$

where x is the distance between the source and the measurement station. The undulations of any surface (like the water table or the self-potential map) can be described using oscillating functions with a distribution of spatial wavelengths. The Fourier transform of the self-potential anomaly is,

$$FT[\psi(P)] = -2\pi |\mathbf{P}| \exp(-kz_0) \sum_{n=1}^{\infty} \frac{(-k)^{n-1}}{n!} FT[h^n(P)], \tag{6.30}$$

where k is the spatial wave number, $h(P)$ is the depth to the interface (positive downwards), and z_0 is the mean depth of the interface. This equation is rearranged to obtain:

$$FT[h(P)] = -\frac{FT[\psi(P)]}{2\pi |\mathbf{P}|} \exp(-kz_0) - \sum_{n=2}^{\infty} \frac{k^{n-1}}{n!} FT[h^n(P)]. \tag{6.31}$$

This expression is used to determine the topography of the water table by means of an iterative procedure. Indeed, we assume the value of the mean depth of the interface, z_0, and the polarity contrast associated with two media, $|\mathbf{P}|$. The first term of Eq. (6.31) is computed by assigning $h(x) = 0$ and its inverse Fourier transform provides the first approximation of the topography interface, $h(x)$. This value of $h(x)$ is then used in Eq. (6.31) to evaluate a new estimate of $h(x)$. This process is continued until an acceptable solution is achieved.

The process is convergent if the depth to the interface is greater than zero and it does not intercept the topography of the ground surface and if the amplitude of the interface relief is less than the mean depth of the interface. Moreover, as the iterative process involving Eq. (6.31) is unstable at high spatial frequencies, a high-cut filter, $HCF(k)$, is included in the inversion procedure to ensure convergence of the series in Eq. (6.31). This filter is defined by:

$$HCF(k) = \frac{1}{2} \left[1 + \cos\left(\frac{k - 2\pi W_H}{2(S_H - W_H)}\right) \right], \tag{6.32}$$

for $W_H < k < S_H$, $HCF(k) = 0$, for $k > S_H$, and $HCF(k) = 1$ for $k < W_H$. This filter is used to restrict the high spatial frequency contents in the Fourier spectrum of the observed self-potential anomaly. The spatial frequency, k can be expressed as $1/\lambda$; λ being the spatial wavelength (in m). The iterative process is terminated when a certain number of iterations has been accomplished or when the difference between

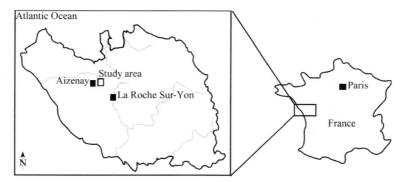

Figure 6.17. Position of the test site used to test the vadose zone model for self-potential data. The test site is located in Vendée, in the west of France, near the city of Aizenay.

two successive approximations to the topography is lower than a pre-assigned value as the convergence criteria.

6.2.4 Field application

The test site investigated in this section is located in the vicinity of the village of Aizenay, in a place called La Proutière in Vendée (France) (Figure 6.17). This site is located on the granitic dome of "La Roche-Sur-Yon" and corresponds to the flank of a small valley along which a river flows. The substratum is made of granite with a cover of arenitic and permeable sand. The topography of the test site, obtained with a differential GPS, varies from 58 to 65 m. The flank of the valley is itself crossed by a small talweg that is slightly asymmetric (Figure 6.18).

The area investigated in this study represents an area of 50 m by 100 m and the geophysical measurements were carried out in March 2002. They comprised electrical resistivity from the EM-31 tool, pole–pole vertical electrical resistivity sounding (PPVS), and self-potential measurements. These methods were applied after a period of intense rain in order to characterize the path of water flow at this site.

We use first the EM-31 and the PPVS resistivity data to characterize the geometry of the test site, especially the interface between the granitic substratum and the arenetic sand. The apparent conductivity map obtained from the EM-31 survey represents mean resistivity values averaged over a depth of 1–5 m. It shows that the axis of the talweg is conductive while on the East flank, the resistivive body reveals the presence of the granitic substratum. The vertical electrical resistivity soundings were used to draw the vertical conductivity profile at the different locations of the test site on a regular grid of 10×10 m (with x varying from 5 to 45 m and y varying in the range 5 to 95 m). The use of the pole–pole array is quite unusual in

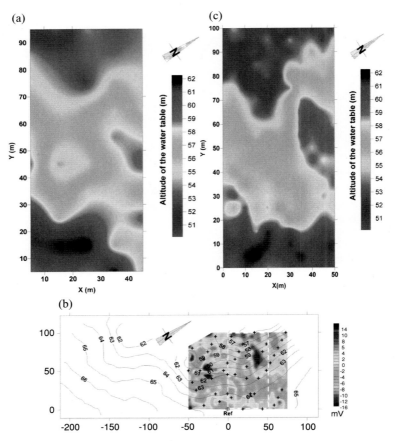

Figure 6.18. (a) Altitude of the water table from vertical electrical resistivity soundings. (b) Self-potential map of the investigated area and position of the measurement stations. Data from Nicolas Florsch. (c) Altitude of the water table using the Parker algorithm ($P = 0.45$ mV m^{-1} and $Z_0 = 4$ m). See also color plate section.

vertical sounding. It was chosen here because a sensitivity analysis demonstrates that the technique is robust to lateral change in the resistivity structure. The PPSV is used to distinguish between the different lithological units; the soil, the dry and saturated arenitic sand, the granitic substratum and so on. Inversion of these data was realized with the software QWSELN (developed by UMR Sisyphe, University of Paris 6, France; data from Nicolas Florsch).

The self-potential survey was performed with a high input impedance (10 MOhm) voltmeter (Metrix MX-20) and two non-polarizing electrodes. The reference electrode is at the point (0,0), and the self-potential map shows both positive and negative anomalies with respect to the electrical potential at the reference

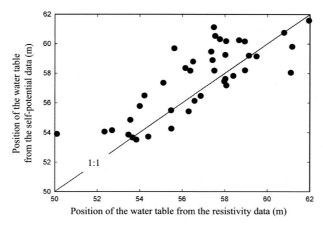

Figure 6.19. Comparison with the altitude of the water table determined from the self-potential data (inverted) versus the altitude of the water table determined from the resistivity (SEV) data.

electrode (Figure 6.18a). We assume here that these anomalies are the footprint signature of the vertical movement of water in the vadose zone. Other self-potential sources could affect the response like flow of the ground water in an unconfined aquifer for instance. Our assumption will be, however, validated by the final result of the inversion of the self-potential signals. Prior to inverting the self-potential data, we used the topographic correction proposed by Zhang and Aubert (2003). Then we apply the Parker algorithm to determine the depth of thickness of the vadose zone. The mean depth of the bottom of the vadose zone is determined from the PPSV inverted data with $z_0 = 4$ m.

Figures 6.18 and 6.19 show that the shape of the water table determined from the self-potential data matches fairly well with the result obtained with the inversion of the PPSV data except over the resistive body on the left side of the investigated area. In this area, however, the granitic substratum is unaltered and therefore very impermeable, and consequently we do not expect large self-potential signals of electrokinetic nature. However, the depths determined from the two methods agree fairly well in the other places where the permeability of the medium is higher.

6.3 Catchments hydrogeology

6.3.1 Position of the problem

In the past two decades, two potential sources of the self-potential signals associated with ground water flow have been identified. The first conceptual model corresponds to slow infiltration of water through the vadose zone; see Zablocki

(1976), Jackson and Kauahikaua (1987) and Aubert and Yéné Atangana (1996). These authors assumed that the self-potential signals are mainly related to the distance along which water percolates vertically through the vadose zone before reaching the water table. This model is referred to as the infiltration model. The second conceptual model is related to ground water flow in unconfined aquifers (see Fournier (1989)). According to this model, the strengths of the observed self-potential signals depend on the contrast of the streaming potential coupling coefficient through the water table (see Fournier (1989) and/or on the contrast of electrical resistivity through the water table (see Revil *et al.* (2003)). This model will be referred to below as the water table model.

The infiltration model

Given that the vadose zone is uniformly polarized, where each point in space acts as a vertically oriented dipole, and that the electrical conductivity distribution in the vadose zone is homogeneous, the resulting self-potential is proportional to the thickness of the vadose zone; see Zablocki (1976) and Jackson and Kauahikaua (1987). Zhang and Aubert (2003) demonstrated that, to the first order, the following simple formula can be used to relate self-potential signals to the thickness of the vadose zone,

$$\psi(x, y, z) - \psi_0(x_0, y_0, z_0) = c_V\{[z(x, y) - h(x, y)] - [z_0(x_0, y_0) - h_0(x_0, y_0)]\},$$
$$(6.33)$$

where $\psi(x, y, z)$ is the electrical potential at a self-potential station located at the ground surface; c_V is an apparent streaming potential coupling coefficient; $z(x, y)$ and $z_0(x_0, y_0)$ are the elevation of the self-potential station and the reference station, respectively; $h(x, y)$ and $h_0(x_0, y_0)$ are the corresponding elevations of the ground water table. This model predicts negative self-potentials at the ground surface if the self-potential reference station is located at the water table. The predicted elevation of the water table (or the elevation of the first impermeable formation) can be obtained from Eq. (6.33) (see Aubert and Yéné Atangana (1996)) as:

$$h(x, y) = z(x, y) - [z_0(x_0, y_0) - h_0(x_0, y_0)] - [\varphi(x, y, z) - \varphi_0(x_0, y_0, z_0)]/c_V.$$
$$(6.34)$$

Valley-like depressions of the predicted elevations correspond to preferential ground water flow pathways, whereas ridges correspond to water divides. The infiltration model has been successfully applied to several case studies (see, for example, Aubert and Yéné Atangana (1996) and Zhang and Aubert (2003)) in which the shape of the piezometric surface was confirmed by borehole data.

However, Fournier (1989) described several field observations that cannot be correctly explained by Eq. (6.34).

The water table model

According to the water table model, self-potential signals originate from contrasts in the coupling coefficient across the water table; see, for example, Fournier (1989). A first-order approximation that relates the self-potential signals to the water table (see Revil *et al.* 2003) is

$$h(x, y) = h_0(x, y) + [\psi(x, y, z) - \psi_0(x, y, z)]/c_W, \qquad (6.35)$$

where $c_W = (C_2 - C_1)$ is an apparent streaming potential coefficient, where C_1 and C_2 are the streaming potential coupling coefficients in the vadose zone and the saturated zone, respectively. Consequently, according to the water table model, the self-potential is linearly related to the piezometric head. In the case where there are capillary fringes, it is possible that there is an insignificant contrast of the coupling coefficient between the aquifer and the vadose zone, so $c_W \approx 0$.

6.3.2 A Bayesian inversion approach

We have reviewed above two alternative conceptual models relating the shape of the water table to self-potential data (Eqs. (6.34) and (6.35)). In this section, we describe how to use these conceptual models to update prior estimates of the water table obtained from kriged piezometric data. The resulting models provide uncertainty estimates that incorporate the number of data points, the measurement errors, the accuracy of the conceptual models, as well as the consistency between different conceptual models.

Bayes' theorem is an appealing framework for the integration of different data types and a priori information. A Bayesian formulation of data integration is more general than co-kriging (see Deutsch and Journel (1998)), because a Bayesian formulation allows arbitrary prior distributions and non-linear relationships between primary and secondary variables. We adapted the method of Chen *et al.* (2001) to the problem of estimating the geometry of the water table given piezometric and self-potential data by incorporating: (1) likelihood functions that have space-varying variance; (2) likelihood functions where the parameter structure is based on conceptual models (Eqs. (6.34) and (6.35)); (3) geophysical data (i.e., the self-potential data) that are kriged (not inverted) to ensure estimates at all locations of interest; and (4) concepts from Bayesian Model Averaging (BMA) to include the uncertainty arising from different possible conceptual models.

When different models are consistent with the available data, it may not be justifiable to rely on a single model. Instead, it might be better to weight the predictions of the various models with weights based on estimated model probabilities. We used concepts from BMA to condition the posterior probabilities to the two conceptual model types $\mathbf{M} = (M_1, M_2)$, where M_1 corresponds to the infiltration model and M_2 corresponds to the water table model.

The estimation method is described in detail in the following section. An estimate of the water table $h(x, y)$ using model M_k can be obtained with Bayes' theorem (see Linde *et al.* (2007b)) as follows

$$p[h(x, y) \mid \psi(x, y), M_k] = AL[\psi(x, y) \mid h(x, y), M_k]p[h(x, y)], \qquad (6.36)$$

where A is a normalizing coefficient, $\psi(x, y)$ is obtained by kriging the self-potential data, $L[\varphi(x, y) \mid h(x, y), M_k]$ is the likelihood of observing $\psi(x, y)$ given $h(x, y)$ and M_k, $p[h(x, y) \mid \psi(x, y), M_k]$ and $p[h(x, y)]$ are the posterior and prior probability density functions (pdfs) of the water table, respectively. It should be noted that only collocated estimates are used in Eq. (6.36), which simplifies the computations without significantly affecting the resulting models.

The prior pdf $p[h(x, y)]$ is obtained by kriging the available piezometric data. The likelihood functions $L[\varphi(x, y) \mid h(x, y), M_k]$ are first estimated at the locations where we have collocated self-potential data φ_{col} and piezometric data $\mathbf{h}_{col}$, where the parameters c_V (Eq. (6.34)) and c_W (Eq. (6.35)) can be estimated using linear regression. The likelihood functions are assumed to have a Gaussian distribution where the estimation variances $(\sigma_k^{reg})^2$ from regressions of the collocated data are used to estimate the variance of the likelihood functions at the collocated locations. The variance of the likelihood functions at other locations $(\sigma_k^{est}(x, y))^2$ are estimated by adding an additional error term that reflects the uncertainty associated with the estimation variance of the kriged self-potential estimates away from the self-potential stations (see Linde *et al.* (2007b)), as follows

$$\left(\sigma_k^{est}(x, y)\right)^2 = \left(\sigma_k^{reg}\right)^2 + \left(\sigma_\varphi^{krige}(x, y)\right)^2 - C_0, \qquad (6.37)$$

where $(\sigma_\varphi^{krige}(x, y))^2$ is the estimation variance obtained from ordinary kriging of the self-potential data and C_0 is the corresponding nugget (i.e., the uncorrelated contribution to the total kriging variance). The contribution of C_0 is already incorporated into $(\sigma_k^{reg})^2$ because the collocated self-potential and piezometric data are in practice separated by a short distance. Equation (6.36) is solved numerically.

The posterior probability that a model M_k is correct is estimated from Bayes' theorem as follows

$$p[M_k \mid \varphi_{col}] = A_1 L[\varphi_{col} \mid M_k]p[M_k], \qquad (6.38)$$

where A_1 is a normalizing coefficient, $p[M_k]$ is the prior probability that M_k is the correct model type, and the likelihood function $L[\varphi_{col} \mid M_k]$ describes the likelihood that the observed collocated self-potential data φ_{col} are observed given that model type M_k is the correct model. It is assumed that the data errors in φ_{col} are normally distributed with a variance that corresponds to the nugget C_0 of the kriged self-potential data. The likelihood function $L[\varphi_{\text{col}} \mid M_k]$ is given by

$$L[\varphi_{\text{col}} \mid M_k] = \frac{1}{\sqrt{2\pi C_0}} \exp\left[-\frac{\sigma_{\text{col}}^2}{2C_0}\right], \tag{6.39}$$

where σ_{col}^2 is the variance of the differences between the predicted and measured self-potential data at the collocated locations and where we have assumed that the errors in the head observations are negligible. In Bayesian Model Averaging, the likelihood is evaluated for all possible set of model parameters that the parameter-ization of the model type can take (Draper, 1995), however, we only calculate the likelihood function (Eq. 6.39) for the least-squares estimates of model M_k.

The sum of the prior probabilities $p(M_k)$ is one:

$$\sum_{k=1}^{K} p(M_k) = 1. \tag{6.40}$$

We have no strong evidence a priori about which model type is preferable and we make the neutral choice that all models are a priori equally likely (Draper, 1995):

$$p(M_1) = p(M_2) = 1/2. \tag{6.41}$$

The posterior mean $E[h(x, y) \mid \psi(x, y)]$ and variance $var[h(x, y) \mid \psi(x, y)]$ of the BMA model are (Draper, 1995):

$$E[h(x, y) \mid \psi(x, y)] = \sum_{k=1}^{K} E[h(x, y) \mid \psi(x, y), M_k] p(M_k \mid \varphi_{\text{col}}), \tag{6.42}$$

$$var[h(x, y) \mid \psi(x, y)] = \sum_{k=1}^{K} var[h(x, y) \mid \psi(x, y), M_k] p(M_k \mid \varphi_{\text{col}})$$

$$+ \sum_{k=1}^{K} \{E[h(x, y) \mid \psi(x, y), M_k]$$

$$- E[h(x, y) \mid \psi(x, y)]\}^2 p(M_k \mid \varphi_{\text{col}}). \tag{6.43}$$

The posterior mean is simply a weighted mean of the different model predictions, but the posterior variance includes an additional term that quantifies the deviations of the individual models from the posterior mean (see Linde *et al.* (2007b)).

6.3.3 Application to the catchment of Roujan (France)

The catchment of Roujan is located in southern France (Figure 6.20). It covers a surface area of approximately 1 km^2. The site is primarily man-made, with terraced

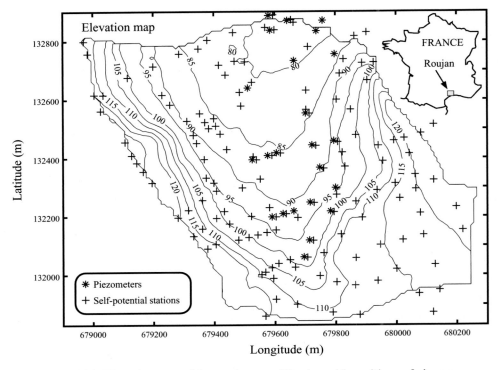

Figure 6.20. Elevation map of the catchment of Roujan with positions of piezometers (*) and self-potential stations (+). Roujan is located in the South of France
near the city of Béziers (modified from Linde *et al.* (2007b)).

slopes and a major network of ditches collecting the runoff water. Its land use
consists mainly of vineyards and shrubs. In geological terms, the substrata derive
from marine, lacustrine or fluvial sediments. The soils of the catchment range from
clayey to loamy. Most parts of the basin exhibit water tables whose depth fluctuates
largely both in space and time. The vadose zone can reach thicknesses of more than
10 m (see Linde *et al.* (2007b)). The Miocene marine deposits form a layer of marls
which can be considered impervious and corresponds to the lower boundary of most
shallow groundwaters of the catchment. Twentyeight piezometers were installed
and the piezometric levels were recorded every week. The electrical conductivity of
the ground water is approximately in the range 0.10 to 0.30 S m^{-1} (at 25 °C). The
soil is rich in clay minerals and its cation exchange capacity varies between 5 and
17 meq g^{-1}. High-resolution electrical resistivity tomography surveys (not shown
here) indicate a mean resistivity of 20 $\pm$ 5 Ωm with no resistivity contrasts between
the vadose zone and the unconfined aquifer which is in agreement with laboratory
measurements (see Table 1 in Suski *et al.* (2006)). However, the electrical resistivity

distribution in the first 50 cm of the soil appears to be highly heterogeneous during the summer (see Linde *et al.* (2007b)).

The self-potential survey was conducted during two days in June 2005 using a Metrix MX-20 multimeters and Cu/CuSO$_4$ non-polarizable electrodes. In order to get some understanding of the scatter of the data due to near-surface effects, we first performed 10–13 self-potential measurements around each piezometer. We determined the mean and then we removed the mean from the data set. Finally, we summed up all the measurements (286 measurements in total) and averaged the data per classes. The result supports the idea that the distribution of the self-potential measurements around a given measurement station is described by a Gaussian probability distribution,

$$p(\psi) = \frac{1}{(2\pi)^{1/2}\sigma_\psi} \exp\left[-\frac{(\psi - \bar{\psi})^2}{2\sigma_\psi^2}\right],$$ (6.44)

where $\bar{\psi}$ is the mean of the distribution ($\bar{\psi} = 0$ here) and σ_ψ^2 is the associated variance. For the self-potential measurements made at Roujan, we obtained $\sigma_\psi = 1.1$ mV. We believe that this variability mainly originates from the heterogeneity of the electrical resistivity in the upper 50 cm of the ground (see Linde *et al.* (2007b)).

We performed laboratory measurements of the streaming potential coupling coefficient for a set of samples extracted from the basin of Roujan. The samples were taken at different depths, sampling both the vadose zone and the aquifer. All the experiments were performed at saturation using water from the aquifer of Roujan ($\sim$0.1 S m^{-1} at 25 °C) using the experimental set-up described in Chapter 3. The streaming potential coupling coefficient varies from -5.6 to -3.2 mV m^{-1} (see Table 1 in Suski *et al.* (2006)). These values are representative of field conditions because the streaming potential coupling coefficient is only very weakly dependent on temperature (see, for example, Revil *et al.* (2005)).

We mapped the self-potential signals over the entire catchment area with 140 self-potential stations, where four to six self-potential measurements have been performed around each self-potential station. The experimental and fitted semi-variograms of the detrended piezometric data in the North–South direction (Figure 6.21a), and the experimental and fitted semi-variograms of the detrended self-potential data in the North–South (Figure 6.21b) and East–West (Figure 6.21c) directions are shown in Figure 6.21. The model used to detrend the data has a cone shape with its apex at the northernmost piezometer. The slope of the trend model for the heads was found by linear regression to be 1.4%. An isotropic spherical semi-variogram model with a sill of 0.8 m^2 and an effective range of 600 m was used to fit the residual piezometric data. The distribution of the piezometers

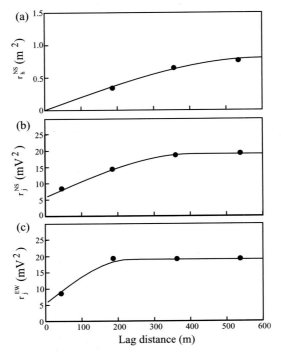

Figure 6.21. Experimental (filled circles) and fitted (solid line) semi-variograms. (a) Semi-variograms for the detrended piezometric levels in the North–South direction. (b) Semi-variograms for the detrended self-potential data in the North–South direction. (c) Semi-variograms for the detrended self-potential data in the East–West direction.

(see Figure 6.20) did not allow reliable estimates of the semi-variogram of the piezometric data in the East–West direction, and the range was assumed to be the same as in the North–South direction. The residual self-potential data were fitted with a nugget of 6 $(mV)^2$ and a spherical semi-variogram model with an effective range of 400 m in the North–South direction and 175 m in the East–West direction.

Ordinary kriging (see Deutsch and Journel (1998)) of the detrended self-potential and piezometric data were performed to obtain maps of the self-potential signals (Figure 6.22a) and water table (Figure 6.23a). The trend model is commonly assumed to be perfectly known and this is a good assumption in the North–South direction in the central part of the study area, but there is no reason to assume that the cone shape used to represent the head data is correct far away from the available head data (i.e., on the western and eastern slopes surrounding the catchment). We believe that the choice of the type of trend model is rather arbitrary, for example, a linear trend model with only a North–South component would essentially provide

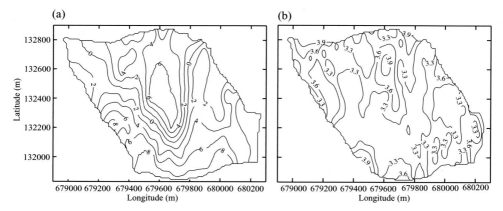

Figure 6.22. Self-potential map. (a) Kriged self-potential map in mV. (b) Kriging errors in mV. The dataset comprises 140 self-potential stations located throughout the catchment of Roujan (see position in Figure 6.20) (see Linde *et al.* (2007b)).

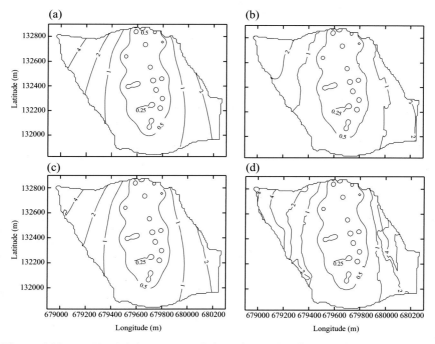

Figure 6.23. (a) The kriging errors of the prior hydraulic head increase outside the basin where the piezometers are located. The estimation errors of the updated models based on the infiltration model (b) and the water table model (c) are lower compared with the prior model, especially outside the basin. The weighted estimate of the infiltration model and the water table model have larger uncertainty estimates (d) compared with the individual models because conceptual uncertainty is included (see Linde *et al.* (2007b)).

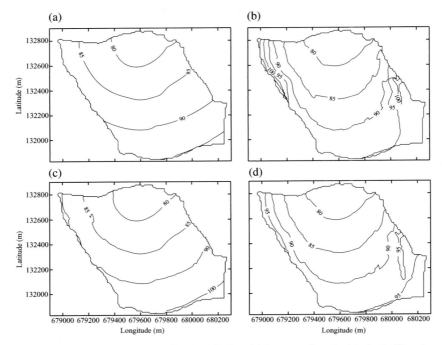

Figure 6.24. The prior model of hydraulic head (a) was updated with the infiltration
model (b) and the water table model (c). The infiltration and water table models
were averaged using the BMA approach (d) (from Linde *et al.* (2007b)).

the same size of the head residuals. We assumed that the difference between the
trend model that assumed a cone shape and a linear trend model in the North–
South direction corresponded to the largest possible error in the choice of the trend
model. This difference was assumed to correspond to two standard deviations of
the assumed normally distributed error of the cone shaped trend model. We can thus
add an additional term to the kriging errors in order to provide error estimates that
are more reasonable away from the available head data. This additional variance
term is scaled by the kriging variance divided by the sill in order to keep the
estimation errors low close to the piezometers. The corresponding kriging errors
are shown in Figures 6.22b and 6.23a, respectively.

The prior model for the hydraulic heads is shown in Figure 6.24. The likeli-
hood functions $L[\varphi(x, y) \,|\, h(x, y), M_k]$ (see Eq. (3.36)) were determined by linear
regression. A scatter plot between the measured thickness of the vadose zone and
the collocated kriged self-potential estimates (Figure 6.25a) reveals a linear trend
with a correlation coefficient of -0.86, where c_V is $-0.95 \,\mathrm{mV\,m^{-1}}$ (see Eq. (3.34)).
A scatter plot between the elevation of the water table and the collocated kriged

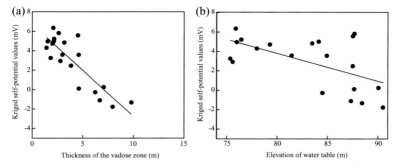

Figure 6.25. Relationship between the self-potential data measured in the field and the water table. (a) Correlation between the self-potential data (in mV) and the thickness of the vadose zone (in m) (correlation coefficient is -0.86). (b) Correlation between the self-potential data (in mV) and the piezometric heads (in meters above sea level) (correlation coefficient is -0.58).

self-potential estimates (Figure 6.25b) reveals a weaker linear trend with a correlation coefficient of -0.58, where c_W is -0.29 mV m^{-1} (see Eq. (3.35)). The apparent coupling coefficients c_V and c_W are unusually low. For example, Revil *et al.* (2003) and Naudet *et al.* (2004) reported values of c_W of -3.2 mV m^{-1} and -10.6 mV m^{-1}, respectively. The low values of c_V and c_W are also indicated by the low values of the streaming potential coupling coefficient C with a mean value of -4.3 mV m^{-1}, which is inversely related to the electrical conductivity of the pore water (Figure 6.26). The low values of the apparent coupling coefficients are attributed both to the relatively high conductivity of the pore water (0.1 S m^{-1}), which results in a relatively low streaming potential coupling coefficient and to the low electrical conductivity of the aquifer materials (i.e., approximately 20 Ohm) that results in lower electrical potentials at the ground surface compared with more resistive aquifer materials.

The resistivity of the ground (vadose zone and aquifer), determined from a geoelectric survey, is rather uniform $\sim 20 \pm 5$ Ohm in agreement with the petrophysical measurements reported by Suski *et al.* (2006). It follows that we can apply the Bayesian estimation method to update the prior water table map with the self-potential map.

For each model (Eqs. (6.34) and (6.35)) and location, we determine the position of the water table (Figures 6.24b, c) and the uncertainty of the estimates using Eq. (6.36) (Figures 6.23b, c). Finally, the posterior mean of the position of the water table incorporating both models is estimated from Eq. (6.42) (Figure 6.24d). The updated models estimate steep gradients associated with the slopes of the

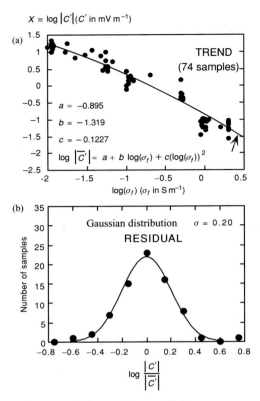

$X = \log |C'|(C' \text{ in mV m}^{-1})$

(a)

TREND
(74 samples)

$a = -0.895$

$b = -1.319$

$c = -0.1227$

$\log |\overline{C'}| = a + b \log(\sigma_f) + c(\log(\sigma_f))^2$

$\log(\sigma_f) \ (\sigma_f \text{ in S m}^{-1})$

(b)

Number of samples

Gaussian distribution $\sigma = 0.20$

RESIDUAL

$\log \dfrac{|C'|}{|\overline{C'}|}$

Figure 6.26. Streaming potential coupling coefficient versus pore water conductivity. (a) A compilation of 74 laboratory and field measurements of the streaming potential coupling coefficient (in mV m^{-1}) (see Revil *et al.* (2003), Naudet *et al.* (2003), Rizzo *et al.* (2005) and Suski *et al.* (2006)) and its dependence on the electrical conductivity of the pore water that is fitted with a second-order polynomial. (b) At a given ground water conductivity, the streaming potential coupling coefficient is given by a log normal distribution.

catchment (see elevation map, Figure 6.20). These steep gradients were not resolved in the prior model because of the distribution of the piezometers. The estimated spatial variations in the water table are larger in the model based on the infiltration model compared with the model based on the water table model. This difference is something that can be explained by the higher sensitivity of the self-potential data to changes in the thickness of the vadose zone (see Figure 6.25a) compared with changes in the elevation of the water table (see Figure 6.25b). The weighted estimates give a large weight to the infiltration model because of its higher probability of being the correct model.

The estimation uncertainties of the prior model (Figure 6.23a), posterior model based on the infiltration model, posterior model based on the water table model

(Figure 6.23c), and the total uncertainty estimated from Eq. (6.43) (Figure 6.23d) are shown. The improvements in the estimation errors for the updated models are most significant away from the piezometers, e.g. the maximum estimation error of the model based on the infiltration model is 2 m compared with more than 4 m for the prior model. The estimation errors are smaller for the models based on the infiltration model compared with the water table model because of the lower variances of the associated likelihood functions. The weighted models have larger estimation errors, because uncertainties regarding the proper model type are included. Areas in the weighted models with much higher estimation errors compared with the individual models indicate areas where the two individual models deviate significantly. It should be noted that the uncertainty arising from applying the calibrated conceptual models outside of the area sampled by the piezometers is not included in the estimation method, and these errors might be considerable.

The estimation error of using the self-potential data only to estimate the elevation of the water table is approximately 2 m at Roujan when using the infiltration model (see Figure 6.23b) and the question arises if future surveys can be conducted such that the estimation errors decrease substantially, say to 1 m. Adding more self-potential data will improve the estimates as seen by the self-potential semi-variograms (see Figures 6.21b, c), but just to a certain extent (i.e., to the estimation error at a given self-potential station). There are three obvious ways to improve the estimates presented in this work: (i) estimate the electrical conductivity of the pore water at each self-potential station; (ii) measure the temperature of the electrolyte in the electrodes; and (iii) monitor the self-potential stations at high-frequencies and transform the signals into the frequency domain to remove distortions from telluric currents and cultural activity. A discussion of these three points follows.

(i) The electrokinetic coupling coefficient is inversely proportional to the electrical conductivity of the pore water (see Revil *et al.* (2003)). Variations in the electrical conductivity of the pore water are therefore partly responsible for the scatter in the estimated likelihood functions and for the nugget in the semi-variograms of the self-potential signals (see Figures 6.21b, c). Variations in the electrical conductivity of the pore water could be incorporated in the estimation, but it would then be necessary to sample the electrical conductivity of the pore water at the self-potential stations.

(ii) The Cu/CuSO$_4$ electrodes have a temperature drift of $\sim$0.8 mV $^\circ$C^{-1} and this effect is considerable given the low amplitudes of the self-potential signals at this site even if the drift is corrected for by repeated measurements at the reference station. A temperature reading of the electrolyte at each self-potential station and continuous readings at the reference station would make it possible to remove this effect. Alternatively, use of the more stable Petiau electrodes should improve our measurements, according to the supplier (SDEC, Reignac sur Indre, France), the temperature drift of the Petiau electrodes are only ~ 0.2 mV $°C^{-1}$.

(iii) The recorded signals are composed by in situ signals and telluric currents, which correspond to the main source of noise in addition to cultural activity (50 Hz and harmonics). For monitoring purposes, the removal of cultural noise and telluric currents can be accomplished by filtering the self-potential measurements using the magnetic variations recorded with a magnetometer and a reference station for the self-potential survey (see Kawakami and Takasugi (1994)).

The first two additions to the survey procedure can be carried out without slowing down the survey significantly, whereas the third addition is probably too time-consuming for this type of application. In addition, the physical models behind the infiltration model could be improved by taking the distribution of the water content into account, something we have not attempted in this work. The apparent coupling coefficients in the Roujan catchment are at the lower end of the range usually observed (-1 to -10 mV m^{-1}). This is probably the consequence of the relatively high electrical conductivity of the pore water and the aquifer materials. Consequently, the signal-to-noise ratio is likely to be higher at sites where the electrical conductivity of the pore water and the geological formation are lower.

It should be noted that the self-potential method can be used, with much lower estimation uncertainty, to estimate temporal variations in the water table below a given station. Suski *et al.* (2006) found in an infiltration experiment carried out within the Roujan catchment that the relative temporal variations of the piezometric levels could be estimated with a precision of 20 cm when using a fixed network of electrodes.

6.4 Contaminant plumes

The Entressen landfill (43°60′ N, 4°93′ E) is located in the South of France near the delta of the Rhône River. This site was studied because it is the largest open air landfill in Europe and it is lying over a very simple geology consisting of a superficial aquifer overlying a marl aquitard. Consequently, the site is well-suited to test the possibility of using self-potential data to delineate the extent of a major leachate plume. More information is provided below in Section 6.4.2. The self-potential map was built using the loop approach discussed in Chapter 1. In the following, we are going to show how the self-potential signals downstream Entressen can be divided into two contributions: one associated with the flow of the ground water and one associated with redox processes occurring in the contaminated aquifer.

6.4.1 Streaming potential versus redox-controlled potential

In the subsurface, self-potential anomalies result from non-equilibrium thermo-dynamic processes that generate polarization of electrical charges throughout the porous materials (see Chapter 2). In the vicinity of equilibrium, the total current density $\mathbf{J}$ (A m^{-2}) can be written as the sum of a conductive current described by Ohm's law and a net source (or driving) current density,

$$\mathbf{J} = -\sigma \nabla \psi + \mathbf{J}_S. \tag{6.45}$$

The electrical potential ψ (V) corresponds to the natural electrical potential gener-ated by non-equilibrium processes (the so-called self-potential), σ is the electrical conductivity (S m^{-1}), and $\mathbf{J}_S$ (A m^{-2}) represents the source current density. We write the total current density as the sum of a redox contribution and an electroki-netic contribution, as discussed in Chapters 2 and 3,

$$\mathbf{J}_S = \sigma(\nabla E_H + C\nabla h), \tag{6.46}$$

where E_H is the redox potential (V), h the hydraulic head (m), and C the streaming potential coupling coefficient (V m^{-1}) (see, for example, Revil *et al.* (2003)). Equations (6.45) and (6.46) imply that, through an electrocell, the difference of electrical potential is equal to the difference of redox potential $\Delta\psi = \Delta E_H$, which is consistent with basic electrochemical models (see Sato and Mooney (1960)).

If the charge carriers of the geobattery model of a contaminant plume are elec-trons, we need to understand the nature of the electromotive force and how the elec-trons are produced and exchanged within the system. In a leachate plume, inorganic elements exist in different valence states and may participate in oxido-reduction reactions. Some of the redox reactions are also mediated by micro-organisms like those involving the decomposition of organic matter (see, for example, Lappin-Scott and Costerton (1995)). The electrons are associated with ionic carriers (e.g. Mg^{4+}, Fe^{3+}) and with the organic matter itself. There is the possibility that bacteria play a direct role in the generation of self-potential anomalies by offering direct pathways for the migration of these electrons between the reducing and oxidizing parts of the system (e.g., through the water table). Bacteria can be interconnected by electronically conducting microbial nanowires (see Childers *et al.* (2002) and Reguera *et al.* (2005)).

The growth of bacteria is closely related both to the presence of organic contaminants and to the oxygen concentration. There is poor micro-organism growth outside the contaminant plume (depleted in nutrients) and in the interior of the contaminant plume (depleted in oxygen) with the exception of anaerobic bacteria that represent, however, only a small fraction of the total biomass associated with a contaminant plume. The main growth of bacteria is therefore likely to be restricted to the water table where contaminated water mixes with oxygen-rich meteoric water. In short, the rapid bacterial growth in a nutrient and oxygen-rich environment, and the strong association of bacteria to the mineral surfaces lead to potential important biomass accumulation in the vicinity of the water table where both organic matter and oxygen are present. Similar situations would also occur at the end-boundaries of the plume but the gradient of the redox potential is expected to be smaller because of the lower amount of available oxygen in the saturated zone compared with the vadose zone.

If hardwired bacteria support migration of electrons through the system (see the recent work of Risgaard-Petersen *et al.* (2012)), every element of the water table behaves like a dipole with cathodic half-cell reactions occurring at the upper, more oxidizing end, and anodic half-cell reactions at the lower, more reducing end. As in geobatteries associated with ore deposits (see Sato and Mooney (1960)), oxygen (electron acceptor) is probably the electro-active species at the cathode, whereas organic matter, reducing metal ions (like sulfate, nitrate, ferric iron), and carbon dioxide are controlling electro-active species at the anode of the system.

In the field, measurements are carried out at the ground surface and a reference station is positioned upgradient where the aquifer is uncontaminated (Figure 6.27a). The flowchart in Figure 6.27b shows the procedure we followed to remove the streaming potential contribution and to relate the residual self-potential signals to the redox potentials throughout the contaminated area. As discussed above, the relationship between the electrokinetic (streaming) component of the self-potentials, and the hydraulic head values often takes the form of a linear relationship for unconfined aquifers (see, for example, Rizzo *et al.* (2004), Suski *et al.* (2006) and Linde *et al.* (2007b)). Naudet *et al.* (2003) showed that the same type of linear relationship occurs between the redox potential and the self-potential. Therefore, the effective self-potential value measured at station P can often be approximated by,

$$\psi(P) \approx c_H \left(E_H - E_H^0 \right) + c'(h - h_0),\qquad(6.47)$$

where h_0 and E_H^0 are the head (m) and the redox potential (V), respectively, measured in a piezometer close to the reference self-potential station, while E_H and

(a) General setting of a contaminant plume

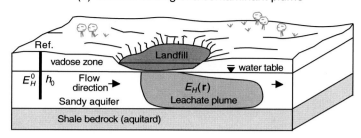

(b) Flow chart used to discriminate between the contributions

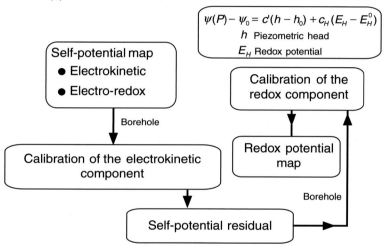

Figure 6.27. Interpetation of self-potential data at a landfill. (a) Sketch of a contaminant plume associated with a landfill. "Ref" represents a reference piezometer taken upgradient and used to determine the reference piezometric head, the reference redox potential, and the reference electrical conductivity of the ground water. The reference for the self-potential (also called the base station in the main text) is also located in the vicinity of this reference piezometer. All the measured self-potentials are measured relative to this base station. (b) Flowchart showing the steps followed to analyze the self-potential data associated with a contaminant plume. The self-potential data measured at the ground surface of the Earth is the sum of a contribution associated with the flow of the ground water and a contribution associated with the redox potential distribution in the contaminated portion of the aquifer.

h are measured at station P, c' (V m^{-1}) and c_H (dimensionless) are the apparent coupling coefficients associated with the streaming potential and electroredox contributions, respectively. Equation (6.47) is a first-order approximation that is based on the assumptions that the source current density is located at the water table and that there are no significant lateral variations in the electrical conductivity structure.

There have also been successful attempts to relate the streaming-potential signals to the depth to the water table by assuming that the vadose zone is uniformly polarized (see Aubert and Yéné Atangana (1996)). The predictive power of Eq. (6.47) depends on the site-conditions and accurate modeling under general conditions can only be done by solving Eq. (6.45).

6.4.2 Application to the Entressen Landfill

The Entressen Landfill (43°60′ N, 4°93′ E) is located in the South of France at a distance of 70 km from the city of Marseille, in the Crau alluvial plain of Provence. This landfill has been active since 1912, it covers about 0.5 km^2, and its maximum elevation is 30 m above the regional land surface elevation. Approximately 460 000 tons of municipal wastes are brought to this landfill every year and it is one of the largest open-air landfills in Europe. Additional details (log data and electrical resistivity tomograms) can be found in Naudet *et al.* (2003, 2004). Borehole and geophysical data indicate that the landfill is located on a shallow aquifer of Quaternary alluvium, which contains calcareous, metamorphic, and endogenous materials. The thickness of the quaternary alluvium is a few meters (see Naudet *et al.* (2004)). Below the aquifer, there is a thick layer of low-permeability Pliocene marls. The hydraulic head data indicate that the depths to the unconfined aquifer vary between 2 m and 12 m below the ground surface. The average hydraulic gradient of the shallow aquifer in the vicinity of the landfill is 0.005 based on hydraulic head data collected within the study area and the yearly fluctuation of the water level is about one meter. The average hydraulic gradient of 0.003 reported by Naudet *et al.* (2004) was based on estimations for the whole Crau area. The contaminated portion of the aquifer is characterized by high electrical conductivities of the ground water, approximately ten times higher than the reference value that was measured upgradient in well #7, and predominantly negative values of the redox potentials. Note that piezometer #7 is outside the map shown in Figure 6.28, approximately 1 km upgradient. These negative values of the redox potential indicate the presence of an anaerobic zone in the contaminated plume (see Naudet *et al.* (2004)).

Figure 6.28 shows a map of the kriged self-potential measurements. The error on self-potential measurement is about 8 mV. The reference station for the self-potential measurements is taken upgradient (see the filled circle of Figure 6.28) and the neighboring well is used as reference station for the hydraulic head data. The self-potential data show a general trend with increasingly positive values in the direction of the ground water flow and a strong (∼−500 mV) anomaly above the contaminant plume. This general trend is expected from the electrokinetic theory as positive charges are dragged along with the water flow.

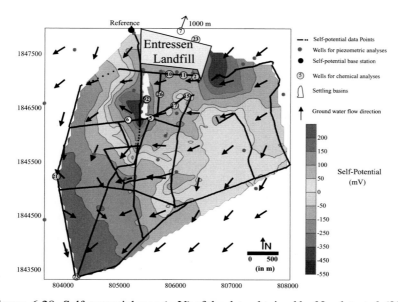

Figure. 6.28. Self-potential map (mV) of the data obtained by Naudet *et al.* (2003) over the Entressen Landfill. Black dots correspond to the 2800 self-potential measurements points made in the field. The locations of the piezometers for the chemical analyses and the measurement of the electrical conductivity of the groundwater. Piezometer #7 is located 1 km upgradient from well 23. See also color plate section.

The self-potential map (Figure 6.28) has two contributions. Before interpreting the self-potential data in terms of redox processes, it is therefore necessary to remove the contribution of streaming potentials. To interpret a self-potential map over a contaminated area, a two-step procedure is proposed (Figure 6.27b). The first step consists of determining the apparent streaming potential coefficient c'. It can be estimated by performing a regression of the hydraulic head data and the kriged collocated self-potential estimates collected at various locations outside the contamination plume. We used a Bayesian method developed by Linde *et al.* (2007b) to compute the streaming potential component of the recorded self-potential signals. This allows us to remove the streaming potential contribution from the self-potential data to conserve only the electro-redox potential contribution. We note the residual self-potential distribution as

$$\psi_r(x, y) = \psi(x, y) - c'[h(x, y) - h_0], \qquad (6.48)$$

where (x, y) are the coordinates of point P. The redox potential data measured in a set of piezometers can then be compared directly with the estimated redox potential, which is obtained as

$$E_H(x, y) = \frac{\psi_r(x, y)}{c_H} + E_H^0.$$ (6.49)

We employed the method of Linde *et al.* (2007b) to calculate the streaming potential contribution to the measured self-potential data. This method is based on the work of Chen *et al.* (2001) who developed a Bayesian method to calculate permeability between boreholes by updating a prior permeability field (based on flowmeter data) with crosshole geophysical data using site-specific relationships between permeability and different geophysical properties. In our application, we first kriged the detrended hydraulic head (our prior model) and the self-potential data collected outside the contaminant plume. In the second step, we use a linear regression model to relate the hydraulic head and self-potential data. In the third step, we calculate the posterior head model at each location (x, y) by solving Bayes' theorem through discretization

$$p[h(x, y) \,|\, \psi(x, y)] = AL[\psi(x, y) \,|\, h(x, y)] p[h(x, y)],$$ (6.50)

where A is a normalizing coefficient, $\psi(x, y)$ is the kriged self-potential data, $L[\psi(x, y) \,|\, h(x, y)_k]$ is the likelihood of observing $\psi(x, y)$ given the hydraulic head $h(x, y)$, and $p[h(x, y) \,|\, \psi(x, y)]$ and $p[h(x, y)]$ are the posterior and prior probability density functions (pdfs) of the water table, respectively. Only collocated estimates are used in Eq. (6.50), which simplifies the computations without significantly affecting the resulting models. It is assumed that both the prior and posterior distributions have a Gaussian distribution. The method uses the kriging errors to calculate space-varying standard deviations of the likelihood function $L[\psi(x, y) \,|\, h(x, y)_k]$ as described in detail by Linde *et al.* (2007b). In the fourth step, we subtract the streaming potential contribution from the measured self-potential values, where the streaming potential contribution is calculated based on the regression model and the posterior head model.

The hydraulic head data of Entressen were detrended with a linear trend model using linear regression. The slope of the trend was 0.005 with increasing values in the direction 47° N. The detrended head data were kriged with ordinary kriging by fitting an isotropic spherical model with a sill of 6.2 m and a range of 1300 m (see Figure 6.29a). The self-potential data collected outside the contaminated area were detrended in the same way as the hydraulic head data. The slope of the trend was 0.062 mV m^{-1} in the direction 235° N. The detrended self-potential data collected outside the contaminated area were kriged with ordinary kriging using a nugget effect of 130 mV and an isotropic spherical model with a sill of

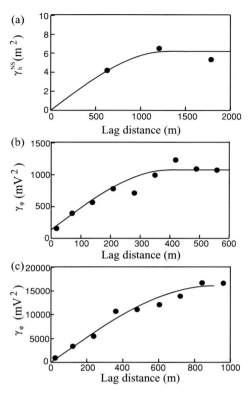

Figure 6.29. Semi-variograms of the heads (a), the self-potentials measurements outside the contaminant plume (b), and the self-potentials measurements at all locations (c) (from Arora *et al.* (2007)).

940 mV and a range of 400 m (see Figure 6.29b). A scatter plot of available head data and the collocated kriged self-potential values with the trend included are shown in Figure 6.30. The correlation coefficient is high (-0.94) and the resulting regression relationship is $\psi \approx -9.67 \, (h - h_0)$. The Bayesian updating was performed to estimate a posterior head map, which was used to estimate the streaming component of the self-potential signals (Figure 6.31). The residual self-potential data were calculated by subtracting this contribution for all self-potential data points.

In a first attempt, we investigated whether the self-potential data correlate with redox measurements collected within the contaminated area. All self-potential data were kriged with a nugget effect of 130 mV and an isotropic spherical model with a sill of $16\,000 \, \text{mV}^2$ and a range of 900 m (Figure 6.32). A scatter plot between the redox values and the collocated self-potential values (Figure 6.32) reveals a

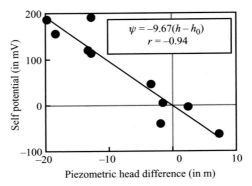

Figure 6.30. Correlation between the measured hydraulic head difference (m) (measured in all the available piezometers in the investigated area) and the kriged self-potential data (mV) at the ground surface of the Earth. The self-potential values are kriged values at the land surface location of each piezometer. The error bar is ± 10 mV for the self-potential measurements (determined from the standard deviation of the measurements) and ± 0.2 m for the heads (from Arora *et al.* (2007)).

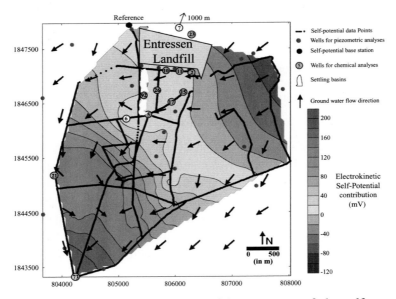

Figure 6.31. Map of the streaming potential component of the self-potential signals (mV). The self-potential base station represents the reference station for the self-potential measurements. Note that in most places the arrows resulting from the interpolation of the piezometric data are normal to the electrical equipotentials, which in principle mimic the hydraulic equipotentials. See also color plate section.

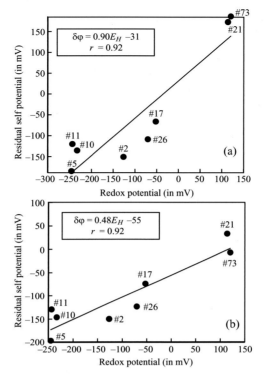

Figure 6.32. (a) The correlation between the kriged self-potential signals and the redox potential data. (b) The correlation between the kriged residual self-potential data (corrected for the streaming potential contribution) and the measured redox potentials. The error bar is ± 10 mV for the self-potential measurements (determined from the standard deviation of the measurements) and ± 25 mV for the redox potentials.

strong correlation (correlation coefficient of 0.92). Piezometer #92 was not used because it is in an area where there is a strong gradient of the contamination and the self-potential measurements are only available downstream, just after the settling basin responsible for the contamination in this area. The slope of the regression curve based on these data is inaccurate because the streaming potential contribution was not removed.

In the next step, we kriged the residual self-potential data ψ_r with a nugget effect of 130 mV and an isotropic spherical model with a sill of 1600 mV^2 and a range of 700 m. A scatter plot between the redox values and the collocated self-potential values (Figure 6.32) reveals a strong correlation (correlation coefficient 0.92). The estimated relationship between residual self-potential data and the redox values is

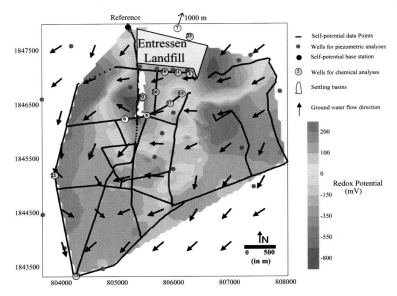

Figure 6.33. Map of the redox potential (mV). The small black circles represent the location of the self-potential stations used for the measurements. The base station represents the reference station for the self-potential measurements. See also color plate section.

$\psi_r = 0.48 \, (E_H - 115)$, where ψ_r is expressed in mV. This analysis suggests that 115 mV is approximately the redox potential in the undisturbed area. This value is consistent with the redox potential recorded outside the contaminant plume. Indeed, in the two reference wells (#7 and #73), the redox potential is in the range 110–140 mV.

These results indicate a strong relationship between the redox potential and self-potential signals. Consequently, self-potential appears to be a very useful method to estimate the redox potential of contaminated plumes, and to constrain mapped distributions of the redox potential. The resulting redox map (Figure 6.33) is improved in comparison with the previously generated redox map of the site (see Naudet *et al.* (2004)). This improvement is due to the fact that the estimates of the hydraulic head were obtained by kriging the residual head data as well as the self-potential data collected outside the survey area following the method of Linde *et al.* (2007b), whereas Naudet *et al.* (2004) interpolated the hydraulic head data linearly. In addition, the residual self-potential data were kriged and not linearly interpolated. The map clearly shows the extent of the contaminated plume which is spreading southerly. We cannot provide a map of the electrical conductivity of the

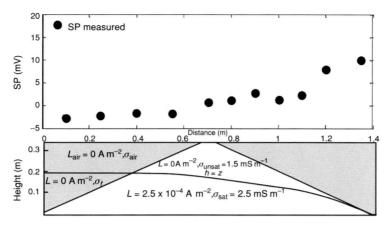

Figure 6.34. Self-potential signal associated with ground water flow through an earth dam.

pore water because of the scarcity of the measurements, which are only available at a few piezometers.

6.5 Conclusions

We have investigated four distinct cases at different scales where the self-potential method can be used to address problems related to water resources. The first case is to use the self-potential method as a non-intrusive source of information regarding ground water flow during pumping tests. The second application was related to vadose zone hydrology to investigate the thickness of the vadose zone. The third application was at the watershed scale and showed the power of the self-potential method as a quick reconnaissance tool at this scale to characterize watershed hydrogeology. The final application was the application of the self-potential method to the delineation of a contaminant plume dowstream a landfill and to investigate the possibility of a biogeobattery running in the aquifer. We think that the next step will be the installation, here again, of permanent sensors and the long-term monitoring of ground water flow during long periods of time.

Exercises

Exercise 6.1. Seepage problem. The response of the self-potential signals due to the seepage of water in dams was studied in the laboratory (Figure 6.34). A study of a homogeneous embankment was constructed in an acrylic tank measuring 14 cm in length, 10 cm in with and 31 cm in height, filled with Ottawa sand. On the dam, a steady-state flow was imposed using two constant reservoirs in up 18 cm and 22.5 cm. The hydraulic conductivity and the electrical properties of the sand are: $K = 4.5 \times 10^{-4}$ m s^{-1}, $L_{\text{sat}} = 2.5 \times 10^{-4}$ A m^{-2} $\sigma = 2.5 \times 10^{-3}$ S m^{-1} in the saturated zone, and $\sigma = 1.5 \times 10^{-3}$ S m^{-1} in the unsaturated zone. The electrical conductivity of the water is $\sigma_f = 2.8 \times 10^{-3}$ S m^{-1}. Use the finite element code to reproduce the self-potential field.

7

Application to hydrothermal systems

Another application of the self-potential method concerns the mapping of ground water flow in hydrothermal and geothermal systems. In Section 7.1, we propose a stochastic inversion of borehole temperature measurements and surface self-potential data to determine the pattern and magnitude of flow in a geothermal field assuming that the lithological boundaries and the position of the faults are independently known (through geology and additional geophysical methods). This approach is applied to the Cerro Prieto geothermal field in Baja California (Mexico) in Section 7.2. In Section 7.3, we develop an alternative gradient-based approach for the self-potential data that can be combined with d.c.-resistivity tomograms to determine the flux of thermal water upwelling through tectonic faults. This approach is applied to two geothermal fields in Colorado and Oregon.

7.1 Stochastic inversion of temperature and self-potential data

Both temperature and self-potential data contain information regarding the flow pattern of water and are therefore suitable to be jointly inverted. This technique can be used to recover the flow of water at various scales, particularly when the information from the two data sets is complementary. In this section, we develop a stochastic inversion scheme of in-situ (borehole) temperature measurements and surface self-potential data to determine the flow pattern in a geothermal field. This approach is tested on a realistic, synthetic case study where we assume that the position of the main lithological bodies and faults are perfectly known by combining geological, tectonic, and geophysical mesurements; the focus is to find the flow pattern of the ground water in this setting.

7.1.1 Field equations

We consider steady state conditions, no source or sink terms for water, and an incompressible pore fluid. With these assumptions, the continuity equation for the flow of the ground water is $\nabla \cdot U = 0$. The seepage (Darcy) velocity U (in m s^{-1}) obeys Darcy's law $U = -K\nabla H$, where K is the hydraulic conductivity (in m s^{-1}) and H the hydraulic head (in m). The distribution of the pressure head is therefore obtained by solving an ellisptic equation using boundary conditions and the distribution of the hydraulic conductivity. The relationship between the hydraulic conductivity and the permeability k (in m^2) is $K = k\rho_w g/\eta_w$, where η_w and ρ_w are the dynamic viscosity (in Pa s) and the mass density of the pore water (kg m^{-3}), respectively, and g denotes the acceleration of gravity (m s^{-2}). The viscosity and the mass density of the pore fluid are both dependent on the temperature. In the following study, we use first-order linear relationships to account for temperature.

In our model, the self-potential response is related to ground water flow through an electrokinetic coupling mechanism. As discussed in Chapters 4 and 5, the self-potential field ψ (in V) is governed by a Poisson equation,

$$\nabla \cdot (\sigma\nabla\psi) = \nabla \cdot (\hat{Q}_V U), \tag{7.1}$$

where σ (in S m^{-1}) is the electrical conductivity of the porous material. The right-hand side of Eq. (7.1) corresponds to the source term associated with the Darcy velocity distribution and the heterogeneity in the distribution of the volumetric charge density $\hat{Q}_V$ (expressed in C m^{-3}). The volumetric charge density $\hat{Q}_V$ is the effective charge density occurring in the pore space of the porous material because of the electrical double layer at the mineral/water interface; see Leroy *et al.* (2007, 2008) and Jougnot *et al.* (2009). The relationship between this volumetric charge density and the more classical streaming potential coupling coefficient C (in V Pa^{-1}) has been investigated in Chapter 3. The coupling coefficient is given by $C = -\hat{Q}_V k/\sigma$, where k and σ have been defined above. As an example for an intact Inada granite sample, Tosha *et al.* (2003) measured $C = -0.02 \times 10^{-6}$ V Pa^{-1}, $k = 9 \times 10^{-19}$ m^2 and $\sigma = 0.008$ S m^{-1} (pore fluid: 0.1 mM KCl). This yields an effective charge density of $\hat{Q}_V = 2 \times 10^8$ C m^{-3}. The relationship between the excess of charge density $\hat{Q}_V$ draged by the pore water flow and the permeability k has been given in Chapter 3 (see Figure 3.11).

The self-potential distribution is therefore obtained by first solving the boundary-value problem for the hydraulic head. The solution is then used to determine the seepage velocity distribution using the distribution of the hydraulic conductivity. Knowing the distribution of the excess of electrical charge per unit volume, the source term of the Poisson equation can be obtained and the Poisson equation can be solved with appropriate boundary conditions.

In our modeling, the ground surface is an insulating boundary and therefore $\hat{n} \cdot \nabla\psi = 0$, where $\hat{n}$ is the unit vector normal to the ground surface. As noted in Chapter 1, the electrical potential is never measured in an absolute sense at the ground surface of the Earth. It is measured relative to a self-potential station called the "reference station." This peculiarity of the self-potential data should be considered both in the forward and inverse numerical modeling (see Jardani and Revil (2009)).

From Eq. (2.48) of Chapter 2 and in steady-state condition (the time derivative of Eq. (2.48) is equal to zero), the heat flow equation is given by

$$\nabla \cdot (\lambda \nabla T) = \nabla \cdot (\rho_w C_w T \mathbf{U}), \qquad (7.2)$$

where T is the average temperature of the porous medium (in K), λ (in W m^{-1} K^{-1}) is the thermal conductivity of the porous material, and C_w (in J kg^{-1} K^{-1}) is the heat capacity of the pore water per unit mass. For water saturated porous rocks, the thermal conductivity typically varies within one order of magnitude. Note that Eq. (7.2) has the same structure as Eq. (7.1) and that the source term in each of these equations (right-hand side) is controlled by the same seepage velocity $\mathbf{U}$. However, the source term of Eq. (7.2) is temperature dependent, making this a non-linear partial differential equation. Because the pore fluid viscosity and density entering the hydraulic problem are also temperature dependent (as mentioned above), the hydraulic and the thermal problems are coupled problems. The partial differential equation governing the electrical problem can be safely decoupled from the primary flow problem where electro-osmosis is negligible; see Sill (1983).

7.1.2 Relationships between the material properties

So far, we have said nothing about the relationships that can be used to connect the material properties entering the hydraulic, electrical, and thermal problems described above. The hydraulic problem, in steady-state conditions, is controlled only by the distribution of the hydraulic conductivity K. The self-potential problem is controlled by the distribution of the electrical conductivity σ, the distribution of the charge density per unit pore volume $\hat{Q}_V$, and the distribution of the seepage velocity. The thermal problem, in steady-state conditions, is controlled by the distribution of the thermal conductivity λ and the distribution of the seepage velocity. It is legitimate to wonder if the material properties entering the field equations are entirely independent or if they are correlated in the field and, if so, what is the degree of these correlations. Are these relationships truly useful to establish prior constraints in an inverse model?

The pore-scale problems defining the permeability, the electrical conductivity, and the thermal conductivity share some common characteristics and therefore these material properties can exhibit some degree of correlation in the field. However, there are no universal laws to formulate general relationships between these three properties. Experimental data and borehole measurements show that such

Table 7.1. *Material properties used for the synthetic model.*
U6 and U7 correspond to the faults. U1 corresponds to the
granitic basement, U2 to a carbonate, U3 to a shale,
U4 and U5 to clayey sandstones.

Unit	k (in m^2)	σ (in S m^{-1})	λ (in W m^{-1} K^{-1})	Log($\hat{Q}_V$) (in C m^{-3})
U1	1×10^{-20}	1×10^{-3}	6	7.2
U2	1×10^{-14}	2×10^{-3}	2.5	2.3
U3	1×10^{-19}	1×10^{-1}	3	6.4
U4	1×10^{-16}	2×10^{-2}	3	3.9
U5	1×10^{-16}	1×10^{-2}	2.5	3.9
U6	5×10^{-13}	1×10^{-2}	4	0.87
U7	5×10^{-13}	1×10^{-2}	4	0.87

universal relationships are unlikely to exist. In the present work we take the conservative assumption that these three petrophysical properties are independent. Note, however, that the approach we follow below is flexible and could be used to incorporate statistical relationships between material properties for a specific environment. These statistical relationships can be field-dependent and may be derived, for example, from borehole data in existing wells using a coded lithological description (see Rabaute *et al.*, 2003).

7.1.3 Forward modeling

The goal of the present section is to understand the self-potential and temperature signatures of ground water flow. We are especially interested in understanding the sensitivity of these signatures to the values of the material properties included in the partial differential equations we are solving, with our end-goal being the inversion of permeability – the Holy Grail of all reservoir characterization. We perform a sensitivity analysis of the three material properties entering the system of equations: the permeability k, the electrical conductivity σ, and the thermal conductivity λ of each "unit" of a geological system. A "unit" is here understood as being a sedimentary layer, the basement, or a fault, for instance (see Jardani and Revil (2009)).

The synthetic model shows a reasonable degree of geological complexity that is similar to real case studies. The synthetic model comprises seven units. Each unit is characterized by a value of the permeability k, a value of the electrical conductivity σ, and a value of the thermal conductivity λ. The "true" values are reported in Table 7.1. The substratum is the less permeable unit while the two fault zones are the most permeable pathways of the system. The geometry of the system is assumed to be deterministically known (such as from seismic data). In the following, we will consider arbitrarily that the properties of the two faults are the same.

To solve the forward problem, we need to specify the values of the material properties and to impose boundary conditions for the temperature (or for the heat flux), the hydraulic head (or the Darcy velocity), and the (self-)potential (or the current density). We first solve the hydraulic problem by imposing the following boundary conditions at the top surface of the system for the hydraulic head: $H = -0.011\,25\,x + 5920$ m for $x \in [0\,\mathrm{m}, 2680\,\mathrm{m}]$, $H = 5900$ m for $x \in [2680\,\mathrm{m}, 2920\,\mathrm{m}]$, $H = 0.005\,x + 5884$ m for $x \in [2920\,\mathrm{m}, 11\,000\,\mathrm{m}]$. These boundary conditions are used to simulate the topography of the water table. We also impose that the hydraulic heads (above hydrostatic levels) along the left and right boundaries of the system are in agreement with the value given at the top boundary and hydrostatic trends. The side and bottom boundaries are considered to be impervious.

The solution of the hydraulic problem is used to infer the distribution of the Darcy velocity $\mathbf{U}$. This distribution can be used in turn to compute the source current density $\mathbf{J}_S = \hat{Q}_V \mathbf{U}$, where $\hat{Q}_V$ is the charge per unit volume of a given formation. We use Eq. (7.3) in a deterministic way. The divergence of the source current density corresponds to the source term used to determine the distribution of the self-potential ψ, see Eq. (7.1). We impose $\psi = 0$ at all the boundaries except at the ground surface, which corresponds to an insulating boundary with air ($\hat{\mathbf{n}} \cdot \nabla \psi = 0$ where $\hat{\mathbf{n}}$ is the unit vector normal to the ground surface). Such boundary conditions should be examined carefully for a real case study. Usually, it is better to use a padded mesh so this boundary condition applies far enough from the primary and secondary self-potential sources in the investigated system (see Jardani and Revil (2009)).

All the self-potential values need to be referenced to a specific point (called the reference) like in a real experiment (see Suski *et al.* (2006) and Jardani *et al.* (2008)). In our case, the reference station $\psi = 0$ is chosen arbitrarily at the position Ref($x = 6000$ m, $z = 0$ m) located at the ground surface in the middle of the profile. This choice has no effect on the results as long as we keep in mind where this reference is located. We use the distribution of the Darcy velocity to compute also the distribution of the advective heat flux $\mathbf{H}_S = \rho_w C_w T \mathbf{U}$ (in W m^{-2}). For the thermal boundary conditions, we use a constant temperature ($T_0 = 273$ K) at the ground surface and we impose the heat flux (60 mW m^{-2}) at the bottom boundary of the system. This generates a temperature gradient of approximately 15 °C km^{-1}.

The partial differential equations discussed above are solved with the finite element code "Comsol Multiphysics 3.4" with a fine triangular meshing. Regarding the mesh, we performed various tests to get a mesh-independent result. The code was also benchmarked against analytical solutions for the ground water flow and

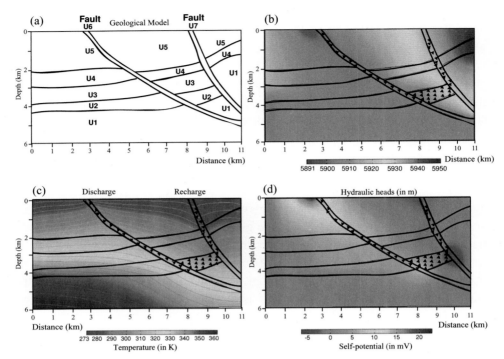

Figure 7.1. Sketch of the synthetic model and forward numerical simulations. (a) The synthetic model comprises seven hydraulic units (including two faults U6 and U7). The material properties of the different formations materials are reported in Table 7.1 (b) Distribution of the hydraulic heads. The arrows represent the direction of the Darcy velocity (the true magnitudes are smaller in unit U2 where the flow is more pervasive, than in the fault planes). (c) Distribution of the temperature in steady-state condition of ground water flow. Note the position of the recharge and discharge areas. (d) Distribution of the self-potential voltages associated with the ground water flow pattern. A negative anomaly is associated with the recharge and a positive anomaly is associated with the discharge area (see Jardani and Revil (2009b)). See also color plate section.

thermal problems. The resulting distributions of the hydraulic heads, the self-potential data, and the temperature data are shown in Figures 7.1b to 7.1d.

The fault U7 on the right-hand side of the Figure 7.2 creates a recharge zone because of its high hydraulic transmissivity and the boundary condition for the hydraulic head at the top surface of the system. At the opposite, the fault U6 is discharging hot water near the ground surface (see Figure 7.1b). These two areas are connected by the permeable unit U2 (see Figure 7.1b). Such a ground water flow pattern is quite common (for example, see, Revil and Pezard (1998)).

Self-potential anomalies recorded at the ground surface of the Earth and the temperature data recorded in boreholes represent useful and complementary

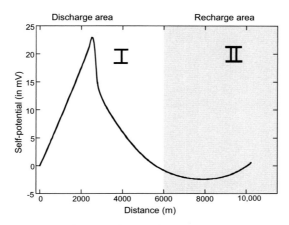

Figure 7.2. Distribution of the self-potential (in mV) at the ground surface in steady-state conditions (see flow pattern in Figure 7.1a). The signature of the discharge area (Zone I) corresponds to a positive self-potential anomaly (with respect to the potential at the position of the reference electrode located at $x = 6000$ m, $z = 0$). The self-potential signature of the recharge area (Zone II) corresponds to a negative self-potential anomaly with respect to the position of the reference (see Jardani and Revil (2009)).

signatures of the flow of the ground water. In the present case, the self-potential profile computed at the ground surface can be subdivided into two anomalies. A positive anomaly (with respect to the value of the potential at the reference electrode) that is associated with the upflow of water in the discharge area and a negative anomaly associated with the recharge area (Figures 7.1b and 7.2). Concerning the temperature anomalies, the recharge area is clearly associated, at depth, with a temperature gradient below the mean geothermal gradient (Figure 7.3). The opposite occurs for the discharge area (Figure 7.3) (see Jardani and Revil (2009)).

We now analyze the sensitivity of the self-potential and thermal signatures to a variation of the material properties of the system. Two stations are monitored. One is located in the recharge area and the second in the discharge area.

Influence of the permeability

We first investigate the influence of the hydraulic transmissivity of the fault zones upon the two self-potential anomalies. The hydraulic transmissivity is equal to the permeability times the thickness of the fault. In the following, we keep the thickness of the faults as constant and we will only discuss their intrinsic permeability. In the discharge area, Figure 7.4a shows that the magnitude of the self-potential anomaly is inversely proportional to the permeability. This is easily explained by the fact that, in Darcy's law, the hydraulic head gradient varies inversely with permeability when the flux is imposed. The volumetric charge density is also

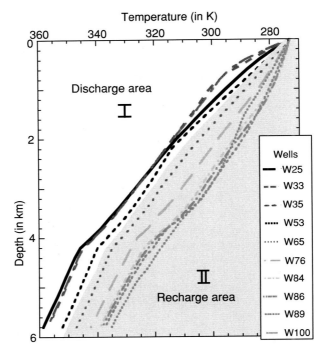

Figure 7.3. Temperature profiles along 10 vertical wells at different positions (the horizontal position of the wells is given on the right-hand side of the figure). The shape of the temperature profile depends on the direction of the ground water flow. The recharge area (zone I) is characterized by a decrease of the temperature with respect to the reference temperature gradient while the opposite situation arises in the discharge area (zone II). The reference temperature profile corresponds to the boundary between the gray and white areas. The names of the vertical wells are derived from their position (e.g., W35 means that the well is located at 3500 m from the beginning of the profile).

dependent on the permeability of the material. For the recharge area, the situation is a bit more complex to interpret. Indeed, two behaviors can be observed depending on the value of the permeability with respect to a critical permeability value that is equal to 10^{-14} m^2. This is explained by the fact that the flow pattern, in the recharge area, is strongly influenced by the ratio of the permeability of the fault zone and the permeability of the surrounding material. Consequently, the critical permeability discussed above is just the permeability of the material in which the fault zone is embedded. When the permeability of the fault zone is smaller than the permeability of the surrounding material, the flow pattern is more horizontal, generating a horizontal electrostatic dipole with a negative pole close to the fault zone. This explains the negative self-potential anomaly observed near the fault in this situation. When the fault is more permeable than the surrounding sediment,

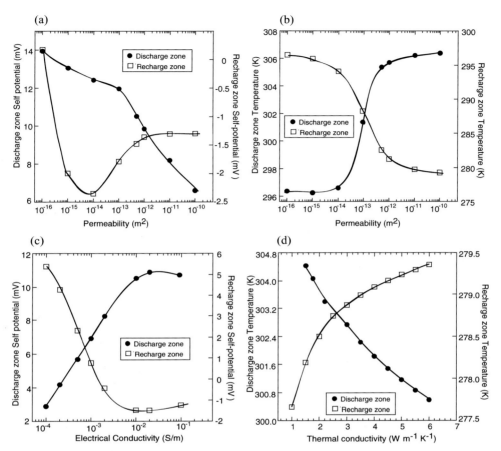

Figure 7.4. Sensitivity analysis in the recharge and discharge areas. (a) Influence of the permeability upon the strength of the self-potential anomaly at the ground surface. (b) Effect of the permeability on the variations of the temperature profile versus depth. (c) Influence of the electrical resistivity upon the strength of the self-potential anomaly at the ground surface. (d) Sensitivity analysis of the temperature profile versus depth. Each point corresponds to a numerical simulation. The lines are guides to the eyes (see Jardani and Revil (2009)).

the fault is a permeable pathway and the strength of the self-potential anomaly is inversely proportional to the permeability of the fault zone. In this case, the self-potential anomaly in the recharge area is not heavily influenced by the permeability of the sediments surrounding the fault due to the boundary conditions imposed on the fault at the top surface of the system. This situation corresponds to a forced hydrogeological regime; see Jardani and Revil (2009).

   The influence of the permeability upon the temperature profiles is exactly the opposite of the influence of the permeability upon the self-potential signals. Figure 7.4b shows that when the permeability increases, the heat flow increases proportionally to it. As is the case of the self-potential anomaly, the temperature anomaly depends on the ratio of the permeability of the fault to the permeability of the surrounding sediment. In the case where the permeability of the fault is smaller than the permeability of the surrounding sediment, the flow pattern does not create a thermal anomaly because the flow is mainly horizontal (this case corresponds to an annihilator for the thermal problem). As faults can act as permeability barriers or conduits, both cases can exist in nature.

Influence of the electrical conductivity

Electrical conductivity is a key-parameter influencing the distribution of the self-potential data, both in terms of polarity and strength. Our sensitivity analysis (Figure 7.4c) reveals that the strength of the two self-potential anomalies (for the recharge and discharge areas) increases with the increase of the electrical conductivity of the fault zones. If the ratio of the conductivity of the fault zone and the conductivity of the surrounding material is very high, the polarity of the self-potential anomalies can also change because the influence of the secondary sources can override that of the primary sources associated with the electrokinetic conversion of the ground water flow. However, the effect of resistivity contrast upon the signature of the self-potential signals depends on the distance between the primary (electrokinetic) sources and the area where there is a drop in the electrical resistivity. It is also important to note, for practical applications, that electrical conductivity can be obtained independently in the field using electrical impedance tomography, low-frequency electromagnetic methods, or borehole data to reconstruct the conductivity of each unit, including the effect of the salinity of the pore water, the cation exchange capacity, and the temperature.

Influence of the thermal conductivity

The distribution of the thermal conductivity controls the distribution of the temperature. Figure 7.4d reveals that the temperature anomalies created by a convective heat flow are better observed when the thermal conductivity is high. However, the domain of variation of the thermal conductivity is relatively small (less than one order of magnitude) by comparison with the domains of variation of permeability and electrical conductivity. Therefore the influence of the heterogeneity of the thermal conductivity is less important. Consequently, for a given formation, the gradient of the temperature with depth depends mainly on the heat flux at the bottom boundary and the value of the thermal conductivity of the formation according to the following relationship:

$$g_G \equiv \frac{\partial T}{\partial z} = \frac{q_G}{\lambda}, \tag{7.3}$$

where g_G is the geothermal gradient (in °C m^{-1}) and q_G is the heat flux (in W m^{-2}). It follows that the isotherms are closer in the region of low thermal conductivity. The heat flux remains roughly the same through all the formations, but distinct thermal conductivities will locally modify the temperature gradient (see Jardani and Revil (2009)).

7.1.4 Stochastic joint inversion

Next we use a Bayesian approach to estimate the material properties $\mathbf{m} = (\mathbf{m}_\sigma, \mathbf{m}_k, \mathbf{m}_\lambda)$ from self-potential and temperature data where $\mathbf{m}_\sigma$ refers to the log of the electrical conductivity distribution, $\mathbf{m}_k$ to the log of the permeability distribution, and $\mathbf{m}_\lambda$ to the log of the thermal conductivity distribution. The Bayesian solution to an inverse problem is based on combining the information coming from the geophysical data with some prior knowledge. The Bayesian analysis considers both the data vector $\mathbf{d}$ and the model parameters $\mathbf{m}$ of a model M as random variables. Several geometrical or petrophysical models M are possible to explain the data. Random variables are characterized with distributions and we assume that all distributions have density functions; see Gelman *et al.* (2004) and Tarantola (2005).

The objective of inverse modeling is to update the information on $\mathbf{m}$, assuming a petrophysical model or a geometrical model M, given the data $\mathbf{d}$ and prior information regarding $\mathbf{m}$. The prior information can come from independent observations and petrophysical relationships. In a probabilistic framework, the inverse problem corresponds to maximize the conditional probability density of m corresponding to the model M given the data vector $\mathbf{d}$. We note $P_0(\mathbf{m} \,|\, M)$ the prior probability density of $\mathbf{m}$ (model M) and $P(\mathbf{d} \,|\, \mathbf{m}, M)$ represents the likelihood corresponding to the data for fixed $\mathbf{m}$ and M. The posteriori probability density $\pi(\mathbf{m} \,|\, \mathbf{d})$ of the model parameters $\mathbf{m}$ given the data $\mathbf{d}$ is obtained by using Bayes' formula:

$$\pi(\mathbf{m} \,|\, \mathbf{d}, M) = \frac{P(\mathbf{d} \,|\, \mathbf{m}, M) P_0(\mathbf{m} \,|\, M)}{P(\mathbf{d} \,|\, M)}, \qquad (7.4)$$

where $P(\mathbf{d} \,|\, M)$ is a normalizing term known as evidence,

$$P(\mathbf{d} \,|\, M) = \int P_0(\mathbf{m} \,|\, M) P(\mathbf{d} \,|\, \mathbf{m}, M) d\mathbf{m}. \qquad (7.5)$$

In the following, we assume that the model M is certain (i.e., the position of the sedimentary layers and faults is known from other geophysical data sets and from borehole or geological information, and the relationship between the charge density and the permeability is considered to be deterministic). Therefore, we drop the term M from Eq. (7.4). The a posteriori probability density $\pi(\mathbf{m}\,|\,\mathbf{d})$ of the model parameters $\mathbf{m}$ given the data $\mathbf{d}$ is written as:

$$\pi(\mathbf{m}\,|\,\mathbf{d}) \propto P(\mathbf{d}\,|\,\mathbf{m})P_0(\mathbf{m}). \tag{7.6}$$

The Bayesian solution of the inverse problem is the whole posterior probability distribution of the material properties. An estimate of the unknown parameters can be computed, e.g., as the expectation value with respect to the posterior distribution (i.e., as the posterior mean value) or as the maximum posterior value. The likelihood function used to assess the quality of a model $\mathbf{m}$ can be considered to follow a Gaussian distribution:

$$P(\mathbf{d}\,|\,\mathbf{m}) = \frac{1}{[(2\pi)^N \det C_d]^{1/2}} \exp\left[-\frac{1}{2}(g(\mathbf{m}) - \mathbf{d})^T C_d^{-1}(g(\mathbf{m}) - \mathbf{d})\right] \tag{7.7}$$

$$\mathbf{d} = (\mathbf{d}_{sp}, \mathbf{d}_T)^T, \tag{7.8}$$

where $g(\mathbf{m})$ is the forward modeling operator for the thermohydroelectrical problem. It connects the generation of a self-potential anomaly and a temperature anomaly to a variation of the material properties of the ground, where $\mathbf{d}$ is an N-vector of the observed surface self-potential data $\mathbf{d}_{sp}$ and temperature data $\mathbf{d}_T$ ("measured" in boreholes if a joint inversion is performed, or just $\mathbf{d}_{sp}$ or $\mathbf{d}_T$ if only the self-potential or temperature data are considered alone). The $(N \times N)$-covariance matrix C_d is a diagonal matrix written as

$$C_d = \begin{bmatrix} \sigma_{sp}^2 & 0 \\ 0 & \sigma_T^2 \end{bmatrix}, \tag{7.9}$$

where σ_{sp} and σ_T are determined from the standard deviations of the self-potential and temperature measurements. The measurements errors for the self-potential and temperature data are usually uncorrelated and Gaussian. The prior probability distribution, if available, is also taken to be Gaussian:

$$P_0(\mathbf{m}) = \frac{1}{[(2\pi)^M \det C_m]^{1/2}} \exp\left[-\frac{1}{2}(\mathbf{m} - \mathbf{m}_{prior})^T C_m^{-1}(\mathbf{m} - \mathbf{m}_{prior})\right] \tag{7.10}$$

where $\mathbf{m}_{prior}$ is the prior value of the model parameters for each unit and C_m is the model diagonal covariance matrix incorporating the uncertainties related of the prior model of material properties.

We now explore the posteriori probability density $\pi(\mathbf{m}\,|\,\mathbf{d})$ expressed by Eq. (7.6). Markov chain Monte Carlo (McMC) algorithms are well-suited for Bayesian inference problems of this type (see Gelman *et al.* (2004)). After an initial period in which the random walker moves toward the highest a posteriori probability regions, the chain returns a number of model vectors sampling the a posteriori probability density $\pi(\mathbf{m} \mid \mathbf{d})$. The characteristic of the probability density $\pi(\mathbf{m} \mid \mathbf{d})$, like the mean and the standard deviation, can be easily determined. The memory mechanisms of the McMC algorithms (making the chain stay in the high a posteriori probability regions of the model space) are responsible for the greater efficiency of the algorithm in comparison to the Monte Carlo methods in which the models are independently chosen and tested against the observations (see Jardani and Revil (2009)).

To improve the performance of the standard Metropolis–Hasting algorithm, Haario *et al.* (2001) introduced recently an efficient McMC algorithm called the Adaptive-Metropolis Algorithm (AMA) to find the optimal proposal distribution (see discussion also in Chapter 4). This algorithm is based on the traditional Metropolis algorithm with a symmetric Gaussian proposal distribution centered at the current model $\mathbf{m}^i$ and with the covariance $\mathbf{C}^i$ that changes during the sampling in such a way that the sampling efficiency increases over time; see Haario *et al.* (2001, 2004). It can be shown that the AMA algorithm, although not Markovian, simulates correctly the target distribution. An important advantage of the AMA algorithm is that the rapid start of the adaptation of the algorithm ensures that the search becomes more effective at an early stage of the simulation. This diminishes the number of function evaluations needed. Let us assume that we have sampled the states $(\mathbf{m}^0, \dots, \mathbf{m}^{i-1})$, where $\mathbf{m}^0$ corresponds to the initial state. Then a candidate point $\mathbf{m}'$ is sampled from the Gaussian proposal distribution q with mean point at the present point $\mathbf{m}^{i-1}$ and with the covariance:

$$\mathbf{C}^i = \begin{cases} \mathbf{C}^0 & \text{if } i \leq n_0 \\ s_n \mathbf{K}^i + s_n \varepsilon \mathbf{I}_n & \text{if } i > n_0 \end{cases} \tag{7.11}$$

where $\mathbf{I}_n$ denotes the n-dimensional identity matrix, $\mathbf{K}^i = \text{Cov}(\mathbf{m}^0, \dots, \mathbf{m}^{i-1})$ is the regularization factor (a small positive number that prevents the covariance matrix from becoming singular), $\mathbf{C}^0$ is the initial covariance matrix that is strictly positive (note that the AMA algorithm is not too sensitive to the actual values of $\mathbf{C}^0$), and $s_n = (2.4)^2/n$ is a parameter that depends only on the dimension of the vector $m \in \Re^n$ (see Haario *et al.* (2001)). According to Gelman *et al.* (1996), this choice of s_n yields an optimal acceptance in the case of a Gaussian target distribution and a Gaussian proposal distribution. The candidate point $\mathbf{m}'$ is accepted with the acceptance probability:

$$\alpha(\mathbf{m}^{i-1}; \mathbf{m}') = \min\left[1, \frac{\pi(\mathbf{m}' \mid \mathbf{d})}{\pi(\mathbf{m}^{i-1} \mid \mathbf{d})}\right]. \tag{7.12}$$

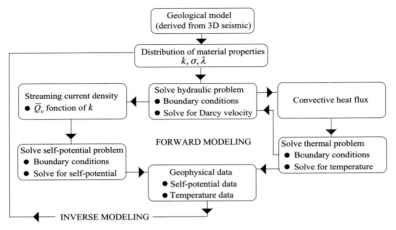

Figure 7.5. Flowchart showing the methodology used to jointly invert the self-potential and temperature data associated with the steady-state flow of ground water in a geothermal field. The inversion of the two types of data yields values of the petrophysical properties (permeability and/or thermal and electrical conductivities) from self-potential measurements at the ground surface and temperature measurements in a set of boreholes. In our case, we assumed that the position (position and shape) of the geological units, including the presence of the faults, are perfectly known. We are only inverting for the mean material properties of each unit.

If the candidate point is accepted, we consider that $\mathbf{m}^i = \mathbf{m}'$, otherwise we choose $\mathbf{m}^i = \mathbf{m}^{i-1}$. The AMA algorithm was written in a Matlab® routine. The flowchart used to solve the forward and inverse problems is summarized in Figure 7.5.

7.1.5 Results

To see the advantage of performing a joint inversion of the temperature and self-potential data, we first start by inverting the self-potential data and the temperature data separately. We used the AMA sampler with 10 000 iterations using the temperature and self-potential data together (both the thermal conductivity and electrical conductivity distributions are unknown and also determined from the algorithm). In Figure 7.6a, we show that the value of the thermal conductivity of each unit inverted from the temperature data is quite good for all geological units. Figure 7.6b shows the inversion of the electrical conductivity of each unit when only the self-potential data are used. In the case where only the temperature data are used, Figure 7.7c demonstrates that the inverted permeabilities are not well-reproduced by the algorithm. This result can be easily explained from the underlying physics as the

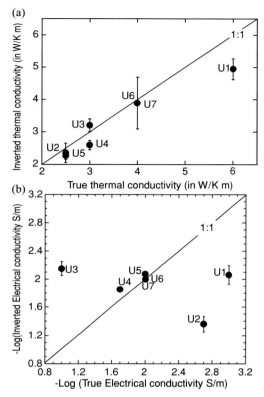

Figure 7.6. Comparison between inverted and true material properties for the six lithological units. (a) Inverted versus true thermal conductivity for the six geological units. The dots correspond to the mean of the posterior probability distribution sampled with the McMC sampler using the Adaptative Metropolis Algorithm (AMA). Only the temperature data are used here. (b) The outputs of the inversion are the thermal conductivity and the permeability of each geological unit. Comparison of the inverted electrical conductivity for the different geological units with the true electrical conductivity data. Only the self-potential data are used here. Only the shallow formations (U4 and U5) and the two faults (U6 and U7) are well-resolved.

temperature distribution depends only on the vertical component of the Darcy velocity. In other words, because the temperature distribution is not sensitive to the horizontal component of the Darcy velocity, the inversion of the temperature data alone yields the poor result shown by Figure 7.7c for the permeability, according to Jardani and Revil (2009).

Figure 7.6b shows a comparison between the inverted electrical conductivities of each unit and their "true" values using the self-potential data. Interestingly, the

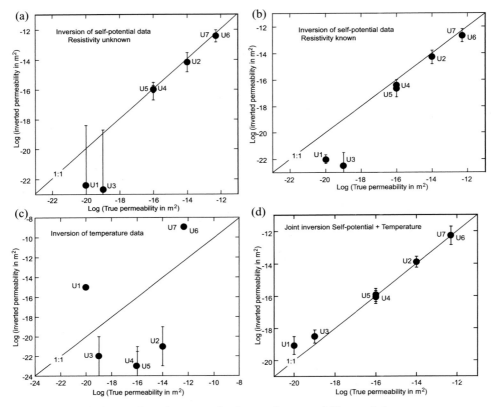

Figure 7.7. Comparison between the inverted permeability and the true permeability. The dots represent the mean of the posterior probability distributions. (a) Result of the inversion of the self-potential data alone. The resistivity is unknown and is also obtained as a result of the inversion. (b) Result of the inversion of the self-potential data alone with the electrical resistivity assumed to be perfectly known. (c) Result of the inversion of the temperature data alone. (d) Result of the joint inversion of the self-potential and temperature data. The joint inversion provides clearly the most reliable estimates of the permeability of each unit. This is a result of the complementarity of the sensitivity maps for the self-potential and temperature data.

self-potential can be used to grossly invert the resistivity value of each unit but the outputs of such an inversion remains rough, especially for the deeper formations. Only the resistivity of the shallow formations and the resistivity of the faults are well-reproduced. After a few kilometers, the effect of secondary sources does not influence the self-potential distribution at the ground surface (see Jardani and Revil (2009)).

In Figures 7.7a and 7.7b, we show that the inversion of the permeability is relatively well-reproduced by considering the self-potential alone except for the permeability of the basement. This is logical if we consider that the permeability of the basement can take a wide range of low values without affecting the self-potential distribution at the ground surface. The knowledge of the electrical conductivity distribution does not really help the algorithm to reach a better result but the results of the inversion are better than by taking the temperature alone (compare Figures 7.7a and 7.7b to Figure 7.7c). Figure 7.7d shows that the joint inversion of the temperature and self-potential data together yields the best estimates of the permeability of each geological unit (compare Figures 7.7a to 7.7c). Therefore, temperature data in a set of boreholes and self-potential data are very complementary data sets in determining the permeability distribution of geological structures. These two sources of information help reduce the non-uniqueness of the inverse problem.

7.2 The Cerro Prieto case study

The Cerro Prieto geothermal field is located in the alluvial plain of the Mexicali Valley, northern Baja California, Mexico, about 35 km south-east of the city of Mexicali (Figure 7.8a, b). Cerro Prieto was extensively studied and drilled (Figure 7.8c) in the 1970s through an international program of collaborative investigations between the Comision Federal de Electricidad (Mexico) and the Department of Energy (USA). It is a suitable target to perform a test of the joint inversion algorithm because there are both surface self-potential data and borehole temperatures.

The geological description of this geothermal field can be found in de la Peña and Puente (1979) and the hydrogeology can be found in Lippmann and Bodvarsson (1982, 1983). The available information was summarized by Jardani *et al.* (2008). The Cerro Prieto geothermal field can be grossly divided into three main lithostratigraphic units (see de la Peña and Puente (1979)) (Figure 7.9). The first unit is made of unconsolidated and semi-consolidated continental deltaic sediments of Quaternary age. These sediments show repeated sequences of clays, silts, sands and gravels, and therefore they are quite hydraulically conductive and characterized by low electrical resistivities.

The second unit corresponds to consolidated continental deltaic sediments of Tertiary age. These sediments are composed of alternating shales, siltstones and sandstones presenting lenticular bedding. This implies that the permeability of this unit is anisotropic. The sandstones are fine-grained, usually well-sorted, varying between graywackes and arkoses (see de la Peña and Puente (1979) and Lyons and van de Kamp (1980)). This formation is discordant on the third unit, the granitic and metasedimentary Upper Cretaceous basement. This basement has suffered tectonic uplift and falls. It is electrically resistive, according to Wilt and Goldstein (1981).

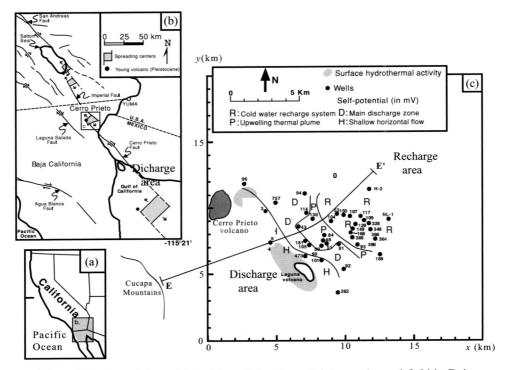

Figure 7.8. Cerro Prieto. (a) Position of the Cerro Prieto geothermal field in Baja California (Mexico). (b) Cerro Prieto is located between the south-east end of the Imperial Fault and the northern end of the Cerro Prieto Fault. (c) Position of the profile EE′ with respect to the recharge and discharge areas of the hydrothermal systems. This profile has been chosen because it is normal to the main geothermal features of the system and crosses the recharge and discharge areas of the system.

A simplified geological cross-section in the central part of profile EE′ (see position in Figure 7.8c) is shown in Figure 7.9.

A map of the self-potential anomalies measured over the Cerro Prieto geothermal field has been presented by Fitterman and Corwin (1982). This map covers an area of ≈ 300 km². A 2D profile is shown on Figure 7.10a. The self-potential anomaly of Cerro Prieto is definitely 3D (see Fitterman and Corwin (1982) and Jardani *et al.* (2008)), but the inversion done below will be performed in 2D. The geometry of the hydrogeological system for the numerical simulations is sketched in Figure 7.10b and is taken from Manon *et al.* (1977) and Wilt and Goldstein (1981). In our nomenclature, unit U1 corresponds to the granitic and metasedimentary Upper Cretaceous basement while unit U2 corresponds to a zone characterized by intense hydrothermal alteration (see Wilt and Goldstein (1981)). In the interpretation of dipole–dipole resistivity data made by Wilt and Goldstein (1981), unit U3 appears as

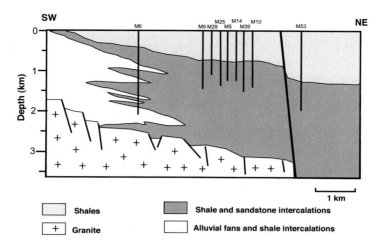

Figure 7.9. Simplified geological description of the central part of the investigated cross-section EE′ in the Cerro Prieto geothermal field with the position of the geothermal wells used for the temperature. The position of wells M6 and M53 can be found in Figure 7.8.

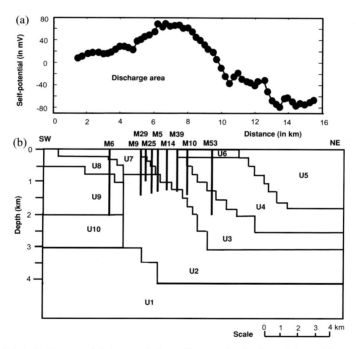

Figure 7.10. Self-potential data and simplified geological model used for the inversion. The model comprises 10 geological units labeled U1 to U10. (a) The self-potential data are taken from Fitterman and Corwin (1982). (b) The hydrogeological model is taken from the resistivity profile EE′ discussed by Wilt and Goldstein (1981). The value of the resistivity of the geological units are taken from Wilt and Goldstein (1981) (their Figure 3) and are reported in Table 7.2. The self-potential anomaly points out that the flow occurs mainly in units U2 and U3 (see Corwin and Hoover (1979) for an extended discussion).

Table 7.2. *Material properties (electrical resistivity and thermal conductivities) used for the model of Cerro Prieto.*

Unit	$\rho^{(1)}$ (in Ohm m)	$\lambda^{(2)}$ (in W m^{-1} K^{-1})
U1	300	1.5
U2	4.0	3.0
U3	1.5	3.0
U4	5.0	3.0
U5	8.5	2.0
U6	2.2	3.0
U7	0.6	2.0
U8	3.5	2.5
U9	1.2	2.5
U10	2.0	3.0

[1] The resistivity data are from Wilt and Goldstein (1981).
[2] The thermal conductivity data are from Lippmann and Bodvarsson (1982).

a thin eastward-dipping conductive body contrasting sharply with the surrounding more resistive formations (units U2 and U4). Unit 7 corresponds to a zone of hot water discharge. Units U4, U5 and U6 are characterized by high clay contents. They are, however, more resistive than the rocks westward by a factor of 5 (see Wilt and Goldstein (1981) and Table 7.2). According to Lyons and van de Kamp (1980), this increase in resistivity is due to the transition from brackish pore water to fresher ground waters associated with a recharge of the system in this part of the EE′ profile. Some recharge also takes place in units U9 and U10.

We assume that the resistivity distribution is known (this is actually the case along profile EE′ as discussed by Jardani *et al.* (2008). Because the thermal conductivity of the encountered lithologies is fairly well-constrained, we also consider that their values are known. The resistivity and thermal conductivity values for each of the 10 formations of Figure 7.10b are summarized in Table 7.2. The unknowns of the inverse problem are the horizontal and vertical effective permeabilities of each geological unit and the constraints are the self-potential data measured at the ground surface (see Figure 7.10a and the temperature profile observed in the nine wells of Figure 7.10b). The temperature data used for the inversion are displayed in Figure 7.11.

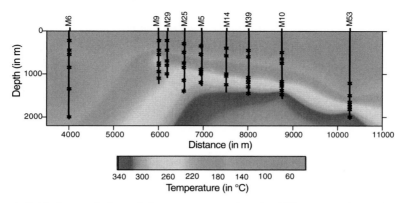

Figure 7.11. Interpolation of the temperature data in different geothermal wells in the central part of the profile EE′. The stars along the borehole indicate the position of the temperature data used for the inversion.

The forward problem is solved with the following boundary conditions. For the flow problem, we impose at the top surface of the system a hydraulic pressure of 7.62×10^6 Pa for x between 11.5 km and 15.5 km. On the left-hand side of the system ($x = 0$ km) and for 0 km $< z < 0.550$ km, we impose an excess fluid pressure of 7×10^6 Pa and for 0.55 km $< z < 2$ km, we impose a hydraulic flux of 8×10^{-9} m s^{-1}. On the right-hand side of the system ($x = 15.5$ km), we impose a hydraulic pressure of 7.5×10^6 Pa for 3.00 km $< z < 4.07$ km. All the other boundaries are impervious. We point out that while the hydraulic boundary conditions are quite well-constrained for the Cerro Prieto geothermal field, this may not be the case elsewhere. In such cases, the boundary conditions should be inverted together with the material properties of each unit; see Jardani and Revil (2009).

Different types of constraints are used for modeling the heat flow, one of which is fixing the temperature distribution in a number of geothermal wells. The boundary conditions used are as follows. At the upper boundary of the model we use $T = 50 °C$ for x between 0 and 6.9 km to account for the discharge of hot water in this area (see Figure 7.8c): for 6.9 km $< x < 15.5$ km, $T = 30 °C$. On the left-hand side of the model ($x = 0$), for $0 < z < 0.55$ km, we impose $T = 220 °C$ because of the observed shallow horizontal flow of hot water in the U7 formation; for 0.55 km $< z < 2$ km ($x = 0$), we use $T = 150 °C$; for 2 km $< z < 3$ km, T is fixed at 220 °C. On the right-hand side of the system (15.5 km), and for 2.6 km $< z < 4.1$ km, we impose $T = 300 °C$. Along the bottom boundary of the system, we impose $T = 640 °C$ to reproduce the mean geothermal gradient along the profile. All the other boundaries are insulating boundaries. For the self-potential

Table 7.3. *Value of the optimized horizontal and vertical permeability for each unit of the Cerro Prieto geothermal field (from Jardani and Revil (2009)).*

Material	Log (k_x) (k_x in m^2)	Log (k_z) (k_z in m^2)
U1	-18.0 ± 1.8	-18.0 ± 1.8
U2	-14.2 ± 0.5	-15.1 ± 1.5
U3	-13.0 ± 0.8	-15.3 ± 0.8
U4	-15.1 ± 0.3	-15.1 ± 0.3
U5	-15.0 ± 0.5	-14.1 ± 1.5
U6	-15.6 ± 0.3	-17.2 ± 1.0
U7	-13.7 ± 2.1	-14.1 ± 1.2
U8	-13.0 ± 1.8	-15.0 ± 0.5
U9	-12.4 ± 2.1	-13.0 ± 1.0
U10	-14.5 ± 2.1	-16.1 ± 0.68

signals, we used insulating boundary conditions at the top and bottom surfaces of the system and the boundary condition $\psi = 0$ on the side boundaries using a padded mesh to avoid the effect of this boundary condition over the investigated area.

The inverse problem is accomplished by performing 5000 realizations using the McMC (AMA) sampler. The optimized vertical and horizontal permeabilities of each unit are listed in Table 7.3 with the standard deviation determined from the last 1000 iterations of the McMC sampler. These optimized values yield the flow model shown in Figure 7.12c with the fit of the self-potential and temperature data shown in Figures 7.12a and 7.12b, respectively.

We point out that we did not use any prior information to constrain the values of the permeabilities. The permeability of the basement unit U1 is found to be very low as expected (the mean value is 10^{-18} m^2). The optimized permeability of unit U3, that contains the transmissive fault, is found to be quite high (mean of 10^{-13} m^2). The mean value we found (10^{-13} m^2) is in excellent agreement with the value used by Lippmann and Bodvarsson (1982; their Table 1) for the isotropic fault material (5×10^{-14} m^2). U2, a unit that is known to be hydrothermally altered, is found to have a lower permeability ($\sim 10^{-14}$ m^2). The permeabilities of formations U4 and U5 are lower by one order of magnitude (10^{-15} m^2) in agreement with the higher clay-content of these formations. Formation U7 has very high horizontal and vertical permeabilities in agreement with field observations; this formation corresponds to a discharge area of the hydrothermal system

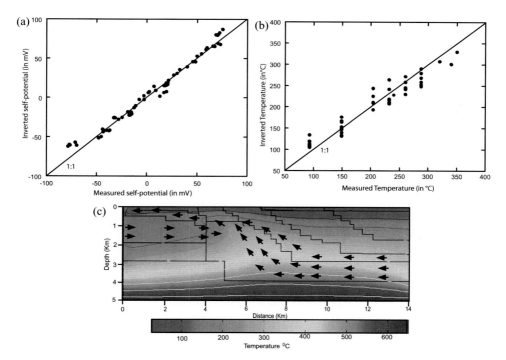

Figure 7.12. Results of the inversion. (a) Best fit of the self-potential data from Figure 7.10a. (b) Best fit of the temperature data from Figure 7.11. (c) Flow pattern associated with the highest posterior probability densities of the permeabilities values of each geological unit. Note the upflow of the hot water in the central part of the system and the horizontal flow in the Western part of the system. The arrows represent the direction of the Darcy velocity with a threshold value of 5×10^{-9} m s^{-1}. See also color plate section.

and horizontal shallow flow of hot water as discussed above (see Figure 7.8c and Jardani *et al.* (2008)). Units U8, U9, and U10 have a high anisotropy of permeability with a very high horizontal permeability. This is consistent with the known lithology of these formations (see Figure 7.9) because of the interbedding of shale and sand geological formations. For the formations U8 and U9, Lippmann and Bodvarsson (1982; their Table 1) used a horizontal permeability of 10^{-13} m^2 and a vertical permeability of 10^{-14} m^2, which is in close agreement with the optimized values reported in Table 7.3 using the self-potential and temperature data as constraints.

We can therefore conclude that the parametric inversion developed below and based on the joint inversion of temperature and self-potential data yields correct estimates of the large-scale permeability of these formations.

7.3 Gradient-based approach applied to hydrothermal fields

7.3.1 Gauss–Newton inversion

The self-potential ψ (in V) is governed by Eq. (7.1). As previously stated, this relationship is obtained by combining the generalized Ohm's law includes the advective drag of the excess of electrical charges of the diffuse layer (coating the surface of the minerals) per unit volume of pore water, $\hat{Q}_V$ (excess charge density in C m^{-3}), and the Darcy or seepage velocity $\mathbf{U}$ (in m s^{-1}). In Eq. (7.1), σ (in S m^{-1}) is the electrical conductivity of the porous material. The right-hand side of Eq. (7.1) corresponds to the self-potential source term associated with the Darcy velocity distribution and the heterogeneity in the distribution of the volumetric charge density $\hat{Q}_V$. The charge density $\hat{Q}_V$ is the effective volumetric charge density occurring in the pore space of the porous material due to the electrical double layer at the mineral/water interface; see, for example, Revil and Leroy (2001). The relationship between this volumetric charge density and the more classical streaming potential coupling coefficient C (in V Pa^{-1}) (see Aizawa *et al.* (2009)) is: $C = -\hat{Q}_V k \rho / \eta_f$ where $\rho = 1/\sigma$ is the electrical resistivity of the porous material (in Ohm m) and η_f denotes the dynamic viscosity of the pore water (in Pa s). For pH ranging from 5 to 8, the relationship between the charge density $\hat{Q}_V$ and the permeability k (in m^2) is given in Chapter 3 (see Figure 3.11). Although slightly above this range, we will use the relationship as a first-order approximation when applying our method at Mount Princeton where the hot springs have a pH between 7.8 and 8.6 (see Table 7.4).

In order to find the source current density responsible for the observed self-potential anomalies, we developed an algorithm based on the Gauss–Newton method (see Revil *et al.* (2011b)). We first need to compute the kernel matrix that represents the relationship between the electrical current density at point M and the measured self-potential signals at a self-potential station P. The relationship between the electrical potential at the observation point P, $\psi(P)$, and the current density at the source position M, $\mathbf{J}_S(M)$, is given by the integral form of the Poisson equation, Eq. (7.1),

$$\psi(P) = \int_\Omega \mathbf{K}(P, M)\mathbf{J}_S(M)dV, \qquad (7.13)$$

where $\mathbf{J}_S = \bar{Q}_V \mathbf{U}$ is the source current density vector associated with ground water flow, dV denotes an integration over the volume of rocks (Ω represents the source rock volume in which fluid flow takes place), and $\mathbf{K}(P, M)$ denotes the kernel connecting the self-potential data measured at a set of non-polarizing electrodes

Table 7.4. *Composition of the thermal waters for Mount Princeton (MP, Hortense Hot Springs, sampling: Oct. 1975; from Dimick (2007)) and Neal Hot Springs (NH, measurements made in 1972; see Mariner et al. (1980)).*

Property	MP	NH
T (°C)	82	87
pH (–)	8.5	7.3
σ_f (10^{-2} S m^{-1}, 25 °C)	4.80	10.10
K$^+$	3.10	16.0
Na$^+$	94	190
Ca^{2+}	4.4	8.8
Mg^{2+}	0.10	0.2
SiO$_2$(aq)	68	180
HCO$_3^-$ (alkalinity)	71	200
SO$_4^{2-}$	100	120
Cl$^-$	10	120
F$^-$	14	9.4

P (with respect to a reference electrode) and the source of current at point M in the conducting ground. The kernel $\mathbf{K}$ depends on the number of measurement stations N at the ground surface, the number of elements M in which the source current density is going to be determined, and the resistivity distribution of the medium, which is taken directly from the electrical resistivity tomogram. For a 2D problem, each element of $\mathbf{K}$ is a Green function. As explained in Chapter 4, the matrix $\mathbf{K}$ depends also on the boundary conditions for the electrical potential or the total current density. The ground surface is considered to be an electrically insulating boundary (Neumann boundary) and therefore the normal component of the current density vanishes at this boundary ($\hat{\mathbf{n}} \cdot \nabla \psi = 0$, where $\hat{\mathbf{n}}$ is the unit vector normal to the boundary and ψ the electrical potential). However, for the rest of the boundaries a null electrical potential is imposed ($\psi = 0$ V). Finally, when computing the elements of $\mathbf{K}$, one has to remember that the electrical potential is determined relative to a reference electrode located somewhere at the ground surface. As explained above, this choice is arbitrary but needs to be consistent between the display of the data and the numerical forward modeling used to compute the kernel (see Jardani *et al.* (2008)).

The inversion of the self-potential data follows a two-step process. The first step is the inversion of the spatial distribution of the source current density $\mathbf{J}_S$. The second step is the determination of $\mathbf{U}$ using the distribution of $\mathbf{J}_S$ and assuming reasonable values for the charge density $\hat{Q}_V$. The self-potential inverse problem is a typical (vector) potential field problem and the solution of such problem is known to be ill-posed and non-unique. It is therefore important to add additional constraints to reduce the solution space. The criteria of the data misfit and the model objective function place different and competing requirements on the models. Using the L_2 norm, these two components of the global cost function ψ are balanced using Tikhonov regularization (see Tikhonov and Arsenin (1977))

$$C = \|\mathbf{W}_d(\mathbf{Km} - \psi_d)\|^2 + \lambda\|\mathbf{W}_m\mathbf{m}\|^2, \qquad (7.14)$$

where λ is a positive regularization constant; $\|\mathbf{Af}\|^2 = \mathbf{f}^t\mathbf{A}^t\mathbf{Af}$ (where the superscript "t" means transpose); N is the number of self-potential stations and M is the number of discretized cells used to represent the ground ($2M$ represents the number of elementary current sources to consider – one horizontal component and one vertical component per cell for a 2D problem); $\mathbf{K} = (\mathbf{K}_{ij}^x, \mathbf{K}_{ij}^z)$ is the kernel ($N \times 2M$) matrix formed from two matrices corresponding to the kernel of the horizontal and vertical vector components of the electrical source density for 2D problems; $\mathbf{m} = (j_i^x, j_i^z)$ is the vector of $2M$ model parameters (source current density); φ_d is a vector of N elements corresponding to the self-potential data measured at the ground surface or in boreholes; $\mathbf{W}_d = \text{diag}\{1/\varepsilon_1, \ldots, 1/\varepsilon_N\}$ is a square diagonal data-weighting $N \times N$ matrix (elements along the diagonal of this matrix are the reciprocals of the standard deviations ε_i of the self-potential data); and $\mathbf{W}_m$ is the $2(M - 2) \times 2M$ model-weighting matrix or regularization matrix (e.g., the flatness matrix or the differential Laplacian operator). The product $\mathbf{Km}$ in Eq. (7.14) represents the predicted (simulated) self-potential data. If a prior model $\mathbf{m}_0$ is considered, $\|\mathbf{W}_m\mathbf{m}\|^2$ is replaced by $\|\mathbf{W}_m(\mathbf{m} - \mathbf{m}_0)\|^2$. This Gaussian assumption on the data is used to set up the matrix $\mathbf{W}_d$. For $\mathbf{W}_m$, we use the smoothness operator (the discrete approximation of the second order derivative). At each inverse iteration step i, we compute a quadratic approximation of C at the current model where $\mathbf{m}_i$ is minimized, yielding a linear system of equations to be solved for a new model update vector $\Delta\mathbf{m}_i$ (Revil *et al.* (2011b)) according to

$$\mathbf{A}_i\Delta\mathbf{m}_i = \mathbf{B}_i, \qquad (7.15)$$

with

$$\mathbf{A}_i = \left[\mathbf{K}^T\left(\mathbf{W}_d^T\mathbf{W}_d\right)\mathbf{K} + \lambda_i\left(\mathbf{W}_m^T\mathbf{W}_m\right)\right], \qquad (7.16)$$

$$\mathbf{B}_i = \mathbf{K}^T\left(\mathbf{W}_d^T\mathbf{W}_d\right)(\psi_d - \mathbf{Km}_i) - \lambda_i\left(\mathbf{W}_m^T\mathbf{W}_m\mathbf{m}_i\right). \qquad (7.17)$$

The update vector $\Delta \mathbf{m}_i$, when added to $\mathbf{m}$, decreases the value of the cost function. The regularization parameter used in our approach is initially set at a large value, λ_0, and it is progressively reduced after each iteration i until it reaches the selected minimum limit, λ_m. In the following example, the minimum value of λ_m is set at one-tenth the value of λ_0. The value of the initial damping factor λ_0 depends on the level of random noise present in the data, with a large value for noisy data. At each iteration step, (i) we compute the inverse solution, (ii) we simulate the self-potential data, and (iii) we compute the data misfit contribution. If the data misfit is larger than that suggested by the self-potential noise, the value of the regularization parameter is reduced and the process repeated until the data are appropriately fitted assuring that we can find the smoothest model that fits the data. The Gauss–Newton method was implemented in a Matlab® routine.

7.3.2 Application to Mount Princeton Hot Springs

The Mount Princeton area represents a complex system where the interaction of faults has resulted in hot springs. These hot springs include the Hortense Hot Springs, which are the hottest springs in Colorado (see Limbach (1975)). The Upper Arkansas Valley is a half-graben located between the Sawatch Range to the west and the Mosquito Range to the east. The dominant faulting corresponds to the Sawatch Range Fault, a north-west trending normal fault bordering the Sawatch Range, which is composed of a relatively young (34–38 Ma) granitic batholith including Mount Princeton (Figure 7.13). The Sawatch fault is segmented in several places by transfer faults and accommodation zones (see Miller (1999)). Here we focus on the geothermal field associated with the Mount Princeton Hot Springs. In this area, the Sawatch normal fault is segmented by a strike slip fault. The surface expression of this segmentation corresponds to the Chalk Cliffs, named for the white color of the highly fractured and hydrothermally altered quartz monzonite (Figure 7.13) (the white color is due to kaolinite replacing feldspar).

Geophysical data collection consisted of a series of nine (~1.2 km long) resistivity profiles and 2500 self-potential measurements performed in May 2008, May 2009 and May 2010. In the following, we will focus on the resistivity and self-potential data obtained along profile P3 crossing the strike-slip fault shown in Figure 7.13c (Fault B). The resistivity data were obtained with an ABEM SAS-4000 resistivity meter using the Wenner-α arrays and 64 stainless steel electrodes with 20 m take-outs. A current of 200 mA was generally injected in the ground for each measurement. Each measurement was repeated until the standard deviation was below 5% of the mean (a maximum of 16 measurements were stacked together). The profile P3 comprises a total of 472 measurements. The self-potential measurements were performed with Pb/PbCl$_2$ Petiau non-polarizing

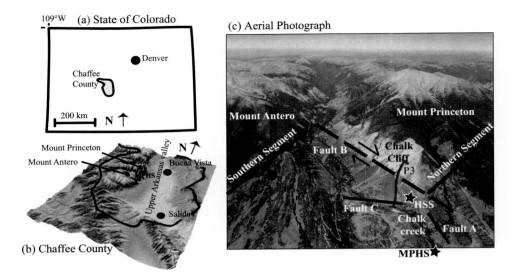

Figure 7.13. Localization of Mount Princeton Hot Springs in Colorado. (a) Sketch of the State of Colorado. (b) Chaffee County. (c) Aerial photograph of the investigated area (courtesy from Jeffrey A. Coe, USGS) showing the position of the Profile P3 and the position of the northern and southern segments of the Sawatch Range fault near Chalk Cliffs. Chalk Cliff results from the alteration of the quartz monzonite of the Mount Princeton batholith. HHS corresponds to the Hortense Hot Springs and MPHS corresponds the Mount Princeton Hot Springs. We have also indicated the position of the dextral strike slip fault zone (Fault B in our nomenclature). Fault A corresponds to an open conduit for the upwelling of the thermal water.

electrodes. We used a high impedance (100 MOhm) calibrated Metrix voltmeter with a sensitivity of 0.01 mV. The mean standard deviation of the self-potential data was 5 mV on average. Temperature was also recorded along profile P3 at a depth of 30±5 cm (see Figure 7.14).

Resistivity data were inverted with the software RES2DINV (see Loke and Barker (1996)) using a Gauss–Newton method and a finite element solver. The inverted resistivity section and the self-potential data are displayed in Figure 7.14. The d.c.-resistivity tomogram on profiles P3 (and other profiles not shown here) show a ~150 m wide, near-vertical, low-resistivity anomaly (named B3) consistent with the presence of a dextral strike slip fault zone in this area (Fault B in Figure 7.13). This anomaly is confirmed by additional profiles performed parallel to P3. Additionally, P3 exhibits a clear, positive self-potential and temperature anomaly associated with the conductive anomaly B3. We interpret this anomaly as being due to the upwelling of thermal waters along this portion of the dextral strike slip fault zone (see Poldini (1938)). The thermal water upwelling along

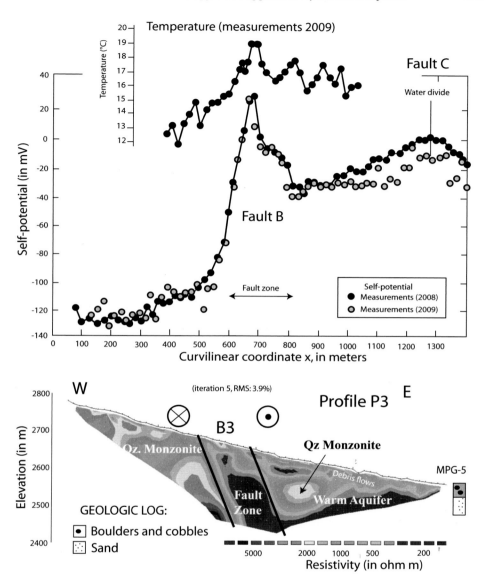

Figure 7.14. Resistivity tomogram (RMS error 4%) and self-potential data along profile P3 (vertical exaggeration factor of the resistivity tomogram: 1.3). The conductive body B3 is consistent with the position of the dextral strike slip zone. The upflow in the dextral strike slip fault zone (Fault B) is associated by a self-potential anomaly of 150 mV in the self-potential signals, low resistivity values (in the range 100–300 Ohm m), and an increase of the temperature at a depth of 30 cm. The positive self-potential anomaly evidences the up-flow of the hydrothermal fluids in this portion of the fault zone. Note the consistency of the self-potential measurements over a year of time interval.

Table 7.5. *Material properties of the geological units used for the numerical modeling.*

Unit	Meaning	Permeability (m^2)	Resistivity (Ohm m)	Charge density $\hat{Q}_V$ (C m^{-3})
U1	Fault	10^{-13}	$200^{(2)}$	$30^{(3)}$
U2	Basement	10^{-20}	$2000^{(2)}$	–
U3	Aquifer	$10^{-12(1)}$	$1000^{(2)}$	$4^{(3)}$

[1] The high permeability of the aquifer agrees with the high permeability of the formation observed in well MPG-5 and composed of boulders, cobbles, aggregates, and sands. The permeability of the aquifer is only used to infer a value for $\hat{Q}_V$.

[2] Resistivity is estimated from the resistivity tomogram.

[3] Using the relationship between the excess charge density and the permeability data.

the fault plane is then flowing downslope in a shallow unconfined aquifer as evidenced by domestic wells and the shape of the self-potential signal, which increases in the direction of Chalk Creek in the center of the valley (see Revil *et al.* (2011b)).

We use the geometry shown in Figure 7.14 with the material properties given in Table 7.5 to set up the geometry for inversion of the self-potential data (Figure 7.15). This geometry comprises three units: unit U1 for the fault, and units U2 and U3 represent the granitic basement and the shallow aquifer, respectively. The presence of this shallow aquifer is confirmed by a drill-hole (MPG-5) where a water temperature of 59 °C and a water table depth of 40 m were measured (see position in Figure 7.14). The boundary conditions are (i) impervious boundaries except at the base of the fault and at the outflow of the aquifer and (ii) insulating boundaries. We inverted the self-potential measurements to determine the magnitude of the Darcy velocity using the deterministic approach proposed above. In the following, we use a constant value for $\hat{Q}_V$ in the fault because we assume a constant temperature, salinity, and lithology in the fluid flow path. Using a reservoir depth of 5 km and a mean geothermal gradient of 28 °C km^{-1}, the permeability of the fault plane is estimated to be on the order of 10^{-13} m^2. This yields an approximate value of $\hat{Q}_V = 30$ C m^{-3} using the relationship discussed earlier. The conductivity of the hydrothermal water is σ_f (25 °C) $= 0.048$ S m^{-1} (Table 7.4). Revil *et al.* (2003) developed the following empirical equation between the value of the coupling coefficient and the value of the conductivity of the pore water at 25 °C: $\log_{10}C = -0.921 - 1.091 \log_{10}\sigma_f$. This gives a streaming potential coupling coefficient of $\sim -3\pm1$ mV m^{-1}. Using C (in V m^{-1}) $= -\hat{Q}_V \, k \, \rho \, \rho_f \, g/\eta_f$ and $k = 10^{-13}$ m^2

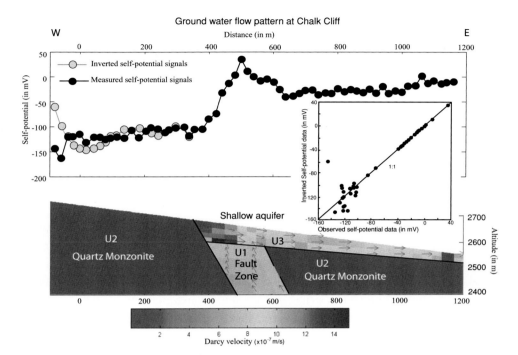

Figure 7.15. Ground water flow pattern as constrained by d.c.-resistivity and self-potential data along profile P3 (data from 2008, the end of the profile has been omitted). The unit U1 corresponds to the dextral strike slip zone, the unit U2 to the quartz monzonite basement, and the unit U3 to the shallow aquifer. The boundary conditions are (i) impervious boundaries except at the base of the dextral strike slip fault and at the outflow of the aquifer and (ii) insulating boundaries. We ignore the possibility of a mix between the thermal water and some cold water that would come from the upper section of Chalk Cliff. This may explained the discrepancy between the model and the data occurs at the top of the profile. The arrows and the colors represent the direction and the amplitude of the Darcy velocity, respectively. Inset: comparison between the measured self-potential data and those resulting from the optimized ground water flow model (RMS error = 1.2%).

(as discussed above), a bulk resistivity of $\rho = 200$ Ohm m (from the resistivity tomogram displayed in Figure 7.14), a mass density for the pore water of $\rho_f = 1000$ kg m^{-3}, a viscosity $\eta_f = 4.6 \times 10^{-4}$ Pa s (water at 60 °C), and a coupling coefficient of $C = -3$ mV m^{-1}, we find $\hat{Q}_V = 7$ C m^{-3}. As $\hat{Q}_V$ can vary over 12 orders of magnitude, this estimate is consistent with the previous estimate (30 C m^{-3}).

In order to have a hydrogeologically reasonable model, we also use the following constraints on the direction and magnitude of the source current density in each unit (see Revil *et al.* (2011b)):

$$\mathbf{m} = \left(j_i^x \leq 0, \, j_i^z \approx 0 \right) \text{ in U3,} \tag{7.18}$$

$$\mathbf{m} = \left(j_i^x < j_i^z \right) \text{ in U1,} \tag{7.19}$$

$$\mathbf{m} = (0, 0) \text{ in U2,} \tag{7.20}$$

which means that the flow is mainly horizontal in the shallow aquifer U3, vertical in the fault zone U1 and null in the basement U2. The minimization of Eq. (7.14) is performed iteratively by the Gauss–Newton method described above. We use the initial value of the regularization parameter equal to $\lambda_0 = 0.08$ on the basis of the noise level in the self-potential data.

The result of the self-potential inversion is shown in Figure 7.15 (61st iteration, RMS error $= 1.2\%$). This small RMS error is due to the fact that the data are relatively smooth and noise free. Note that the value of the RMS in itself does not convey important information because better RMS can be obtained with models that have no hydrogeological basis (i.e., not using the constraints imposed by Eqs. (7.7) to (7.9)). Using the constraints described above, a converged solution gives a mean Darcy velocity in the fault of $7 \pm 2 \times 10^{-7}$ m s^{-1}. Taking a fault zone thickness of 150 m as suggested from the resistivity tomograms and an open pathway of 500 m along the dextral strike slip fault zone, a rough estimate of the water flux is $4 \pm 1 \times 10^3$ m^3 per day of thermal water upwelling along the fault plane at Chalk Cliff at a temperature of roughly 60 °C.

From the pattern of hot and cold domestic wells mapped in the area, we know that the upwelling thermal water coming from a portion of the A- and B-faults shown in Figure 7.13c is channeled in a shallow unconfined aquifer flowing toward the Mount Princeton Hot Springs. Therefore, the previous upflow estimate ($4 \pm 1 \times 10^3$ m^3 day^{-1}) based on the inversion of the geophysical data can be compared with the Mount Princeton hot water production. This production is about $(4.3$–$4.9) \times 10^3$ m^3 day^{-1} at $\sim$60–65 °C. This production rate does not account for six fractures leaking directly into Chalk Creek below the pool to the west end of the Mount Princeton property. It is remarkable that the two independent estimates are so close to each other.

7.3.3 Application to Neal Hot Springs

Neal Hot Springs are located within the tectonically complex area of eastern Oregon, in Malheur County, about 150 km northwest of Boise. This area is dominated by extension-related processes with horst–graben structures. This spring is located along a region of north-striking normal faults related to the Oregon–Idaho Graben

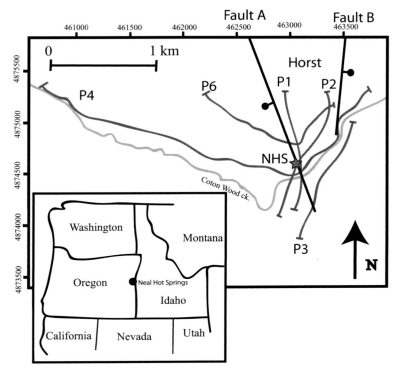

Figure 7.16. Position of Neal Hot Springs and position of the resistivity and self-potential profiles. Inset: position of the Neal Hot Springs (NHS) in Oregon close to the border with Idaho.

and north-west-striking normal faults of the western Snake River Plain. Heat flow averages 71 mW m^{-2} within the plain and ranges to over 105 mW m^{-2} along the margins. This geothermal field is presently in development by US Geothermal Inc. They have drilled two production wells to the West of the hot springs. The first well demonstrated significant flow and a temperature of 141 °C at a depth of 702 m. A second production well, completed 200 m from the first well, intercepted a large aperture fracture possibly associated with Fault "A" discussed below (see Figure 7.16), and with a temperature of 141 °C at a depth of 882 m. The electrical conductivity of the hot springs is slightly higher than at Mount Princeton Hot Springs (see Table 7.6). Temperature can reach 97 °C. The composition of the spring water can be found above in Table 7.4.

In May 2011, we acquired five self-potential profiles (538 measurements in total) and five d.c. electrical resistivity profiles through Neal Hot Springs (Figure 7.16, some profiles used an electrode roll). Figure 7.17 shows the hot springs with a maximum self-potential anomaly of ~40 mV and maximum ground

278

Application to hydrothermal systems

Table 7.6. *Properties of the thermal water of Neal Hot Springs.*

Study	2011[1]	1972[2]
T (°C)	93	87
pH (–)	7.0	7.3
σ_f (10^{-2} S m^{-1}, 25 °C)	9.92	10.10

[1] This work (sampled in May 2011).
[2] From Mariner *et al.* (1980).

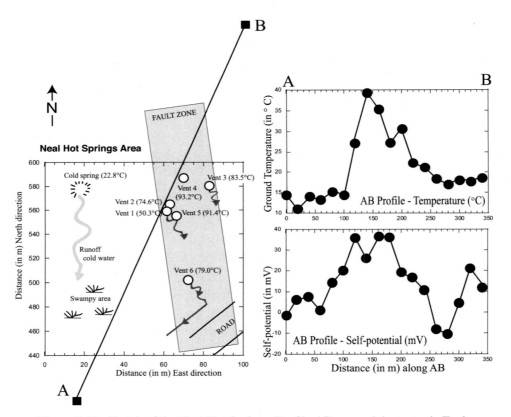

Figure 7.17. Sketch of the Neal Hot Springs. Profile AB crossed the tectonic Fault A associated with the horst graben system. The temperature is measured at a depth of 30 cm. The self-potential anomaly associated with the upflow of water along the open portion of the fault amounts to 40–50 mV. Note the presence of a cold spring (temperature of 22.8 °C) located 50 m east from the A-fault.

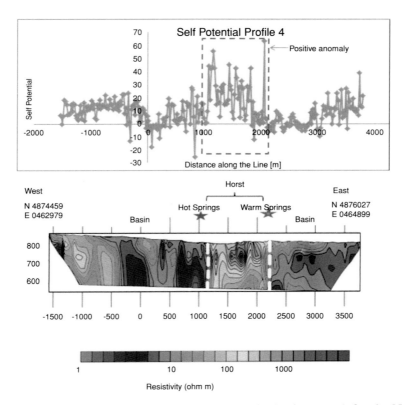

Figure 7.18. Large-scale resistivity tomogram (altitudes in meters) for the Neal Hot Springs area and associated self-potential anomalies (expressed in mV). The self-potential anomaly is mostly symmetric with respect to the position of the fractured horst (located between 1.2 to 2.2 km). The sedimentary basins flanking the horst are associated with negative anomalies corresponding to areas of recharge. The two faults bording the horst (A-fault on the West and B-fault on the East) are associated with the Neal Hot Springs on the West and a warm spring on the East side. They are both associated with positive anomalies showing the discharge of thermal waters. See also color plate section.

temperature anomalies of ~40 °C. As the pore water conductivity at Neal Hot Springs (at 40 °C) is approximately twice the pore water conductivity at Mount Princeton Hot Springs (at 25 °C) and the self-potential anomaly is approximately half, this may imply that the Darcy velocity along the fault plane is similar for the two sites (see Revil *et al.* (2011b)).

Figure 7.18 shows a large-scale self-potential and resistivity survey crossing a basaltic horst along profile 4 (5 km long). This self-potential profile P4 includes 278 stations. The d.c.-resistivity measurements were acquired using a Wenner array configuration with a distance of 20 m between the take-outs and 64 electrodes with

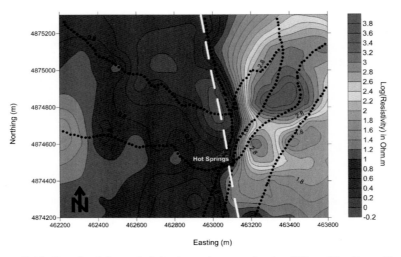

Figure 7.19. Result of the resistivity inversion at a depth of 50 m. The East side of this figure shows high resistivities in the range 100–5000 Ohm m corresponding to the horst. The West side is more conductive and corresponds to the sedimentary infilling of the graben. The filled black circles correspond to the position of the electrodes for the d.c.-resistivity tomogram. Fault A (dashed line) bounds the resistive volcanic rocks of the horst on the West side. See also color plate section.

roll-over of the electrodes along the profile. The resistivity of the basaltic horst is dependent upon fracture density and has a range of resistivities from 50 to 5000 Ohm m. It is generally characterized by a positive self-potential anomaly indicating the upward flow of water through a network of fractures. The horst is bounded by two springs: the Neal Hot Springs ($\sim$93 °C) on the West and a warm unnamed spring ($\sim$40 °C) on the East, suggesting that the horst is bounded by faults on both sides. The sedimentary infillings in the two basins on each side of the horst have low resistivities (1 to 10 Ohm m) due to the presence of clays (including smectite). The basins are characterized by negative anomalies that may correspond to downward infiltration of water and slow recharge of the thermal reservoir at depth, but the residence time of the water in the geothermal system is unknown.

Figure 7.19 shows the resistivity distribution at a depth of 50 m where kriging was applied to all the inverted resistivity profile data. This resistivity map clearly shows the contacts between the horst and the more conductive sediments filling the basin on each side of the horst (see the dashed white lines in Figure 7.19).

The self-potential data of profile 4 have been inverted using the Gauss–Newton approach described above. We use 147 cells, so the number of unknowns in 2D is 294. For the initial model, we consider that the vertical component of the current density is 1×10^{-5} A m^{-2} and the horizontal component is null between $x = 1350$ m and 2400 m (for all depths, in the horst) and we take zero for the two

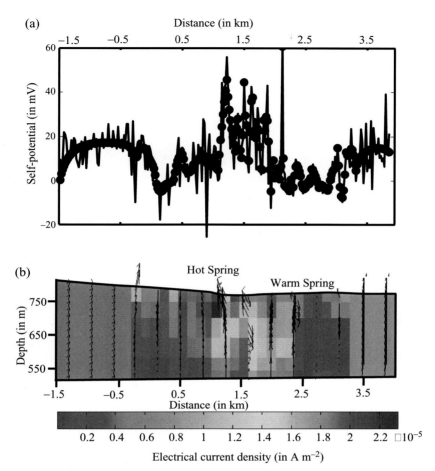

Figure 7.20. Inversion of the self-potential data in terms of the source current density distribution, which is in turn can be related to the ground water flow. (a) Fit of the self-potential data (RMS data misfit 8%). The line corresponds to the (noisy) data while the black filled circles corresponds to the reconstructed self-potential profile based on the source current model distribution shown in Figure 7.20b using the resistivity distribution shown in Figure 7.18. (b) Tomogram of the source current density distribution showing the focus of the flow along Fault A. See also color plate section.

components of the source current density elsewhere. The inversion converged in five iterations. The RMS fit of the measured self-potential data is 8%. This higher value by comparison with the Mount Princeton case study is due to the noise present in the data as shown on the self-potential profile (see Figure 7.18). The results are displayed in Figure 7.20. They show both a comparison between the measured self-potential data and the fitted data (Figure 7.20a), as well as the inverted source current density distribution (Figure 7.20b). Figure 7.20a shows that the predicted self-potential data follow the trend of the measurement, but the higher frequency components are attributed to the heterogeneous resistivity of the top soil. Figure 7.20b shows very well that the main upflow area corresponds to the Neal Hot Springs. A rough estimate of the flux can be obtained as follows. Because we have inferred previously that the Darcy velocity is roughly the same with the shear zone at Mount Princeton Hot Springs, we assume a mean volumetric charge density on the order of 10 C m^{-3} (similar to the previous case study). Using this volumetric charge density and using the maximum current density determined by the inversion of the self-potential data (2×10^{-5} A m^{-2}) yields a Darcy velocity of 2×10^{-6} m s^{-1}, which is very close (as expected) to the Darcy velocity obtained at Mount Princeton Hot Springs (see Figure 7.15). Taking an upwelling area of 100 by 100 meters (100 m is the size of the cell and the lateral extension is determined by the extension of the Hot Springs, see Figure 7.17), we obtained a flux of hot water of 2×10^{-2} m^3 s^{-1} (1700 m^3 per day).

7.4 Conclusions

The inversion of self-potential data remains an ill-posed and underdetermined problem to solve, especially to recover the flow pattern in geothermal fields. The advantage of stochastic, fully coupled methods is to account for the physics of the flow in inverting the self-potential, which helps in reducing the non-uniqueness of the inverse problem. In the case of the gradient-based approach, it is necessary to start with a rough idea of the flow pattern and its associated self-potential field in order to correctly invert for the flow pattern. We have shown in three cases how the self-potential data provide an opportunity to non-intrusively acquire information about the flow pattern and the Darcy velocity. This is a unique feature of the self-potential data over many other geophysical methods. The equipment is not heavy and can be used in rugged terrain. However, the self-potential method is still complementary to other geophysical methods (e.g., gravity, magnetic and seismic data) and does require a distribution of the resistivity which can be obtained by d.c.-resistivity tomography, airborne or ground-based TDEM, or magnetotelluric or controlled source EM-based methods.

Exercises

Exercise 7.1. Application to the vertical infiltration of water in a volcano. In active volcanic areas, the self-potential method has been used to identify the upwelling flow in geothermal zones with a positive signature, and a negative signature for the infiltration of the water through the fractures (Figure 7.21). We propose a synthetic case of a volcano with a geometry inspired from the

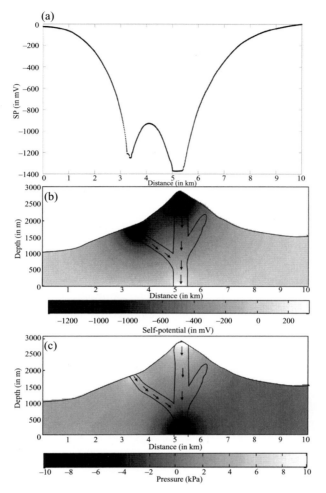

Figure 7.21. The infiltration of rain water along some preferential conduits is responsible for negative self-potential anomalies at the ground surface of the system.

example discussed in the Comsol software to simulate the streaming potential created by the water infiltration through permeable conduits as shown in the figure. Use the geometry and mesh distribution in the attached file, using the hydraulic and electrical properties of the ground to solve the forward problem. Values are reported in Figure 7.21. For the hydraulic boundary conditions, we have imposed a hydraulic entry pressure $p = -10$ kPa on the two conduits, and a pressure of $p = -10$ kPa in the bottom of the conduit. For the electrical problem, use a Neumann boundary condition at the surface, and $\psi = 0$ mV for the other.

8

Seismoelectric coupling

In this chapter, we provide an extension of the electrokinetic theory to the frequency domain accounting for both inertial effects and the partial saturation of the water phase. In Section 8.1, we provide a short history of the seismoelectric (seismic-to-electric conversion) and we discuss the method. In Section 8.2, we discuss the physics of the coupling for fully water saturated conditions. In Section 8.3, we show how to implement this geophysical method and, in Section 8.4, we show an example of numerical application. We develop an extension of the theory in Section 8.5 to unsaturated conditions. In Section 8.6, we provide a numerical example in two-phase flow conditions. This numerical example is related to the water flooding of an oil reservoir and to the computation of the electromagnetic disturbances associated with the passage of the seismic waves at the oil/water encroachment front. This approach may offer a completely new way of monitoring change in saturation in both near-surface and deep applications. In Section 8.7, we show that the passive record of electrical signals can be used to track hydromechanical disturbances in a cement block during the rupture of a seal associated with a fracking experiment. In Section 8.8, we show briefly an approach of using beamforming to extract the electrokinetic properties of the point where the seismic field is focused.

8.1 Position of the problem

The electroseismic (electric-to-seismic) and seismoelectric (seismic-to-electric) phenomena correspond to coupling between electromagnetic disturbances and seismic disturbances in a porous material; see Frenkel (1944) and Pride (1994). Any mineral in contact with water is the setting of electrochemical reactions at its surface. A surface charge is formed, which is fixed in a Lagrangian framework attached to the solid phase. This charge is partly shielded by the sorption of counterions in the Stern layer coating the surface of the minerals (see Stern (1924), Revil and

Leroy (2001) and Leroy and Revil (2004)). Global electroneutrality at the scale of a representative elementary volume requires an excess of electrical charges located in the pore water, in the vicinity of the mineral/water interface, in the so-called electrical diffuse layer (see Pride (1994)). The diffuse and Stern layers form the electrical double layer; the Stern layer is part of the total fixed charge attached to the solid phase, while the diffuse layer can be dragged by the flow of the pore water relative to the mineral framework. The electroseismic and seismoelectric couplings are therefore controlled by the relative displacement between the charged solid phase (with the Stern layer attached to it) and the pore water (with its diffuse layer and consequently an excess of electrical charges per unit pore volume).

In this book, we are interested only in the seismoelectric coupling corresponding to electromagnetic conversion of mechanical energy during the propagation of seismic waves in poroelastic media. We want to know how the coupled seimic and seismoelectric information can be used, for instance, to determine the permeability of an oil reservoir and to locate and track hydromechanical disturbances provoked during the mechanical ruptures and earthquake events. The same type of methodology could be applied to image NAPL/DNAPL plumes for shallow ground water remediation problems.

When seismic waves propagate in a linear poroelastic porous material, two types of electrical disturbances are observed. The propagation of seismic waves (P- and S-waves) generates an electrical current associated with the displacement of the electrical diffuse layer in a Lagrangian framework attached to the solid phase. These co-seismic electrical signals travel at the same speed as the seismic waves (see Pride (1994)). The amplitudes of the co-seismic electromagnetic signals are controlled by the properties of the porous material (the formation factor) and by the properties of the pore fluid/solid interface (the zeta potential in the theory of Pride (1994), the excess charge per unit pore volume in the formulation developed below).

In addition to the co-seismic signals, another phenomenon occurs when a seismic wave moves through a sharp interface characterized by a change in the textural properties or a change in salinity or clay content/mineralogy. In this situation, a fraction of the mechanical energy is converted into electromagnetic energy and a dipolar electromagnetic excitation is produced. The resulting electromagnetic disturbances diffuse very quickly through the geological strata and can be recorded, nearly instantaneously, by electrodes or antennas located at the ground surface, in boreholes (see Mikhailov *et al.* (1997)), or at the sea floor, for instance. This seismoelectric conversion can be used as a tool to diagnose and locate oil reservoirs and this information can be used in principle to determine remotely their permeability and their electrical conductivity (hence possibly the saturation of oil in the reservoir) as shown below.

Thompson and Gist (1993) presented a case study for the exploration of oil and gas using seismoelectric converted electrical signals. They used adapted data processing and common mid-point (CMP) techniques to produce a seismoelectric image of the subsurface to depths on the order of a few hundred meters. They concluded that seismoelectric conversions could be detected from a depth of 300 m. Thompson *et al.* (2007) suggested that these methods could be used for much greater depths (several kilometers in the case of the electroseismic method). The seismoelectric method has also been used for a variety of applications in near-surface geophysics; see Migunov and Kokorev (1977), Fourie (2003) and Kulessa *et al.* (2006). Mikhailov *et al.* (2000) described crosshole seismoelectric measurements in a small-scale laboratory experiment with vertical and inclined fractures located between the source and the receivers. They recorded not only the co-seismic electric signals generated by the seismic wave arriving at the receivers but also the EM-wave associated with the Stoneley wave excited in the fracture. They claimed that a tomography image with the traveltimes extracted from the seismoelectric measurements could be possibly constructed.

The model developed by Pride (1994) couples Biot's theory to the Maxwell equations via a source current density of electrokinetic origin (see Chapter 2). This model has opened the door to numerical modeling of both the co-seismic and seimoelectric conversions using finite-difference or finite element methods and to assess the usefulness of these methods for application to the field. Various authors have used a finite difference algorithm to simulate the 2D seismoelectric response of a heterogeneous medium taking into account all the poroelastic waves modes (fast waves, slow waves and shear waves) and their co-seismic electrical signals plus the seismoelectric conversions. Pain *et al.* (2005) presented a 2D mixed finite-element algorithm to solve the poroelastic Biot equations including the electrokinetic coupling in order to study the sensitivity of the seismoelectric method to material properties like porosity and permeability of geological formations surrounding a borehole.

8.2 Seismoelectric theory in saturated media

8.2.1 The wave equations in a poroelastic body

Biot's theory (see Biot (1956a, b) and Biot and Willis (1957)) provides a starting framework to model the propagation of seismic waves in linear poroelastic media. The theory predicts the existence of an additional compressional wave by comparison with the P- and S-waves found for elastic materials. The existence of this slow P-wave was first confirmed by Plona (1980). According to Biot's theory, the equations of motion in a statistically isotropic, fully saturated, heterogeneous, porous medium are given, in the frequency domain, by Revil and Jardani (2010b) as

$$-\omega^2(\rho\mathbf{u} + \rho_f\mathbf{w}) = \nabla \cdot \mathbf{T} + \mathbf{F}, \tag{8.1}$$

$$\mathbf{T} = [\lambda_u\nabla \cdot \mathbf{u} + C\nabla \cdot \mathbf{w}]\mathbf{I} + G[\nabla\mathbf{u} + \nabla\mathbf{u}^T] \tag{8.2}$$

$$-\omega^2(\rho_f\mathbf{u} + \tilde{\rho}_f\mathbf{w}) - ib\omega\mathbf{w} = -\nabla p, \tag{8.3}$$

$$-p = C\nabla \cdot \mathbf{u} + M\nabla \cdot \mathbf{w} + S, \tag{8.4}$$

where $i^2 = -1$, $\mathbf{u}$ is the displacement vector of the solid, $\mathbf{w}$ is the displacement vector of the fluid relative to the solid (called the filtration displacement), $\mathbf{T}$ is the stress tensor, $\mathbf{I}$ is the identity matrix, $\mathbf{F}$ denotes the body force on the elastic solid phase, S is a pressure source acting on the pore fluid, ρ represents the mass density of the saturated medium, ρ_f and ρ_s are the mass density of the fluid and the solid, respectively, $\tilde{\rho}_f$ is an apparent density of the pore fluid, p is the fluid pressure, $\lambda_u = K_u - (2/3)G$ is the undrained Lamé modulus of the porous material, b is the mobility of the fluid, G is the shear modulus of the porous frame, and C and M are Biot moduli. Equation (8.1) corresponds to Newton's law. Equation (8.2) represents a constitutive expression for the total stress tensor as a function of the displacement (Hooke's law). This constitutive equation comprises the classical term of linear elasticity plus an additional term related to the expansion/contraction of the porous body to accommodate the flow of the pore fluid relatively to a Lagrangian framework attached to the solid phase. Equation (8.3) denotes the Darcy constitutive equation in which the bulk force acting on the fluid phase has been neglected and Eq. (8.4) is one of the classical Biot constitutive equations of poroelasticity. The mass density of the porous material is given as $\rho = \phi\rho_f + (1 - \phi)\rho_s$.

The material properties entering Eqs. (8.1)–(8.4) are given by Pride (1994) and Rañada Shaw et al. (2000) as

$$b = \frac{\eta_f}{k_0}, \tag{8.5}$$

$$\tilde{\rho}_f = \frac{\rho_f\phi}{a}, \tag{8.6}$$

$$\alpha = 1 - K_{fr}/K_s \tag{8.7}$$

$$K_u = \frac{K_f(K_s - K_{fr}) + \phi K_{fr}(K_s - K_f)}{K_f(1 - \phi - K_{fr}/K_s) + \phi K_s}, \tag{8.8}$$

$$C = \frac{K_f(K_s - K_{fr})}{K_f(1 - \phi - K_{fr}/K_s) + \phi K_s}, \tag{8.9}$$

$$M = \frac{C}{\alpha} = \frac{K_f K_s}{K_f(1 - \phi - K_{fr}/K_s) + \phi K_s}, \tag{8.10}$$

where K_u (in Pa) is the bulk modulus of the porous medium, K_{fr} is the bulk modulus of the dry porous frame (skeleton), K_f is the bulk modulus of the fluid, K_s is the bulk modulus of the solid phase, and α is the Biot–Willis coefficient. Equation (8.8) is the Gassman equation; η_f is the dynamic viscosity of the pore fluid, k_0 the permeability of the medium, ϕ the porosity, and a is the tortuosity. The ratio a/ϕ corresponds to the electrical formation factor F also defined by Archie's law $F = \phi^{-m}$ where m is called the cementation exponent. In the following, we consider the tortuosity equal to $\phi^{-1/2}$, which is equivalent of taking a cementation exponent equal to 1.5.

The classical formulation described above in Eqs. (8.1)–(8.4) is based on solving partial differential equations for two unknown fields, $\mathbf{u}$ and $\mathbf{w}$. For a two-dimensional discretized problem, four degrees of freedom per node are therefore present. All the papers dealing with the modeling of the seismoelectric problem use this type of formulation. Atalla *et al.* (1998) introduced an alternative approach using $\mathbf{u}$ and p as unknowns. This implies three unknown parameters (u_1, u_2 and p) to solve at each node. After some algebraic manipulations described in Appendix B, the equations of motion can be written in terms of the two new unknown fields ($\mathbf{u}$, p) (see Jardani *et al.* (2010)) as

$$-\omega^2 \rho_\omega^s \mathbf{u} + \theta_\omega \nabla p = \nabla \cdot \hat{\mathbf{T}} + \mathbf{F}, \tag{8.11}$$

$$\hat{\mathbf{T}} = \lambda (\nabla \cdot \mathbf{u})\mathbf{I} + G[\nabla \mathbf{u} + \nabla \mathbf{u}^T], \tag{8.12}$$

$$\mathbf{T} = \hat{\mathbf{T}} - \alpha p \mathbf{I}, \tag{8.13}$$

$$\frac{1}{M}(p + S) + \nabla \cdot \{k_\omega [\nabla p - \omega^2 (\rho_f + \alpha)\mathbf{u}]\} = 0. \tag{8.14}$$

Equation (8.11) corresponds to Newton's law applied to the solid skeleton of the porous material. This equation is similar to Newton's equation of elastic bodies except for the coupling term, $\theta_w \nabla p$, which represents the coupling between the solid and fluid phases. The stress tensor defined by Eq. (8.12) corresponds to the stress tensor with the porous material in vacuum (i.e., it corresponds to the stress acting on the solid phase if the pore fluid is replaced by vacuum). Equation (8.13) describes the relationship between the total stress tensor and the effective stress tensor. The material properties entering into Eqs. (8.11)–(8.14) are given by Jardani *et al.* (2010) as

$$k_\omega = \frac{1}{\omega^2 \tilde{\rho}_f + i\omega b}, \tag{8.15}$$

$$\lambda = K - \frac{2}{3}G, \tag{8.16}$$

$$\rho_\omega^s = \rho + \omega^2 \tilde{\rho}_f^2 k_\omega, \tag{8.17}$$

$$\theta_\omega = \alpha + \omega^2 \rho_f k_\omega, \tag{8.18}$$

where k_ω is not the dynamic permeability of the porous material (the dynamic permeability is given in Appendix B), $\tilde{\rho}_f$ is an effective fluid density, λ is the Lamé coefficient, and ρ_ω^s corresponds to the apparent mass density of the solid phase at a given frequency ω.

8.2.2 Description of the seismic source

In the following example, we use a source generating only P-waves. This force creates a net force on the solid phase of the porous rock. Because the source generates a displacement of pore water relative to the grain framework, it creates an electromagnetic disturbance. This disturbance diffuses nearly instantaneously to all receivers and has a quite strong amplitude. Because this contribution can be easily removed from the electrograms, we will not model it below.

Using the Fourier transformation of the first time derivative of the Gaussian function for such a source yields the following expressions for the bulk force acting on the solid phase:

$$\mathbf{F}(x, y, \omega) = F(\omega)\nabla[\delta(x - x_0)\delta(y - y_0)] \tag{8.19}$$

$$F(\omega) = \text{FT}[(t - t_0)\exp\{-[\pi f_0(t - t_0)]^2\}] \tag{8.20}$$

where $\text{FT}[f(t)]$ is the Fourier transform of the function $f(t)$, t_0 is the time delay of the source and f_0 is its dominant frequency. This force is a source term acting on the right-hand side of Eq. (8.11). In the following, we will neglect the pore fluid pressure source term S, which is equivalent to neglect the electromagnetic effects associated with the seismic source itself.

8.2.3 The Maxwell equations in saturated conditions

The local Maxwell equations can be volume-averaged to obtain the macroscopic Maxwell equations (see Pride (1994) and also Revil and Linde (2006) who used the Donnan model discussed in Chapter 2). With the Donnan model, the Maxwell equations are written (see Revil and Jardani (2010b)) as

$$\nabla \times \mathbf{E} = -\dot{\mathbf{B}}, \tag{8.21}$$

$$\nabla \times \mathbf{H} = \mathbf{J} + \dot{\mathbf{D}}, \tag{8.22}$$

$$\nabla \cdot \mathbf{B} = 0, \tag{8.23}$$

$$\nabla \cdot \mathbf{D} = \phi \hat{Q}_V, \tag{8.24}$$

where $\mathbf{H}$ is the magnetic field, $\mathbf{B}$ is the magnetic induction, and $\mathbf{D}$ is the displacement vector. These equations are completed by two electromagnetic constitutive equations: $\mathbf{D} = \varepsilon\mathbf{E}$ and $\mathbf{B} = \mu\mathbf{H}$, where ε is the permittivity of the medium and μ is the magnetic permeability. If the porous material does not contain magnetized grains, these two material properties are given by Pride (1994) as

$$\varepsilon = \frac{1}{F}(\varepsilon_f + (F-1)\varepsilon_s), \tag{8.25}$$

$$\mu = \mu_0, \tag{8.26}$$

where F is the electrical formation factor, ε_f and ε_s are the dielectric constants of the pore fluid and the solid, respectively, and μ_0 is the magnetic permeability of free space. Note that only two textural properties, Λ and F, are required to describe the influence of the topology of the pore network upon the material properties entering the transport and electromagnetic constitutive equations.

The coupling between the mechanical and the Maxwell equations occurs in the current density, which can be written in the time domain (see Revil and Jardani (2010b)) as

$$\mathbf{J} = \sigma\mathbf{E} + \mathbf{J}_S, \tag{8.27}$$

$$\mathbf{J}_S = \hat{Q}_V\dot{\mathbf{w}}. \tag{8.28}$$

The electromagnetic problem can be solved in its quasi-static limit. To simplify the problem, we consider that the reservoir is close enough to the sensors (antennas, non-polarizing electrodes, and/or magnetometers). In this case, we can neglect the time required by the electromagnetic disturbances to diffuse between the reservoir and the electromagnetic sensors. With this additional assumption, we can model the problem by solving only the quasi-static electromagnetic problem (rather than the low-frequency diffusive problem) for the electrical potential and the magnetic field (see Revil and Jardani (2010b)):

$$\nabla \cdot (\sigma\nabla\psi) = \nabla \cdot \mathbf{J}_S, \tag{8.29}$$

$$\nabla^2\mathbf{B} = -\mu_0\nabla \times \mathbf{J}_S, \tag{8.30}$$

$$\mathbf{J}_S = \hat{Q}_V\dot{\mathbf{w}} = -i\omega\hat{Q}_V k_\omega(\nabla p - \omega^2\rho_f\mathbf{u}), \tag{8.31}$$

where ψ is the electrostatic potential, σ is the electrical conductivity of the medium, $\mathbf{J}_S$ is the source current density of electrokinetic nature, and $\hat{Q}_V$ is the excess of charge (of the diffuse layer) per unit pore volume (in C m^{-3}). For saturated rocks, $\hat{Q}_V$ can be directly computed from the d.c.-permeability k_0 through a semi-empirical formula derived by Jardani *et al.* (2007b), as discussed in Chapter 3. Therefore, we consider below that $\hat{Q}_V$ and k_0 are not independent parameters.

8.3 Numerical modeling

We use the finite element method with COMSOL Multiphysics package. The field equations are written as (Revil and Jardani, 2010b):

$$\nabla \cdot \overline{\overline{\Gamma}} = \bar{F}, \tag{8.32}$$

$$\overline{\overline{\Gamma}} \equiv \begin{bmatrix} \hat{T}_{xx} & \hat{T}_{xz} \\ \hat{T}_{zx} & \hat{T}_{zz} \\ w_x & w_z \end{bmatrix}, \tag{8.33}$$

$$\bar{F} \equiv \begin{bmatrix} -\omega^2 \rho_\omega^s u_x + \theta_\omega \dfrac{\partial p}{\partial x} \\ -\omega^2 \rho_\omega^s u_z + \theta_\omega \dfrac{\partial p}{\partial z} \\ -\dfrac{1}{M}(p + S) \end{bmatrix}, \tag{8.34}$$

$$w_x = k_\omega \left[\dfrac{\partial p}{\partial x} - \omega^2 (\rho_f + \alpha) u_x \right], \tag{8.35}$$

$$w_z = k_\omega \left[\dfrac{\partial p}{\partial z} - \omega^2 (\rho_f + \alpha) u_z \right], \tag{8.36}$$

where w_x and w_z are the both components of relative velocity of the fluid $\dot{\mathbf{w}}$. This form is suitable for the implementation into a finite element code.

Equations (8.11), (8.12) and (8.14) described the propagation of seismic waves in an infinite unbounded medium. However, when one performs numerical and laboratory simulations, the domain investigated is always bounded. A common approach to limit reflection at the boundaries of the domain is to use the one-way wave equation based on the paraxial approximations of the seismic wave equations; see Clayton and Engquist (1977). The Perfectly Matched Layer (PML) method was proposed later by Berenger (1994), first for electromagnetic problems. With PML boundary layers, no reflection is expected to occur at the interface between the physical domain and the absorbing layer for any frequency and any angle of incidence of the seismic waves.

In this book, we use the Convolution-Perfectly Matched Layer (C-PML) approach. The C-PML method, for first-order systems of partial differential equations, was developed for electromagnetic waves by Roden and Gedney (2000) and in simulations of elastic waves propagation by Bou Matar et al. (2005). It has never been used for the problem of seismoelectric waves so we adapted this method for this type of problem. This method is extended for second-order systems written in terms of displacements. The main advantages of the C-PML approach

over the classical PML approach concerns its numerical satbility and its high efficiency. Using the concept of complex coordinates (see Chew and Weedon (1994))
in the frequency domain (with a time dependence of $e^{-i\omega t}$), the complex coordinate
stretching variables (see Jardani *et al.* (2010)) are

$$\tilde{x}_j = \int_0^{x_j} s_{x_j}(x')dx' \quad j = 1, 2, \tag{8.37}$$

$$s_{x_j} = k_{x_j}(x_j) + \frac{\sigma_{x_j}(x_j)}{\alpha_{x_j} + i\omega}, \tag{8.38}$$

where α_{x_j}, σ_{x_j} are positive real damping coefficients and k_{x_j} are real and positive-
definite numbers that are equal or larger than unity. In this book, we consider
$k_{x_j} = 1$ to keep the waves continuous. To determine the value of the two other
damping coefficients, we use the following formula:

$$\sigma_j = \begin{cases} \dfrac{3c}{2L_0} \log\left(\dfrac{1}{R}\right) \left(\dfrac{x_{j\,min} - x_j}{L_0}\right)^3, & \text{as } x_{j\,min} \geq x_j, \\[2ex] 0, \text{ as } x_{j\,min} \leq x_j \leq x_{j\,max}, \\[2ex] \dfrac{3c}{2L_0} \log\left(\dfrac{1}{R}\right) \left(\dfrac{x_j - x_{j\,max}}{L_0}\right)^3, & \text{as } x_j \geq x_{j\,max}, \end{cases} \tag{8.39}$$

$$\alpha_i = \begin{cases} \pi f_0 \left(\dfrac{x_{j\,min} - x_j}{L_0} + 1\right), & \text{as } x_{j\,min} \geq x_j, \\[2ex] \pi f, \text{ as } x_{j\,min} \leq x_j \leq x_{j\,max}, \\[2ex] \pi f_0 \left(\dfrac{x_j - x_{j\,max}}{L_0} + 1\right), & \text{as } x_j \geq x_{j\,max}, \end{cases} \tag{8.40}$$

where c is the highest of all of the velocities in the domain, $R = 1/1000$ represents
the amount of reflected energy at the outer boundary of the PML layer, L_0 is the
thickness of the PML, and f_0 is the dominant frequency of the source. The derivative
$\partial(\cdot)/\partial\tilde{x}_i$ can be expressed in terms of the regular coordinate stretching variables,
$\partial(\cdot)/\partial\tilde{x}_j = (1/s_{x_j})(\partial(\cdot)/\partial x_j)$. Finally, after replacing $\partial(\cdot)/\partial x_j$ by $\partial(\cdot)/\partial\tilde{x}_j$ and
after some algebraic manipulations, the reduced set of equations for the modified
poroelastic formulation (see Jardani *et al.* (2010)) is

$$-\omega^2 \tilde{\rho}_w^s \mathbf{u} + \tilde{\theta}_w \nabla p = \nabla \cdot \tilde{\mathbf{T}}, \tag{8.41}$$

$$\tilde{\theta}_w = \begin{pmatrix} \theta_w s_{x_2} & 0 \\ 0 & \theta_w s_{x_1} \end{pmatrix} \tag{8.42}$$

$$\tilde{\rho}_w^s = \rho_w^s s_{x_1} s_{x_2}, \tag{8.43}$$

$$\tilde{\mathbf{T}} = \begin{pmatrix} \tilde{T}_{11} & \tilde{T}_{12} \\ \tilde{T}_{21} & \tilde{T}_{22} \end{pmatrix}, \tag{8.44}$$

$$\tilde{T}_{11} = (\lambda + 2G)\frac{s_{x_2}}{s_{x_1}}\frac{\partial u_1}{\partial x_1} + \lambda \frac{\partial u_2}{\partial x_2}, \tag{8.45}$$

$$\tilde{T}_{22} = \lambda \frac{\partial u_1}{\partial x_1} + (\lambda + 2G)\frac{s_{x_1}}{s_{x_2}}\frac{\partial u_2}{\partial x_2}, \tag{8.46}$$

$$\tilde{T}_{12} = G\left(\frac{s_{x_1}}{s_{x_2}}\frac{\partial u_1}{\partial x_2} + \frac{\partial u_2}{\partial x_1}\right), \tag{8.47}$$

$$\tilde{T}_{21} = G\left(\frac{\partial u_1}{\partial x_2} + \frac{s_{x_2}}{s_{x_1}}\frac{\partial u_2}{\partial x_1}\right), \tag{8.48}$$

$$\frac{s_{x_1}s_{x_2}}{M}p + \nabla \cdot [\tilde{\mathbf{k}}_{1\omega}\nabla p - \tilde{\mathbf{k}}_{2\omega}\omega^2(\rho_f + \alpha)\mathbf{u})] = 0, \tag{8.49}$$

$$\tilde{\mathbf{k}}_{1\omega} = \begin{pmatrix} k_\omega s_{x_2}/s_{x_1} & 0 \\ 0 & k_\omega s_{x_1}/s_{x_2} \end{pmatrix}, \tag{8.50}$$

$$\tilde{\mathbf{k}}_{2\omega} = \begin{pmatrix} k_\omega s_{x_2} & 0 \\ 0 & k_\omega s_{x_1} \end{pmatrix}. \tag{8.51}$$

We first solve the poroelastodynamic wave equations (see Eqs. (8.41) and (8.49)) and the electrostatic Poisson equation for the electrical field in the frequency domain. We use the Comsol Multiphysics 4.2a package and the stationary parametric solver PARDISO (http://www.computational.unibas.ch/cs/scicomp/ software/pardiso/); see Schenk *et al.* (2008). The problem is solved as follow: (i) we first compute for the poroelastic and electric properties distribution for the given porosity, fluid permeability and each saturation profile, then (ii) we solve for the displacement of the solid phase and the pore fluid pressure, and (iii) finally we compute the electrical potential by solving the Poisson equation using the solution of poroelastodynamic problem. The solution in the time domain is computed by using an inverse-Fourier transform of the solution in the frequency wave number domain.

8.4 Application in saturated conditions

The applications of seismoelectric methods either in the oil industry or in the hydrogeophysics field are too limited owing to the difficulty of recording the seismoelectric conversion with adequate quality. However, the sensitivity of existing tools is high enough to record this electrical signature. Given that, is it a preventable issue?

Crespy *et al.* (2008) showed that electroencephalographic equipment can be used to provide reliable electrograms with a sensitivity better than 0.1 μV up to a frequency of few kHz. In marine controlled source electromagnetics (CSEM), the measurement of the electrical field is often made with sensitivity of the order of 1 nV once the spurious electromagnetic effect of external origin has been filtered out. A new generation of electrodes exists, developed for EM studies that couple with the ground or the sea floor by capacitance rather than resistance (see http://www.quasargeo.com). These electrodes usually show noise that is lower by a factor of five than non-polarizing Ag/AgCl in the frequency range 1–10 Hz. So we believe that, although difficult, accurate measurements of seismoelectric conversions produced at a depth of a few hundred meters below the sensors, or for cross-hole investigations, is possible. Consequently, seismoelectric investigations made on shore could benefit from all the recent developments made in measuring small electrical fields at the sea floor for Controlled Source EM methods (CSEM) (see Ellingsrud *et al.* (2008)).

Concerning the direct application of the present method on real data, we have recently proposed different strategies across several studies to interpret the seismoelectric data. These processes have also been discussed and applied to interpret the streaming potential signature in the previous chapters of this book. We now introduce two approches: the first approach consists to use in the joint inversion the seismolectric and seismic data to determine the hydromechanical properties of the reservoirs. The second protocol is based to the localizations of the hydrome-chanicals events responsible of the electroseimsic signature from the inversion of this signature.

To check the usefulness of a joint inversion of seismic and seismoelectric data, we test this approach using a numerical case study. We consider two flat layers plus a rectangular reservoir embedded in the second layer (Figure 8.1). The geophones and the electrodes are located at the top surface of the system to simulate an on-shore acquisition. The take-out for the electrodes and the geophones is 10 m. The source wavelet is a first-order derivative of a Gaussian with dominant frequency $f_0 = 30$ Hz and with a time delay factor $t_0 = 0.1$ s, see Eq. (8.20). The seismic source is located at a depth of 20 m (Figure 8.1). The true values of the material properties used for the simulation are reported in Table 8.1. With these values, the velocity of the P-fast wave is 1972 m s^{-1} in the first layer L1, 2188 m s^{-1} in the second layer (labeled L2), and 3118 m s^{-1} in the reservoir (labeled R). The receiver located with an offset of $x = 150$ m from the source, the time required for the P-wave to reach this receiver is therefore 0.076 s in agreement with the numerical simulations (see below) (see Jardani *et al.* (2010)).

In Figure 8.2d, the electrograms show two types of seismoelectric signals. The first type of signals corresponds to the co-seismic electrical signal associated with

Table 8.1. *True values of the material properties used for the synthetic model shown in Figure 8.1. L1 and L2 stand for the two layers and R for the reservoir (see Figure 8.1). Layer L1 corresponds to a clean sand, L2 to a clayey sand, and R to a sand reservoir partially filled with oil (see Jardani et al. (2010)).*

Parameter	Units	Unit L1	Unit L2	Unit R
k_0	m^2	10^{-12}	10^{-16}	10^{-11}
ϕ	–	0.25	0.10	0.33
K_s	GPa	36.50	6.90	37.00
K_f	GPa	0.25	0.25	2.40
K_{fr}	GPa	2.22	6.89	9.60
G	GPa	4.00	3.57	5.00
ρ_s	kg m^{-3}	2650	2650	2650
ρ_f	kg m^{-3}	1040	1040	983
η_f	Pa s	1×10^{-3}	1×10^{-3}	8×10^{-1}
σ	S m^{-1}	0.01	0.1	0.001
Log $\hat{Q}_V$	log C m^{-3}	0.203	3.49	3.2

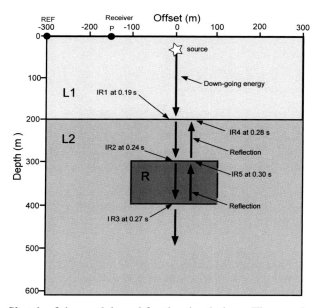

Figure 8.1. Sketch of the model used for the simulations. The geophones and the electrodes are collocated at the top surface of the system. All the electrodes are assumed to be connected to a reference electrode. REF corresponds to the position of the reference electrode. L1 and L2 correspond to the two layers of sediments and R stands for the reservoir. Arrows indicate ray path for seismic energy creates seismoelectric interface response labeled IRi. In addition to this energy, the direct field and reflected seismic arrivals are recorded as coseismic electric field. The seismic source is located at a depth of 20 m below the top surface (see Jardani *et al.* (2010)).

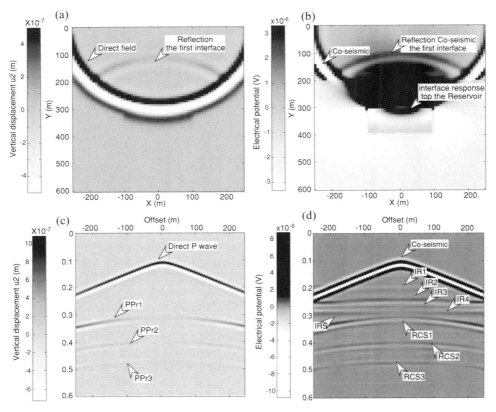

Figure 8.2. Snapshots of seismic and seismoelectric phenomena. (a, b) Snapshots of the seismic and electrical fields at time $t = 0.24$ s. This corresponds to the time when the P wave reaches the top of the reservoir, which acts as a dipole radiating electromagnetic energy. (c) The seismograms reconstructed by the geophones (with a take out of 10 m) present direct field and different reflection of the P-waves: the reflection PPr1 on the interface L1–L2, the reflection PPr2, on the interface L2–top of the reservoir, and the reflection PPr3 on the interface L2–bottom of the reservoir. (d) The electrograms show the co-seismic electrical potential field associated with the direct wave and with the reflections of the P-wave (labeled RCS1, RCS 2, RCS 3) and the seismoelectric conversions with a smaller amplitude and a flat shape (labeled IR1, 2, 3, 4, 5) (see Jardani *et al.* (2010)).

the propagation of the P-wave. The co-seismic electric field related to the direct field wave from the source to the receiver is labeled CS. Other co-seismic signals are associated with the reflected P-waves at the various interfaces like the L1–L2 interface, the L2–reservoir interface, and the reservoir–L2 interface. These co-seismic signals are labeled RCS1, RCS2 and RCS3, respectively. A co-seismic signal occurs when a seismic wave travels through a porous material, creating a relative displacement between the pore water and the solid phase. The associated

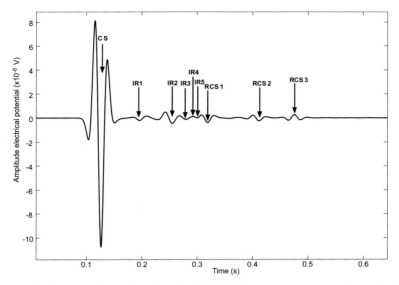

Figure 8.3. Time series of the electrical field (electrogram) at an electrode (the position corresponds to receiver P in Figure 8.1) with a horizontal offset of 150 m. CS stands for the co-seismic disturbance associated with the direct seismic wave. The terms RCS1 and RCS2 stand for the co-seismic disturbances associated with the reflected P-waves (see Figure 8.1). The term IRi stand for the various seismo-electric disturbances associated with the seismoelectric conversions at the different interfaces of the subsurface shown in Figure 8.1.

current density is balanced by a conduction current density. It results an electrical field traveling at the same speed as the seismic wave. Because shear waves are equivoluminal, they are not responsible for any source current density in a homogeneous medium and therefore they have no co-seismic electric field associated with them.

The second type of seismoelectric signals correspond to converted seimoelectric signals associated with the arrival of the P-waves at each interface (between the two layers and at the surface of the reservoir). These converted seimoelectric signals are labeled IR1, IR2, IR3, IR4 and IR5 (see Figures 8.1 to 8.3). When crossing an interface between two domains characterized by different properties, a seismic wave generates a time-varying charge separation, which acts as a dipole radiating electromagnetic energy. In our approach, we neglect the time used by this electromagnetic energy to diffuse from the geological interface, where it is generated, to the receiver (quasi-static field approximation). This assumption is very good for investigations to the first kilometers below the ground surface. Owing to constructive interferences, a significant portion of the first Fresnel zone acts as a disk of electric dipoles oriented normal to the interface. These dipoles oscillate with the waveform of the seismic wave (Figure 8.2b). Because the electromagnetic diffusion of the electrical disturbance is very fast, the seismoelectric conversions are observed nearly at the same time by all the electrodes but with different amplitudes. The seismoelectric conversions appear therefore as flat lines in the electrograms. Note also that the polarity of the converted seismoelectric signals depends on the contrast of electrical material properties (volumetric charge density and electrical conductivity) at the interface where they are generated. On the contrary, the polarity of the co-seismic electrical signals depends on the value of the streaming potential coupling coefficient at the position of the electrode and the polarity of the seismic waves.

Figure 8.3 shows the electric potential for a given electrode. In this figure, we can clearly discriminate the co-seismic signals from the seismoelectric conversions. Note also that the amplitudes of the signals are small. However, they can easily be measured in the field by using the type of ultrasensitive equipment discussed by Crespy *et al.* (2008). This equipment can be used to record electrical potential using up to 256 simultaneous channels at several kHz and with a sensitivity of 10 nV.

We use the AMA algorithm detailed in Chapter 4 to generate 25 000 realizations of the 21 parameters of the material properties of the different geological units, using the data recorded in 60 geophones and 60 electrodes at four frequencies only (25, 30, 35 and 40 Hz). The position and the characteristic of the source are assumed to be perfectly known. The posterior probability distribution functions of the material properties of the three units (layers L1 and L2 and the reservoir R) are shown in Figure 8.4 using the last 5000 iterations. We observe that, except for the porosity, our algorithm is performing a very good job in properly inverting the seismic and seismoelectric data in terms of getting the mean value of the material properties. We believe that a better estimate of the porosity can be obtained through the use of additional petrophysical relationships between the porosity, the electrical conductivity and the bulk modulus of the skeleton.

8.5 Seismoelectric theory in unsaturated media

8.5.1 Generalized constitutive equations

In this section, we develop a new theory of the seismoelectric effect in unsaturated porous material. This theory can be used to understand the electromagnetic signals associated with seismic wave propagation in saturated and unsaturated porous media. We consider below an isotropic representative elementary volume of a porous material with connected pores. The surfaces of the minerals in contact with the pore water are negatively charged. We consider therefore an excess of (positive) charges in the pore space of the porous material in the electrical double layer. As

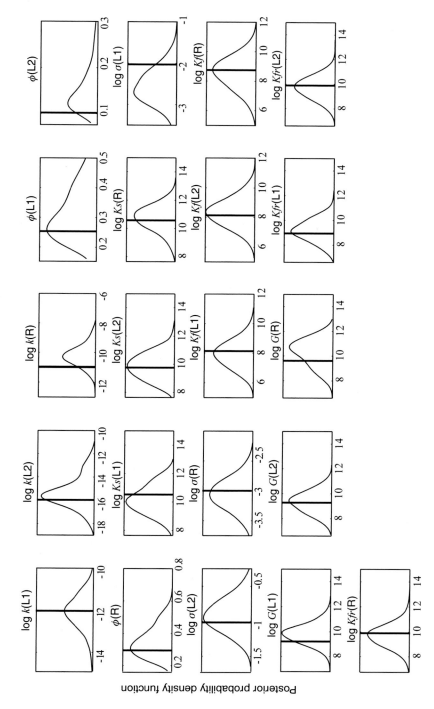

Figure 8.4. Posterior probability density functions of the material properties for the three geological units (the two layers L1 and L2 and the reservoir R) using a McMC sampler (the Adaptative Metropopolis Algorithm, AMA). The vertical bars indicate the real value of the material properties (see Table 8.1) (see Jardani et al. (2010)).

discussed in Chapter 1, the electrical double layer coating the surface of the grains is made of two layers: (1) a layer of (counter) ions sorbed onto the mineral surface and (2) a diffuse layer where the electrostatic (Coulomb) force prevails. In the following, we will use the subscript a to describe the properties of air (the non-wetting phase), and the subscripts w and s will be used to describe the properties of the water and solid phases, respectively. Water is assumed to be the wetting phase. The term skeleton will be used to describe the assemblage of grains alone without the two fluid phases in the pore network.

Another set of assumptions used below pertains to the capillary pressure curve. Hysteretic behavior will be neglected and therefore the porous material will be characterized by a unique set of hydraulic functions (for example, see Brooks and Corey (1964)). We work also with the Richards model, assuming that the air pressure is constant and equal to the atmospheric pressure. This implies, in turn, that the air phase is infinitely mobile and connected to the atmosphere. The capillary pressure p_c (in Pa) is defined as the pressure of the non-wetting phase minus the pressure of the wetting phase (see Bear and Verruijt (1987))

$$p_c = p_a - p_w, \tag{8.52}$$

where p_a and p_w denote the average air and water pressures (in Pa), respectively. The capillary head (suction) Ψ is defined as

$$\Psi = -\frac{p_c}{\rho_w g} = \frac{p_w - p_a}{\rho_w g}. \tag{8.53}$$

In unsaturated flow conditions, the gradient of the capillary head is given by

$$\nabla \Psi = \frac{1}{\rho_w g} \nabla p_w. \tag{8.54}$$

This assumption is used to avoid dealing with the flow of the air phase. In unsaturated conditions, the capillary pressure is positive, the capillary head is negative, and the pressure of the water phase is smaller than the atmospheric pressure. The total head h includes the gravity force and is defined by $\psi + z$, where z denotes the elevation head. Our model will be restricted to the capillary regime, that is for saturation above the irreducible water saturation. There are also many mechanisms of electrical polarization in porous media. At low-frequencies (<1 MHz), the so-called α-polarization prevails (see Revil (2012)); this is due to the polarization of the Stern layer. Finally, attenuation of the seismic waves associated with squirt-flow dissipation mechanisms will be neglected despite the fact that this mechanism is known to control the attenuation of seismic waves in the frequency band usually used in the field for seismic investigations (see Rubino and Holliger (2012)).

In saturated conditions, the (averaged) filtration displacement is defined (see Morency and Tromp (2008)) as

$$\mathbf{w} = \phi(\mathbf{u}_w - \mathbf{u}), \tag{8.55}$$

where ϕ denotes the connected porosity, $\mathbf{u}_w$ and $\mathbf{u}$ correspond to the averaged displacement of the water and solid phases, respectively. All the disturbances considered below will be harmonic $\exp(-i\omega t)$, where $\omega = 2\pi f$ denotes the angular frequency (in rad s^{-1}) and $f = \omega/(2\pi)$ is the frequency (in hertz).

In saturated conditions, the Darcy velocity (filtration velocity) is defined as the time derivative of the filtration displacement. In water-saturated conditions, the generalized Darcy's law is given by Jardani *et al.* (2010) as

$$\dot{\mathbf{w}} = -\frac{k_0}{\eta_f}\left(\nabla p + \rho_f\ddot{\mathbf{u}} + \rho_f\frac{\alpha}{\phi}\ddot{\mathbf{w}} - \mathbf{F}_f\right), \tag{8.56}$$

where k_0 denotes the quasi-static permeability of the porous material (in m^2), $F = \alpha/\phi$ denotes the electrical formation factor (dimensionless), which is the ratio of the bulk tortuosity α of the pore space to the connected porosity and $\mathbf{F}_f$ denotes the body force applied to the pore water phase (in N m^{-3}, e.g. the gravitational body force or the electrical force acting on the excess of electrical charges of the pore water). To keep the notation as light as possible, we will not distinguish below the variables expressed in the time domain or in the frequency domain, but it is easy enough to recognize if the equations are written in the frequency domain or in the time domain, the switch from one domain to the other being done by a simple Fourier transform or its associated inverse Fourier transform.

In unsaturated conditions, the filtration displacement and the mass density of the fluid (subscript f) phase are given by

$$\mathbf{w}_w = s_w\phi(\mathbf{u}_w - \mathbf{u}), \tag{8.57}$$

$$\rho_f = (1 - s_w)\rho_g + s_w\rho_w, \tag{8.58}$$

respectively. In these equations s_w denotes the water saturation ($s_w = 1$ at saturation). The mass density of the gas phase can be neglected and therefore $\rho_f \approx s_w\rho_w$. From Eq. (8.55) the porosity can be replaced by $s_w\phi$ (of course, terms in $(1 - \phi)$, dealing with the solid phase, remain unchanged). The Darcy velocity associated with the water phase is given by,

$$\dot{\mathbf{w}}_w = -\frac{k_r k_0}{\eta_w}\left(\nabla p_w + \rho_w s_w\ddot{\mathbf{u}} + \rho_w s_w\frac{\alpha_w}{\phi}F\ddot{\mathbf{w}}_w - \mathbf{F}_w\right), \tag{8.59}$$

where k_r denotes the relative permeability (dimensionless), η_w denotes the dynamic viscosity of the pore water (in Pa s) and P_w is the pressure of the water phase (it will be replaced later by the suction head defined above). The ratio α_w/ϕ corresponds to the bulk tortuosity of the water phase, α_w, divided by the connected porosity. This ratio can be replaced by Archie's second law $F s_w^n$ where n is called the second Archie's exponent ($n > 1$, dimensionless) (see, for example, Archie (1942) and Revil *et al.* (2007)). From now, the constitutive equations are described in the frequency domain. Therefore $\ddot{\mathbf{w}}_w$ is replaced by $-i\omega\dot{\mathbf{w}}_w$ and so on. The Darcy equation, Eq. (8.59), can be rewritten as

$$-i\omega\mathbf{w}_w = -\frac{k^*(\omega)}{\eta_w}(\nabla p_w - \omega^2\rho_w s_w \mathbf{u} - \mathbf{F}_w),\qquad(8.60)$$

where $k^*(\omega)$ is a complex-valued permeability given by

$$k^*(\omega) = \frac{k_r k_0}{1 - i\omega\tau_k},\qquad(8.61)$$

and where the relaxation time is given by,

$$\tau_k = \frac{k_r k_0 \rho_w F}{\eta_w} s_w^{1-n}.\qquad(8.62)$$

The relaxation time τ_k represents the transition between the viscous laminar flow regime and the inertial laminar flow regime. The critical frequency associated with this relaxation time is given by,

$$f_k = \frac{1}{2\pi\tau_k} = \frac{\eta_w}{2\pi k_r k_0 \rho_w F} s_w^{n-1}.\qquad(8.63)$$

Note that because $n \geq 1$ (see Archie (1942)), $n - 1 \geq 0$. Note the measurement of this relaxation frequency can be used to estimate therefore the permeability of the material.

The two flow regimes in porous media are sketched in Figure 8.5. Low frequencies ($\omega\tau_k \ll 1, f \ll f_k$) corresponds to the viscous laminar flow regime where the flow in a cylindrical pore obeys the Poiseuille law (Figure 8.5c). High frequencies ($\omega\tau_k \gg 1, f \gg f_k$) corresponds to the inertial laminar flow regime for which the pore water flow is a potential flow problem. Some authors (such as Biot in his earlier works) prefer to include the inertial effect in an apparent (or effective) dynamic viscosity $\eta^*(\omega)$ rather than in defining an apparent (or effective) permeability $k^*(\omega)$. This choice (that is arbitrary) has, however, been abandoned more recently.

In poroelasticity, it is customary to define the following two variables (see Morency and Tromp (2008));

$$b = \frac{\eta_w}{k_0},\qquad(8.64)$$

$$\tilde{\rho}_w = \rho_w F,\qquad(8.65)$$

(a) Thick double layer

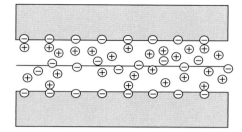

(b) Thin double layer

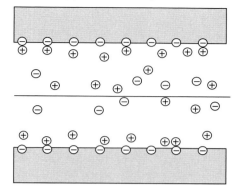

(c) Flow in the viscous laminar regime

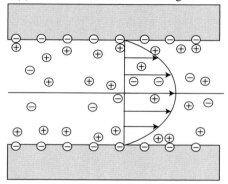

(d) Flow in the inertial laminar regime

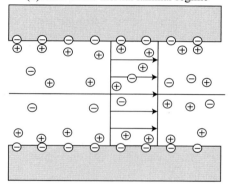

Figure 8.5. Sketch of the charge distribution and flow regime. There are four cases to consider depending on the pore size with respect to the double layer thickness and depending on the frequency. (a) Thick double layer. The counterions of the diffuse layer are uniformly distributed in the pore space. (b) Thin double layer (the thickness of the diffuse layer is much smaller than the size of the pores). (c) Viscous laminar flow regime occurring at low frequencies. (d) Inertial laminar flow regime occurring at high frequencies.

where $\tilde{\rho}_w$ denotes an apparent pore water mass density. The relationship between the permeability and the water saturation can be expressed with the Brooks and Corey (1964) relationship,

$$k_r = s_w^{\frac{2+3\lambda}{\lambda}},\tag{8.66}$$

where λ is termed the Brook and Corey exponent. Using Eq. (8.62) and Eqs. (8.64)–(8.66), the following relationship between the relaxation time and the saturation is obtained,

$$\tau_k = s_w^{\frac{2+3\lambda}{\lambda}+1-n}\frac{\tilde{\rho}_w}{b}.\tag{8.67}$$

In order to write a hydrodynamic equation coupled with the electrical field, the body force $\mathbf{F}_w$ entering Eq. (8.60) should be expressed by Coulomb's law,

$$\mathbf{F}_w = \hat{Q}_V{}^*(\omega)\mathbf{E}, \tag{8.68}$$

where $\hat{Q}_V{}^*(\omega)$ denotes the frequency dependent (effective) excess of charge that can be dragged by the flow of the pore water through the pore space of the material (dynamic excess charge density of the pore space, see Chapter 3, Figure 3.12) and $\mathbf{E}$ denotes the electrical field, in V m^{-1}. The charge density $\hat{Q}_V{}^*(\omega)$ is frequency dependent because there is more charge dragged in the inertial laminar flow regime than in the viscous laminar flow regime, in agreement with the model of Pride (1994). In the following, the parameters $\hat{Q}_V^0$ and $\hat{Q}_V^\infty$ are the volumetric charge density dragged in the low ($\omega\tau_k \ll 1$) and high ($\omega\tau_k \gg 1$) frequency regimes, respectively. Because the transition between low and high frequency regimes is governed by the relaxation time τ_k, the following functional can be used to compute the effective charge density as a function of the frequency:

$$\frac{1}{\hat{Q}_V{}^*(\omega)} = \frac{1}{\hat{Q}_V^\infty} + \left(\frac{1}{\hat{Q}_V^0} - \frac{1}{\hat{Q}_V^\infty}\right)\frac{1}{\sqrt{1 - i\omega\tau_k}}. \tag{8.69}$$

The form of this function is derived and explained further in Revil and Mahardika (2013). We need to find expressions for the low and high frequency charge densities, $\hat{Q}_V^0$ and $\hat{Q}_V^\infty$, respectively. We note the following.

(1) At low frequencies, only a small fraction of the counterions of the diffuse layer are dragged by the flow of the pore water and therefore $\hat{Q}_V^0 \ll \hat{Q}_V^\infty$. An expression to compute $\hat{Q}_V^0$ from the low frequency permeability k_0 is discussed further below.

(2) At high frequencies, all the charge density existing in the pores is uniformly dragged along the pore water flow and therefore the charge density $\hat{Q}_V^\infty$ is also equal to the volumetric charge density of the diffuse layer. An expression to compute $\hat{Q}_V^\infty$ from the cation exchange capacity is discussed further below.

Depending on the size of the electrical double layer with respect to the size of the pores, two cases can be considered.

(1) In the thick double layer approximation (see Figure 8.5a), $\hat{Q}_V^\infty \approx \hat{Q}_V^0$ (all the counterions of the diffuse layer are dragged by the flow whatever the frequency) and, therefore,

$$\hat{Q}_V{}^*(\omega, s_w) \approx \frac{\hat{Q}_V^0}{s_w}. \tag{8.70}$$

(2) In the thin double layer approximation (see Figure 8.5b), one can expect $\hat{Q}_V^\infty \gg \hat{Q}_V^0$ (see examples in Jougnot et al. (2012)). Therefore, we have,

$$\hat{Q}_V{}^*(\omega, s_w) \approx \hat{Q}_V^0 s_w^{-1}\sqrt{1 - i\omega\tau_k(s_w)}. \tag{8.71}$$

Introducing Coulomb's law, Eq. (8.68), into the Darcy equation, Eq. (8.60), yields the following form of Darcy's law,

$$-i\omega\mathbf{w}_w = -\frac{k^*(\omega)}{\eta_w}(\nabla p_w - \omega^2\rho_w s_w\mathbf{u}) + \frac{k^*(\omega)\hat{Q}_V{}^*(\omega)}{\eta_w}\mathbf{E}. \tag{8.72}$$

This equation shows the influence of three forcing terms on the Darcy velocity: (i) the pore fluid pressure gradient, (ii) the displacement of the solid framework, and (iii) the electrical field through electro-osmosis.

We investigate now the macroscopic electrical current density $\mathbf{J}$. The first contribution to $\mathbf{J}$ is the conduction current density given by Ohm's law,

$$\mathbf{J}_e = \sigma^*(\omega)\mathbf{E}, \tag{8.73}$$

where the conductivity $\sigma^*(\omega)$ denotes the complex conductivity. For clayey materials, this conductivity has been modeled recently by Revil (2012).

The second contribution to the total current density corresponds to the advective drag of the excess of charge of the pore space by the flow of the pore water (contribution of advective nature). If the Darcy velocity associated with the poromechanical contribution is written as $\dot{\mathbf{w}}_w^m$, the second contribution to the current density is given by

$$\mathbf{J}_m = -i\omega\hat{Q}_V{}^*(\omega)\mathbf{w}_w^m. \tag{8.74}$$

The mechanical contribution to the filtration displacement is given by the generalized Darcy's law derived above,

$$-i\omega\mathbf{w}_w^m = -\frac{k^*(\omega)}{\eta_w}(\nabla p_w - \omega^2\rho_w s_w\mathbf{u}). \tag{8.75}$$

The total current density is given by the sum of the conductive and advective contributions, which yields the following generalized Ohm's law,

$$\mathbf{J} = \sigma^*(\omega)\mathbf{E} - \frac{k^*(\omega)\hat{Q}_V{}^*(\omega)}{\eta_w}(\nabla p_w - \omega^2\rho_w s_w\mathbf{u}). \tag{8.76}$$

A model for the complex conductivity is now required. The complex conductivity is first written as

$$\sigma^* = \sigma' + i\sigma'' = |\sigma|\exp(i\varphi), \tag{8.77}$$

where $|\sigma^*| = (\sigma'^2 + \sigma''^2)^{1/2}$ denotes the magnitude of the conductivity and $\varphi = \text{atan}(\sigma'/\sigma'')$ denotes the phase lag between the electrical current and the resulting electrical field. For clayey materials, the frequency dependence of the complex conductivity is usually very small and can be neglected in the frequency range 0.1 Hz to 10 kHz (see Vinegar and Waxman (1984) and Revil (2012), for extensive discussions). This is important for field applications as this frequency range is typically the frequency range used in the field. The following linear model is used to describe the in-phase electrical conductivity as a function of the pore water conductivity σ_f (see Chapter 3):

$$\sigma' = \frac{1}{F}s_w^n\sigma_f + \sigma_S, \tag{8.78}$$

where F denotes the formation factor introduced above (ratio of the pore space tortuosity by the connected porosity), n is the second Archie's exponent, and σ_S denotes the surface conductivity. If a linear model between the conductivity of the material and the pore water conductivity is used, the surface conductivity is given by the model developed recently by Revil (2012):

$$\sigma_S = \frac{A(\phi, m)}{F}s_w^{n-1}\beta_{(+)}\hat{Q}_V^\infty, \tag{8.79}$$

$$\sigma_S = \frac{A(\phi, m)}{F}s_w^{n-1}\beta_{(+)}(1-f)Q_V, \tag{8.80}$$

where $A(\phi, m)$ is defined by,

$$A(\phi, m) = m(F-1)\frac{2}{3}\left(\frac{\phi}{1-\phi}\right). \tag{8.81}$$

Equation (8.80) means that the surface conductivity is controlled by the diffuse layer with a fraction of counterions $(1-f)$ (see Appendix A) and a mobility of the counterions $\beta_{(+)}$ that is equal to the mobility of the cations in the bulk pore water; see Revil (2012). Equations (8.78)–(8.81) can be seen as a variant of the Waxman and Smits (1968) model, which is known to be pretty efficient in analyzing downhole resistivity measurements in shaly sand reservoirs.

Following Revil (2012), the quadrature conductivity can be expressed as

$$\sigma'' = -\frac{A(\phi, m)}{F}s_w^p\beta_{(+)}^S f\, Q_V, \tag{8.82}$$

$$\sigma'' = -\frac{A(\phi, m)}{F}s_w^p\beta_{(+)}^S\left(\frac{f}{1-f}\right)\hat{Q}_V^\infty, \tag{8.83}$$

where $p = n - 1$ and $\beta_{(+)}^S$ denotes the mobility of the counterions in the Stern layer. These equations provide a simple and accurate model to describe the complex conductivity of shaly sands and soils and generalize the Vinegar and Waxman

(1984) model. As noted by Vinegar and Waxman (1984) and Revil (2012), the frequency dependence of the quadrature conductivity is not explicit in Eq. (8.83). It should be mentioned that for quasi-static conditions, the quadrature conductivity should go to zero in d.c. conditions,

$$\lim_{\omega \to 0} \sigma''(\omega) = 0. \tag{8.84}$$

The typical frequency below which the quadrature conductivity becomes frequency dependent is typically smaller than 0.1 Hz (see Revil (2012)) and is therefore not relevant to the seismoelectric problem.

The two constitutive equations for the generalized Ohm and Darcy laws are written as:

$$\begin{bmatrix} \mathbf{J} \\ -i\omega\mathbf{w}_w \end{bmatrix} = \begin{bmatrix} \sigma^* & L^*(\omega, s_w) \\ L^*(\omega, s_w) & \dfrac{k^*(\omega, s_w)}{\eta_w} \end{bmatrix} \begin{bmatrix} \mathbf{E} \\ -(\nabla p_w - \omega^2 \rho_w s_w \mathbf{u}) \end{bmatrix}, \tag{8.85}$$

where the coefficient $L^*(\omega)$ is defined as

$$L^*(\omega, s_w) = \frac{k^*(\omega, s_w)\hat{Q}_V^*(\omega, s_w)}{\eta_w}. \tag{8.86}$$

The generalized streaming potential coupling coefficient is defined by the following equations in the quasi-static limit of the Maxwell equations:

$$\nabla \times \mathbf{E} = 0 \Longrightarrow \mathbf{E} = -\nabla\psi, \tag{8.87}$$

$$C_p^*(\omega, s_w) = \left(\frac{\partial\psi}{\partial p_w}\right)_{\mathbf{j}=0;\ddot{\mathbf{u}}=0} = -\frac{L^*(\omega, s_w)}{\sigma^*(\omega, s_w)}, \tag{8.88}$$

$$C_p^*(\omega, s_w) = -\frac{k^*(\omega, s_w)\hat{Q}_V^*(\omega, s_w)}{\eta_w \sigma^*(\omega, s_w)}. \tag{8.89}$$

More explicitly, in the thin double layer approximation, the streaming potential coupling coefficient is given by

$$C_p^*(\omega, s_w) \approx \frac{C_0(s_w)}{\sqrt{1 - i\omega\tau_k}}, \tag{8.90}$$

$$C_0 = \lim_{\omega \to 0} C_p(\omega, s_w) \approx \frac{k_r k_0 \hat{Q}_V^0}{\eta_w s_w \left(\dfrac{1}{F} s_w^n \sigma_f + \sigma_S\right)}. \tag{8.91}$$

Similarly, the generalized electro-osmotic coupling coefficient is defined in the quasi-static limit of the Maxwell equations by the change of pore fluid pressure when the skeleton is at rest and in absence of pore fluid flow by the ratio between the gradient of pore fluid pressure divided by the gradient in electrical potential. For the thin double layer case, this yields

$$C_{os}^{*}(\omega, s_w) = \left(\frac{\partial p_w}{\partial \varphi} \right)_{\dot{w}=0; \ddot{u}=0} = -\frac{L^{*}(\omega, s_w) \eta_w}{k^{*}(\omega, s_w)}, \tag{8.92}$$

$$C_{os}^{*}(\omega, s_w) = -\hat{Q}_V^{*}(\omega), \tag{8.93}$$

$$C_{os}^{*}(\omega, s_w) \approx -\hat{Q}_V^{0} s_w^{-1} \sqrt{1 - i\omega \tau_k}. \tag{8.94}$$

For the thick double layer case, the electro-osmotic coupling coefficient is given by $C_{os}^{*}(\omega) \approx -\hat{Q}_V^{0}/s_w$. Therefore, the electro-osmotic coupling coefficient is simply a measure of the excess of charges that can be moved by the flow of the pore water. In unsaturated flow conditions, it is customary to use the capillary head gradient $\nabla \Psi = \nabla p_w / \rho_w g$ instead of the pore water pressure gradient. Below the relaxation frequency separating the low frequency viscous laminar flow regime from the high frequency inertial laminar flow regime, Eq. (8.85) can be rewritten in the time domain using the pressure head and including the gravitational field in the hydraulic driving force. This yields

$$\begin{bmatrix} \mathbf{J} \\ \dot{\mathbf{w}}_w \end{bmatrix} = - \begin{bmatrix} \sigma^{*} & L^{*}(s_w) \\ L^{*}(s_w) & \dfrac{k_r(s_w)k_0}{\eta_w} \end{bmatrix} \begin{bmatrix} \nabla \psi \\ \rho_w g \nabla(\Psi + z) + \rho_w s_w \ddot{\mathbf{u}} \end{bmatrix}, \tag{8.95}$$

where the quasi-static approximation coupling coefficient L^{*} is given by

$$L^{*}(s_w) = \frac{k_r(s_w)k_0 \hat{Q}_V^{0}}{\eta_w s_w}. \tag{8.96}$$

Equations (8.95) and (8.96) therefore provide a simple model for the occurrence of streaming potential and electro-osmosis in porous media.

8.5.2 Description of the hydromechanical model

The starting point is the following Biot constitutive equation in saturated conditions:

$$-p = C_s \nabla \cdot \mathbf{u} + M_s \nabla \cdot \mathbf{w}. \tag{8.97}$$

Equation (8.97) is also often written as

$$-p = C_s \varepsilon_{kk} - M_s \zeta, \tag{8.98}$$

where $\varepsilon_{kk} = \nabla \cdot \mathbf{u} = \delta V / V$ (where V denotes the volume of the representative elementary volume) denotes the volumetric strain of the porous body and $\zeta = -\nabla \cdot \mathbf{w} = \phi(\nabla \cdot \mathbf{u} - \nabla \cdot \mathbf{u}_w)$ denotes the linearized increment of fluid content (see

Lo et al. (2002)). The parameter ζ represents the fractional volume of water flowing in or out of the representative volume of skeleton in response to an applied stress. The bulk moduli M_s and C_s, in saturated conditions, are defined as

$$\frac{1}{C_s} \equiv - \left(\frac{d\varepsilon_{kk}}{dp_w} \right)_\zeta = \frac{1 + \Delta}{K_f + K_S \Delta}, \tag{8.99}$$

$$\frac{1}{M_s} \equiv \left(\frac{d\zeta}{dp_w} \right)_{\varepsilon_{kk}} = \phi \left(\frac{1 + \Delta}{K_f} \right), \tag{8.100}$$

where $\alpha \equiv 1 - K_{fr}/K_S$ denotes the Biot coefficient in the saturated state and Δ is defined by $\Delta = (K_f/\phi K_S^2)[(1 - \phi)K_S - K_{fr}]$. The Biot modulus M corresponds to the inverse of the poroelastic component of the specific storage and is defined as the increase of the amount of fluid (per unit volume of rock) as a result of a unit increase of pore pressure, under constant volumetric strain. The following relationship is obtained: $C_s = \alpha M_s$.

To extend these equations to the unsaturated case, we apply the classical change of variables discussed above, that is $\rho_f \Rightarrow s_w \rho_w$, $p \Rightarrow p_w$, $\phi \Rightarrow s_w \phi$, and $\mathbf{w} \Rightarrow \mathbf{w}_w = s_w \phi (\mathbf{u}_w - \mathbf{u})$. This yields

$$\frac{1}{C} = \frac{1 + \Delta}{K_f + K_S \Delta}, \tag{8.101}$$

$$\frac{1}{M} = \theta \left(\frac{1 + \Delta}{K_f} \right), \tag{8.102}$$

where $\theta = s_w \phi$ denotes the water content (dimensionless) and where, in unsaturated conditions, the fluid increment is defined by

$$\zeta = -\nabla \cdot \mathbf{w}_w = \phi s_w (\nabla \cdot \mathbf{u} - \nabla \cdot \mathbf{u}_w). \tag{8.103}$$

Note that the term $\Delta = (K_f/\phi K_S^2)[(1 - \phi)K_S - K_{fr}]$ depends on the saturation because the compressibility of the fluid is given by the Wood formula $(1/K_f) = (1 - s_w)/K_a + s_w/K_w$ (see Wood (1955) and Teja and Rice (1981)), where K_a and K_w represent the bulk moduli of air and water, respectively ($K_a = 0.145$ MPa and $K_w = 2.25$ GPa; see Lo et al. (2005)). In unsaturated condition, from Eqs. (8.101) and (8.102), $C = \alpha_w M$ and the Biot coefficient α_w is given by $\alpha_w = s_w \alpha$, where α denotes the Biot coefficient in saturated conditions. That said, there should be no exchange of water below the irreducible water saturation and therefore the correct scaling should be $\alpha_w = s_e \alpha$ rather than $\alpha_w = s_w \alpha$, where s_w denotes the reduced water saturation s_e is the effective saturation, that is related to the relative saturation of the water phase by $s_e = (s_w - s_r)/(1 - s_r)$, where s_r denotes the irreducible water saturation. In other words $(1/M) \to 0$ as $s_w \to s_r$.

Generalizing Eq. (8.97) to unsaturated conditions yields,

$$\left(\frac{1}{M} + \frac{\partial \theta}{\partial p_w}\right)(p_a - p_w) = \frac{C}{M}\nabla \cdot \mathbf{u} + \nabla \cdot \mathbf{w}_w, \tag{8.104}$$

where $\partial \theta / \partial p_w$ denotes the specific moisture capacity which is determined from the derivative of the capillary pressure with respect to the water content (in unsaturated flow, the air pressure is kept constant). The filtration displacement of the water phase is given by

$$-i\omega \mathbf{w}_w = -\frac{k^*(\omega)}{\eta_w}(\nabla p_w - \omega^2 \rho_w s_w \mathbf{u} - \mathbf{F}_w). \tag{8.105}$$

Therefore the filtration displacement is given by

$$\mathbf{w}_w = \frac{k^*(\omega)}{i\omega \eta_w}(\nabla p_w - \omega^2 \rho_w s_w \mathbf{u} - \mathbf{F}_w). \tag{8.106}$$

Equations (8.104) and (8.106) yield

$$\left(\frac{1}{M} + \frac{\partial \theta}{\partial p_w}\right)(p_a - p_w) = \alpha_w \nabla \cdot \mathbf{u} + \nabla \cdot \left[\frac{k^*(\omega)}{i\omega \eta_w}(\nabla p_w - \omega^2 \rho_w s_w \mathbf{u} - \mathbf{F}_w)\right], \tag{8.107}$$

where the relationship $C/M = \alpha_w$ has been used. Equation (8.107) is a non-linear diffusion equation for the fluid pressure. For this to be obvious, the terms of this equation need to be reworked. Multiplying all the terms by $(i\omega \eta_w / k^*(\omega))$, separating the pressure terms in the left-hand side from the source term in the right-hand side and taking into consideration that the air pressure is constant (unsaturated flow assumption), the following non-linear hydraulic diffusion equation is obtained:

$$\nabla \cdot \left[\frac{k^*(\omega)}{\eta_w}\nabla p_w\right] + i\omega \left(\frac{1}{M} + \frac{\partial \theta}{\partial p_w}\right)(p_w - p_a)$$

$$= \nabla \cdot \left[\frac{k^*(\omega)}{\eta_w}(\omega^2 \rho_w s_w \mathbf{u})\right] - i\omega \alpha_w \nabla \cdot \mathbf{u} + \nabla \cdot \left(\frac{k^*(\omega)}{\eta_w}\mathbf{F}_w\right). \tag{8.108}$$

This equation may be written in the time domain using an inverse Fourier transform. Assuming that the permeability is given by its low frequency asymptotic limit (which is correct below 10 kHz) and using Coulomb law plus the gravity force as body force (the frequency-dependent volumetric charge density is also taken in its low frequency limit too) yields

$$
\nabla \cdot \left[\frac{k_r k_0}{\eta_w} \nabla p_w \right] - \left(\frac{1}{M} + \frac{\partial \theta}{\partial p_w} \right) \dot{p}_w
$$

$$
= -\nabla \cdot \left[\frac{k_r k_0 \rho_w s_w}{\eta_w} \ddot{u} \right] + \alpha_w \nabla \cdot \dot{u} + \nabla \cdot \left[\frac{k_r k_0}{\eta_w} \left(\frac{\hat{Q}_V^0}{s_w} \mathbf{E} + \rho_w g \right) \right], \quad (8.109)
$$

where $\dot{\Theta} = \partial \Theta / \partial t$ denotes the time derivative of the parameter Θ and t represents time. The origin of the three forcing terms on the right-hand side of this equation is now clearly established: the first term is related to the acceleration of the seismic wave acting on the skeleton of the material, the second term is due to the velocity of the seismic wave, and the third term (at constant gravity acceleration) corresponds to the pore water flow associated with the electro-osmotic forcing associated with the drag of the pore water by the electromigration of the excess of charge contained in the pore space of the material.

Another possibility is to write a generalized Richards equation (see Richards (1931)) showing the influence of the forcing term in this equation (we assume again that the air pressure is constant). Starting with Eq. (8.109) and replacing the water pressure by the capillary head defined by $\Psi = (p_w - p_a)/\rho_w g$ (in m), the following Richards equation is obtained

$$
\left(\frac{\rho_w g}{M} + \frac{\partial \theta}{\partial \Psi} \right) \dot{\Psi} + \nabla \cdot [-K_h \nabla (\Psi + z)]
$$

$$
= \nabla \cdot \left[K_h \frac{s_w}{g} \ddot{u} \right] - \alpha_w \nabla \cdot \dot{u} - \nabla \cdot \left(\frac{K_h}{\rho_w g} \frac{\hat{Q}_V^0}{s_w} \mathbf{E} \right), \quad (8.110)
$$

$$
C_e \frac{\partial \Psi}{\partial t} + \nabla \cdot [-K_h (\nabla \Psi + 1)] = -\alpha_w \nabla \cdot \dot{u} + \nabla \cdot \left[K_h \left(\frac{s_w}{g} \ddot{u} - \frac{\hat{Q}_V^0}{s_w \rho_w g} \mathbf{E} \right) \right],
$$

$$
(8.111)
$$

$$
K_h = \frac{k_r k_0 \rho_w g}{\eta_w} = k_r K_s, \quad (8.112)
$$

where $\alpha_w = s_e \alpha = s_e (1 - K_{fr}/K_S)$ and where $K = K_{fr}$ denotes the bulk modulus of the skeleton (drained bulk modulus), K_h denotes the hydraulic conductivity (in m s^{-1}), K_s denotes the hydraulic conductivity at saturation, z denotes the elevation above a datum, and C_e denotes the specific storage term. This storage term is the sum of the specific moisture capacity (in m^{-1}) (also called the water capacity function) and the specific storage corresponding to the poroelastic deformation of the material. This yields

$$
C_e = \frac{\partial \theta}{\partial \Psi} + \frac{\rho_w g}{M}. \quad (8.113)
$$

Usually in unsaturated conditions, the poroelastic term is much smaller than the specific moisture capacity but the poroelastic term should be kept to have a formulation that remains consistent with the saturated state. The hydraulic conductivity is related to the relative permeability k_r and K_0, the hydraulic conductivity at saturation. With the Brooks and Corey (1964) model, the porous material is saturated when the fluid pressure reaches the atmospheric pressure ($\psi = 0$ at the water table). The effective saturation, the specific moisture capacity, the relative permeability, and the water content are defined by

$$
s_e = \begin{cases} (\alpha_b \Psi)^{-\lambda}, & \Psi \leq 1/\alpha_b \\ 1, & \Psi > 1/\alpha_b \end{cases}
\tag{8.114}
$$

$$
\frac{\partial \theta}{\partial \Psi} = \begin{cases} -\lambda \alpha_b (\phi - \theta_r)(\alpha_b \Psi)^{-(\lambda+1)}, & \Psi \leq 1/\alpha_b \\ 0, & \Psi > 1/\alpha_b \end{cases}
\tag{8.115}
$$

$$
k_r = \begin{cases} s_e^{\frac{2+3\lambda}{\lambda}} = (\alpha_b \Psi)^{+(2+3\lambda)}, & \Psi \leq 1/\alpha_b \\ 1, & \Psi > 1/\alpha_b \end{cases}
\tag{8.116}
$$

$$
\theta = \begin{cases} \theta_r + s_e (\phi - \theta_r), & \Psi \leq 1/\alpha_b \\ \phi, & \Psi > 1/\alpha_b \end{cases},
\tag{8.117}
$$

respectively, where α_b denotes the inverse of the capillary entry pressure related to the matric suction at which pore fluid begins to leave a drying soil water system, λ is called the pore size distribution index (a textural parameter), and θ_r represents the residual water content ($\theta_r = s_r \phi$). Sometimes the residual water saturation is not accounted for and the capillary pressure curve and the relative permeability are written as

$$
s_w = \begin{cases} \left(\dfrac{p_c}{p_e}\right)^{-\lambda}, & p_c > p_e \\ 1, & p_c \leq p_e \end{cases}
\tag{8.118}
$$

$$
k_r = \begin{cases} s_w^{\frac{2+3\lambda}{\lambda}}, & p_c > p_e \\ 1, & p_c \leq p_e \end{cases}.
\tag{8.119}
$$

Because $\alpha_w = s_e \alpha = s_e(1 - K_{fr}/K_S)$, when the water saturation reaches the irreducible water saturation, the two source terms on the right-hand side of Eq. (8.111) are null. Therefore there is no possible excitation below the irreducible water saturation. In reality, this is not necessarily true, and the model should be completed by including film flow below the irreducible water saturation.

The hydromechanical equations are defined in terms of an effective stress tensor. As explained in detail in Jardani *et al.* (2010) and Revil and Jardani (2010b), there is a computational advantage in expressing the coupled hydromechanical problem in terms of the fluid pressure and displacement of the solid phase (four unknowns in total) rather than using the displacement of the solid and filtration displacement (six unknowns in total).

In saturated conditions, Newton's law (which is a momentum conservation equation for the skeleton partially filled with its pore water) is written as

$$\nabla \cdot \overline{\overline{T}} + \mathbf{F} = -\omega^2(\rho\mathbf{u} + \rho_f\mathbf{w}), \tag{8.120}$$

where $\overline{\overline{T}}$ is the total stress tensor (positive normal stress implies tension; see Detournay and Cheng (1993)) and $\mathbf{F}$ denotes the total body force applied to the porous material. In unsaturated conditions, Newton's law is written as

$$\nabla \cdot \overline{\overline{T}} + \mathbf{F} = -\omega^2(\rho\mathbf{u} + s_w\rho_w\mathbf{w}_w), \tag{8.121}$$

where the filtration displacement of the pore water phase is given by

$$\mathbf{w}_w = \frac{k^*(\omega)}{i\omega\eta_w}(\nabla p_w - \omega^2\rho_w s_w\mathbf{u} - \mathbf{F}_w). \tag{8.122}$$

Combining Eqs. (8.121) and (8.122) yields

$$\nabla \cdot \overline{\overline{T}} + \mathbf{F} = -\omega^2\left[\rho\mathbf{u} + s_w\rho_w\frac{k^*(\omega)}{i\omega\eta_w}(\nabla p_w - \omega^2\rho_w s_w\mathbf{u} - \mathbf{F}_w)\right], \tag{8.123}$$

$$\nabla \cdot \overline{\overline{T}} + \mathbf{F} = -\omega^2\rho_\omega^S\mathbf{u} - \omega^2\left[s_w\rho_w\frac{k^*(\omega)}{i\omega\eta_w}(\nabla p_w - \mathbf{F}_w)\right], \tag{8.124}$$

where

$$\rho_\omega^S = \rho - (s_w\rho_w)^2\frac{k^*(\omega)}{i\omega\eta_w}\omega^2. \tag{8.125}$$

The effective stress in unsaturated conditions is taken as

$$\overline{\overline{T}}_{eff} = \overline{\overline{T}} + p_a\overline{\overline{I}} + s_e\alpha(p_w - p_a)\overline{\overline{I}}, \tag{8.126}$$

which is consistent with the Bishop effective stress principle in unsaturated conditions and the Biot stress principle in saturated conditions. The confining pressure and the effective confining pressure are defined as

$$P = -\frac{1}{3}\mathrm{Trace}\overline{\overline{T}}, \tag{8.127}$$

$$P_{eff} = -\frac{1}{3}\mathrm{Trace}\overline{\overline{T}}_{eff}, \tag{8.128}$$

respectively. This yields $P_{eff} = P - p_a - s_e\alpha(p_w - p_a)$. Equations (8.124) and (8.126) yield

$$\nabla \cdot \overline{\overline{T}}_{eff} + \mathbf{F} - \omega^2 s_w \rho_w \frac{k^*(\omega)}{i\omega\eta_w} \mathbf{F}_w = -\omega^2 \rho_\omega^S \mathbf{u} + \theta_\omega \nabla p_w, \qquad (8.129)$$

where the hydromechanical coupling term θ_ω is defined by

$$\theta_\omega = s_w \left(\alpha - \omega^2 \rho_w \frac{k^*(\omega)}{i\omega\eta_w} \right). \qquad (8.130)$$

The last fundamental constitutive equation needed to complete the hydromechanical model in unsaturated conditions is a relationship between the total stress tensor (or the effective stress tensor) and the displacement of the solid phase and filtration displacement of the pore water phase. This equation is Hooke's law, which, in linear poroelasticity and for saturated conditions, is given by

$$\overline{\overline{T}} = (\lambda_u \nabla \cdot \mathbf{u} + C\nabla \cdot \mathbf{w})\overline{\overline{I}} + G(\nabla\mathbf{u} + \nabla\mathbf{u}^T), \qquad (8.131)$$

where $\overline{\overline{\varepsilon}} = (1/2)(\nabla\mathbf{u} + \nabla\mathbf{u}^T)$ denotes the deformation tensor, $G = G_{fr}$ denotes the shear modulus that is equal to the shear modulus of the skeleton (frame), and $\lambda_u = K_u - (2/3)G$ denotes the Lamé modulus in undrained conditions (K_u denotes the undrained bulk modulus). In unsaturated conditions and accounting for the air pressure, Eq. (8.129) can be written as

$$\overline{\overline{T}} + p_a\overline{\overline{I}} = (\lambda_u \nabla \cdot \mathbf{u} + C\nabla \cdot \mathbf{w}_w)\overline{\overline{I}} + G(\nabla\mathbf{u} + \nabla\mathbf{u}^T). \qquad (8.132)$$

From Eq. (8.104) above, the linearized increment of fluid content is given by

$$-\nabla \cdot \mathbf{w}_w = \frac{1}{M}(p_w - p_a) + \frac{C}{M}\nabla \cdot \mathbf{u}. \qquad (8.133)$$

Combining Eqs. (8.132) and (8.133) yields

$$\overline{\overline{T}} + p_a\overline{\overline{I}} = \left[\left(K_u - \frac{2}{3}G \right) \nabla \cdot \mathbf{u} + C \left(-\frac{1}{M}(p_w - p_a) - \alpha_w \nabla \cdot \mathbf{u} \right) \right] \overline{\overline{I}}$$
$$+ G(\nabla\mathbf{u} + \nabla\mathbf{u}^T), \qquad (8.134)$$

$$\overline{\overline{T}} + p_a\overline{\overline{I}} = \left[\left(K_u - \alpha_w C - \frac{2}{3}G \right) \nabla \cdot \mathbf{u} - \alpha_w(p_w - p_a) \right] \overline{\overline{I}}$$
$$+ G(\nabla\mathbf{u} + \nabla\mathbf{u}^T), \qquad (8.135)$$

where the following expression, derived above, has been used for the Biot coefficient in unsaturated conditions:

$$\frac{C}{M} = \alpha_w = s_e\alpha. \qquad (8.136)$$

In addition, the bulk modulus is given by $K = K_u - \alpha_w C$ and the Lamé modulus is given by $\lambda = K - (2/3)G$. Equation (8.135) yields the following Hooke's law for the effective stress given by

$$\bar{\bar{T}} + p_a \bar{\bar{I}} + \alpha_w (p_w - p_a)\bar{\bar{I}} = \lambda (\nabla \cdot \mathbf{u})\bar{\bar{I}} + G(\nabla \mathbf{u} + \nabla \mathbf{u}^T), \qquad (8.137)$$

$$\bar{\bar{T}}_{eff} = \lambda (\nabla \cdot \mathbf{u})\bar{\bar{I}} + G(\nabla \mathbf{u} + \nabla \mathbf{u}^T), \qquad (8.138)$$

where the effective stress is given by Eq. (8.126). This model generalizes the effective stress concept developed by Lu *et al.* (2010).

8.5.3 Maxwell equations in unsaturated conditions

Pride (1994) volume-averaged the local Maxwell equations to obtain a set of macroscopic Maxwell equations in the thin double layer limit. The same equations were obtained by Revil and Linde (2006) for the thick double layer case. The general form of these macroscopic Maxwell equations is

$$\nabla \times \mathbf{E} = -\dot{\mathbf{B}}, \qquad (8.139)$$

$$\nabla \times \mathbf{H} = \mathbf{J} + \dot{\mathbf{D}}, \qquad (8.140)$$

$$\nabla \cdot \mathbf{B} = 0, \quad \mathbf{B} = \nabla \times \mathbf{A}, \qquad (8.141)$$

$$\nabla \cdot \mathbf{D} = \phi \hat{Q}_V^\infty, \qquad (8.142)$$

where $\mathbf{B}$ is the magnetic induction vector, $\mathbf{H}$ is the magnetic field (in A m^{-1}), and $\mathbf{D}$ is the current displacement vector (in C m^{-2}), $\mathbf{A}$ is the magnetic potential vector, and $\mathbf{E} = -\nabla \psi - \dot{\mathbf{A}}$, where ψ denotes the electrostatic potential (in V). These equations are completed by two electromagnetic constitutive equations: $\mathbf{D} = \varepsilon \mathbf{E}$ and $\mathbf{B} = \mu \mathbf{H}$, where ε is the permittivity of the material and μ denotes its magnetic permeability. In absence of magnetized grains, $\mu = \mu_0$, where μ_0 denotes the magnetic permeability of free space.

When the harmonic electrical field is written as $\mathbf{E} = \mathbf{E}_0 \exp(-i\omega t)$, and ω is the angular frequency with $\mathbf{E}_0$ is a constant electrical field magnitude and direction, the displacement current density vector is given by $\mathbf{J}_d = \dot{\mathbf{D}} = -i\omega \varepsilon \mathbf{E}$. The total current density $\mathbf{J}_T$ entering Ampère's law,

$$\nabla \times \mathbf{H} = \mathbf{J}_T, \qquad (8.143)$$

is given by,

$$\mathbf{J}_T = \mathbf{J} + \mathbf{J}_d, \tag{8.144}$$

$$\mathbf{J}_T = \sigma^*(\omega)\mathbf{E} - \frac{k^*(\omega)\hat{Q}_V}{\eta_w}(\nabla p_w - \omega^2 \rho_w s_w \mathbf{u}) - i\omega\varepsilon\mathbf{E}, \tag{8.145}$$

$$\mathbf{J}_T = (\sigma^*(\omega) - i\omega\varepsilon)\mathbf{E} - \frac{k^*(\omega)\hat{Q}_V}{\eta_w}(\nabla p_w - \omega^2 \rho_w s_w \mathbf{u}), \tag{8.146}$$

and an effective conductivity can be introduced, such as $\sigma_{eff}^* = \sigma^*(\omega) - i\omega\varepsilon$, where $\sigma_{eff}^* = \sigma_{eff} + i\omega\varepsilon_{eff}$ is the effective or apparent complex conductivity and σ_{eff} and ε_{eff} are real scalars dependent upon frequency. These effective parameters are the parameters that are measured during an experiment in the laboratory or in the field, but these terms contain both electromigration and true dielectric polarization effects. They are given by $\sigma_{eff} = \sigma'$ and $\varepsilon_{eff} = \sigma''/\omega - \varepsilon$.

8.5.4 Determination of the volumetric charge density

The goal of this section is to provide a way to estimate the two charge densities $\hat{Q}_V^0$ and $\hat{Q}_V^\infty$ used in the previous sections. We first start by the low frequency charge density and then we provide a model to estimate the high frequency charge density. For a fully water saturated material, the streaming potential coupling coefficient is defined as

$$C_0 = \lim_{\omega \to 0} C(\omega, s_w = 1) = \frac{\hat{Q}_V^0 k_0}{\eta_w \sigma_0}. \tag{8.147}$$

This equation provides a way to estimate the charge density $\hat{Q}_V^0$ from the measurements of the low frequency streaming potential coupling coefficient, the low frequency electrical conductivity and permeability using

$$\hat{Q}_V^0 = \frac{C_0 \eta_w \sigma_0}{k_0}. \tag{8.148}$$

The estimate of the low frequency volumetric charge density $\hat{Q}_V^0$ is reported as a function of the permeability in Figure 3.12 for experiments performed with a broad range of porous materials at near neutral pH values (pH 5–8). According to Jardani *et al.* (2007b), $\hat{Q}_V^0$ can be directly estimated from the quasi-static (saturated) permeability by (see Figure 3.12):

$$\log_{10} \hat{Q}_V^0 = -9.23 - 0.82 \log_{10} k_0. \tag{8.149}$$

This equation provides therefore a way to estimate directly the volumetric charge density from the low frequency permeability reducing the number of material properties to consider in the simulations.

We seek now a way to estimate the high frequency charge density $\hat{Q}_V^\infty$. At full saturation, the surface conductivity is given by,

$$\sigma_S = \frac{A(\phi, m)}{F} \beta_{(+)} \hat{Q}_V^\infty. \tag{8.150}$$

Therefore the determination of the surface conductivity and the formation factor (from the measurements of the conductivity of the porous material at different pore water conductivities). Therefore the high frequency excess charge density $\hat{Q}_V^\infty$ can be estimated to the surface conductivity and the formation factor by

$$\hat{Q}_V^\infty = \frac{\sigma_S F}{\beta_{(+)} A(\phi, m)}. \tag{8.151}$$

Indeed, at high frequencies, all the charge density existing in the pores is uniformly dragged along the pore water flow and therefore the charge density $\hat{Q}_V^\infty$ is also equal to the volumetric charge density of the diffuse layer (see Revil and Florsch (2010))

$$\hat{Q}_V^\infty = (1 - f)Q_V, \tag{8.152}$$

where f is the fraction of counterions in the Stern layer and Q_V denotes the total charge density in the pore space including the contribution of the Stern layer. This total charge density (Stern plus diffuse layers) is related to the cation exchange capacity CEC (in mol kg^{-1}) (see Waxman and Smits (1968)) by

$$Q_V = \rho_S \left(\frac{1 - \phi}{\phi} \right) \text{CEC}, \tag{8.153}$$

where ρ_S denotes the mass density of the grains. The CEC is another fundamental parameter describing the electrochemical properties of the porous material, more precisely the amount of active surface sites on the mineral surface at a given pH value. In SI units, the CEC is expressed in C kg^{-1}, but is classically expressed in meq g^{-1} (with 1 meq = 1 mmol equivalent charge, e.g., 1×10^{-3} e N, where $e = 1.6 \times 10^{-19}$ C and N is the Avogadro constant, 6.022×10^{23} mol^{-1}, 1 meq g^{-1} = 96320 C kg^{-1}). The fraction of counterions f can be determined from electrical double layer theory (see Revil and Florsch (2010)). For material with different types of clay minerals, the average CEC is determined from the respective exchange capacities of the constituent clay types (see Woodruff and Revil (2011)) using

$$\text{CEC} = \chi_K \text{CEC}_K + \chi_I \text{CEC}_I + \chi_S \text{CEC}_S, \tag{8.154}$$

where χ represents the mass fraction in the porous material and the subscripts K, I and S stand for kaolinite, illite and smectite, respectively. If there is only one type of clay minerals, the CEC is given by CEC = φ_W CEC$_C$, where φ_W denotes the

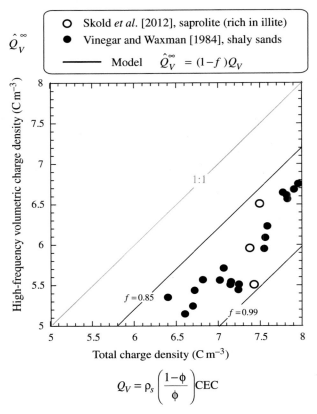

Figure 8.6. High frequency volumetric charge density versus the total charge density. The high frequency volumetric charge density is determined from electrical conductivity measurements at various salinities (from the surface conductivity and the formation factor) while the total charge density is determined from the porosity and the cation exchange capacity.

clay fraction (in weight, unitless) of the porous material and $CEC = \varphi_W \, CEC_C$ denotes the CEC_C of the various clay minerals present in the porous material.

The data from Vinegar and Waxman (1984) were analyzed using the linear conductivity model described above. The results (not shown here) are actually quite close to the results of the differential effective medium model used by Revil (2012). In Figure 8.6, the high frequency charge density determined from surface conductivity is plotted as a function of the total charge density estimated from the measured CEC using a titration method. From this graph, the fraction of counterions in the Stern layer is between 0.85 (85%) and 0.99 (99%). According to Revil (2012), the maximum partition coefficients are reached at high salinities with $f(\text{kaolinite}) = 0.98$, $f(\text{illite}) = 0.90$, and $f(\text{smectite}) = 0.85$. Revil (2012) provides a way of computing the salinity dependence of f based on a simple isotherm for the sorption in the Stern layer.

According to Figure 8.6, $\hat{Q}_V^\infty$ is in the range 10^5 to 10^7 C m^{-3} when the permeability is in the range 10^{-16} to 10^{-12} m^2. For this permeability interval and according to Figure 3.12, the volumetric charge density $\hat{Q}_V^0$ is in the range 1 to 1000 C m^{-3}. Therefore the assumption $\hat{Q}_V^0 \ll \hat{Q}_V^\infty$ (with the exception of shales or formations/soils extremely rich in clays) is justified.

8.6 Application in two-phase flow conditions

There are many potential applications of the previous model and its extension to two fluid phases to understand electrokinetic properties in unsaturated conditions and reservoir conditions. In order to demonstrate how a seismoelectric model works in two-phase flow conditions, we use the approach of Rubino and Holliger (2012) for solving the saturated case, but with the bulk modulus, the viscosity and the density of the fluid mixture given as a fucntion of the properties of the two fluids and the water saturation. This approach is applied to the detection of a moving oil/water encroachment front during the production of an oil reservoir. We used the approach described by Karaoulis *et al.* (2012) to generate a 2D heterogeneous aquifer in terms of porosity and permeability with an isotropic semi-variogram. The result is shown in Figure 8.7. This heterogeneous aquifer is assumed to be initially saturated with 75% of oil and therefore the initial water saturation is uniform and equal to 0.25. Water flooding of this reservoir is performed by injecting water in well A and extracting the oil in the well B. Well B is 250 m away from well A. The simulations are performed by solving to the phase flow equations. The water flooding of the oil reservoir generates the saturation profiles shown in Figure 8.8. We can observe a clear oil/water encroachment front moving from well A to well B over time. Our goal here is to see if this oil/awater encroachment front would generate a seismoelectric conversion, that is if a fraction of the mechanical energy associated with the propagation of the P-waves is converted into electromagnetic energy at this interface.

Figure 8.8 shows a set of six snapshots (denoted as T1 to T6 below) in terms of the water saturation in the aquifer. For each of these snapshots, a seismoelectric acquisition survey between the two wells was simulated. Further information regarding this modeling can be found in Revil and Mahardika (2013). The geometry of the model is a 350×350 m rectangle (see descripion in Figure 8.9). The seismic sources are located in well A and consist of explosive-like sources (denoted as So; see Araji *et al.* (2012) for further descritpions regarding the seismic sources used in our modeling). Regarding the source time function, we choose a Gaussian source time function that is generated with delay time of 30 ms and dominant

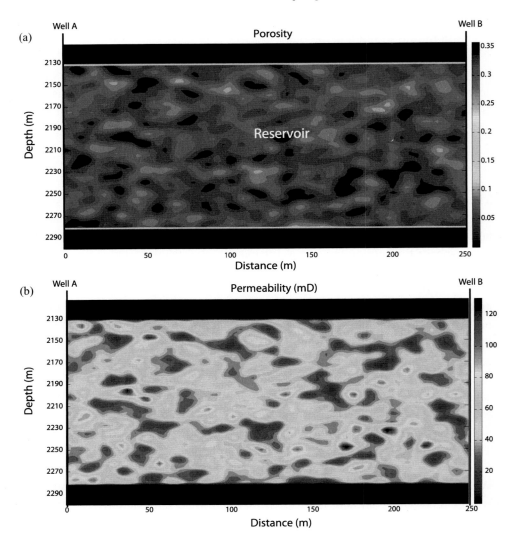

Figure 8.7. Porosity and permeability map of an oil contaminated aquifer or oil reservoir between two wells for a water flood simulation.

frequency equals to 160 Hz. The receivers comprise 28 pairs of seismic stations and electrodes (denoted as Cr1 to Cr28 from bottom to top). All the receivers are located in well B. The separation between receivers in well B is 4 m. The reference position, O(−80,30) for the coordinate system is located at the upper left corner of this domain.

We first solve the poroelastodynamic wave equations and the electrostatic Poisson equation for the electrical field in the frequency domain taking into

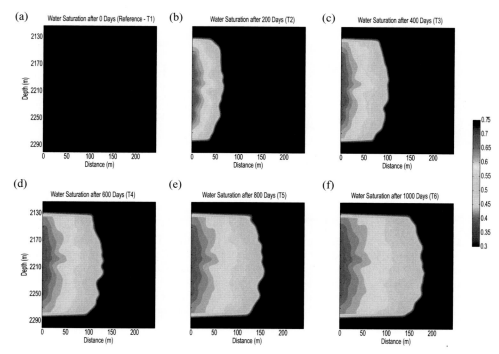

Figure 8.8. Six snapshots showing the evolution of the water saturation over time in a 150-m-thick oil aquifer or reservoir. The initial water saturation in the aquifer is 0.25 (which correspond to 0.75 oil saturation). In this study oil is considered to be the non-wetting phase (simulations by Junwei Zhang and Harry Mahardika, Colorado School of Mines).

account the variable saturation of the water phase. These equations are all described in the previous section. We use the multi-physics modeling package Comsol Multiphysics 4.2a (based on the finite element approach) and the stationary parametric solver PARDISO discussed above in Section 8.3 (see Schenk *et al.* (2008). The problem is solved as follow: (i) we first compute for the poroelastic and electric properties distribution for the given porosity, fluid permeability and each saturation profile, then (ii) we solve for the displacement of the solid phase and the pore fluid pressure, and (iii) finally we compute the electrical potential by solving the Poisson equation using the solution of poroelastodynamic problem. The solution in the time domain is computed by using an inverse Fourier transform of the solution in the frequency wave number domain.

In the frequency domain, we use the frequency range 8–800 Hz, because the appropriate seismic wave in this setting operates in this range and the associated electrical field occurs in the same frequency range. Then using this frequency range,

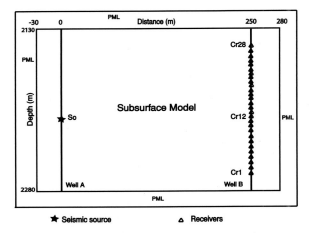

Figure 8.9. Sketch of the domain used for the modeling. The domain is a 350 m × 350 m square. The shooting well, A, is located at $x = 0$ m and the measurement well, B, is located at $x = 250$ m. The discretization of the domain comprises a finite element mesh of 125 × 125 rectangular cells. We put the seismic sources So(0 m, 2216 m) in well A and 71 receivers (Cr1 to Cr71 at the top) located in well B. PML (perfect match layer) boundary conditions are used at the boundaries of the domain.

we compute the inverse fast Fourier transform to get the time series of the seismic displacements u_x and u_z, and the time series of the electric potential response ψ. We use a rectangular mesh of size 2 m × 2 m, smaller than the smallest wavelength of the seismic wave. We have checked that this corresponds to the smallest mesh for which the solution of the partial differential equation is mesh independent. The seismic source is located at position So($x_s = 0$ m, $z_s = 2216$ m) with a magnitude of 1.0×10^4 N m. For theoretical discussion on this type of source, see Araji *et al.* (2012) and Mahardika *et al.* (2012). At the four external boundaries of the domain we apply the 20 m thick convoluted perfect matched layer (CPML) described above. The sensors located at well B are 30 m away from the right-hand side of the CPML layer and therefore the solution is not influenced by the PML boundary condition. This receiver arrangement mimics the acquisition that would be obtained with triaxial geophones and dipole antennas. CPML boundary conditions consist of a strip simulating the propagation of the seismic waves into free space without any reflections going back inside the domain of interest (see Jardani *et al.* (2010) for further details on the implementation of this approach).

Evolution of the seismic displacement and the electrical potential time series recorded at station Cr12 for each saturation profile (T1 to T6) are shown in Figure 8.10. The conductivity of the brine is 5 S m^{-1} and we used $m = n = 1.8$.

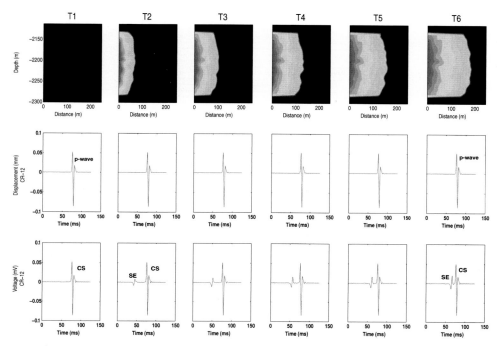

Figure 8.10. Evolution of the seismic displacement and the associated electric potential time series from receiver Cr12 due to water front changes (T1 to T6). Here we show the P-wave arrivals in the seismic time series and the seismoelectric conversions (SE) at interfaces (Type I electrical signal), and CS: co-seismic P-wave (Type II electrical signal) on the electrical counterpart.

The seismic displacements are created from the isotropic-mechanical source which allows only for generation of P-waves. In our case the P-wave velocity of profiles T1 to T6 is almost constant with the average velocity of 5200 m s^{-1}. This "unperturbed" P-wave velocity cause the P-wave arrivals tend to be the same from profiles T1 to T6 (Figure 8.10). Figure 8.10 shows that a variety of seismoelectric (SE) conversions occur for each snapshot. These conversions arrive later as the water front approaches well B. There is therefore a clear conversion mechanism at the oil/water encroachment front for each of the five snapshots T2 to T6. For snapshot T1, since there is no saturation contrast we do not see any strong seismoelectric conversion effect. However, co-seismic seismoelectric signal (CS) associated with the arrival of the P-wave does occur at this snapshot and also at all the other snapshots.

We consider now the snapshot T4. Figure 8.11 shows that both the seismoelectric conversion (SE, generated at the oil water encroachment front) and the co-seismic (CS) electrical signals are shown in all receiving stations Cr1 to Cr28. Figure 8.12

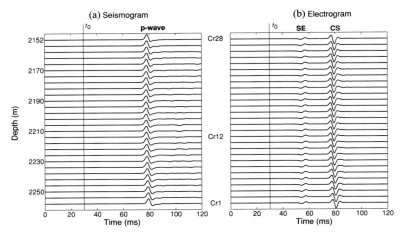

Figure 8.11. Seismogram and electrograms from implementing saturation profile T4. (a) The seismograms reconstructed by the geophones show the P-wave propagation from the seismic source in well A to recording well B. (b) The electrograms show the co-seismic electrical potential field associated with the P-wave (CS) and the seismoelectric conversions with a smaller amplitude and a flat shape (SE).

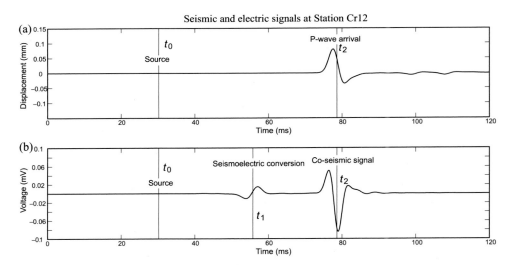

Figure 8.12. Seismogram and electrogram at an electrode (receiver Cr12 in Figure 8.9). Here t_0 denotes the time of source ignition, or the delay time (30 ms). Seismoelectric signals and the co-seismic/P-wave arrival time we denoted as t_1 and t_2, respectively. The strongest signal on the electrogram corresponds to the co-seismic disturbance associated with the P-wave propagation. The seismoelectric signal occurs only on the electric time series and this is associated with the seismoelectric conversions at saturation changes on profile T4 (see Figure 8.8).

shows the seismic displacement and the electrical time series for station Cr12 with information on the delay time t_0, seismoelectric conversion arrival t_1, and the similarities of co-seismic and P-wave arrival times t_2. In these simulations, two types of electromagnetic signals are therefore generated.

(i) Type I denotes the electrical disturbances generated at heterogeneities in the electrical, mechanical and transport properties during the propagation of the seismic waves themselves (see, for example, Huang and Liu (2006)). These conversions are therefore sensitive to the level of heterogeneities of the subsurface and, in our model, these seismoelectric conversions can be associated with sharp change in the water saturation especially at the oil/water encroachment front. When crossing an interface between two domains characterized by different properties (in this case the saturation contrast), a seismic wave generates radiating electromagnetic energy. Because the electromagnetic diffusion of the electrical disturbance is very fast (instantaneous in our simulations), the seismoelectric conversions are observed nearly at the same time by all the sensors (electrodes) but with different amplitudes depending on the resistivity distribution and the distance between the receivers the the position of the radiating source. The seismoelectric conversions therefore appear as flat lines in the electrograms shown in Figure 8.11b.

(ii) Type II corresponds to the co-seismic electromagnetic signals. They are electromagnetic signals propagating at the same speed as the P- and S-seismic waves (see Ivanov (1939) and Frenkel (1944) for example). They are the result of the local fluid flow associated with the passage of the seismic wave. This flow generates a source current density that is locally compensated by the conduction current density. The Type II electrical field is proportional to the acceleration of the seismic wave with a local transfer function that depends on the local electrical properties of the porous materials. It results an electrical field traveling at the same speed as the seismic wave (Figure 8.11).

The next step will be to see if we can invert the sesismoelectric signals obtained in the second well. We could use the approach that we have published recently in Araji *et al.* (2012) to invert the self-potential signals recorded in the second well to localize the causative conversion responsible for these signals in the reservoir. This would open the door to a new way to monitor oil and gas reservoirs or to monitor the remediation of NAPLs and DNAPLs in the shallow subsurface. In Section 8.7, we show an example of localization of a hydromechanical disturbance using the time-lapse record of the electrical field at a set of electrodes.

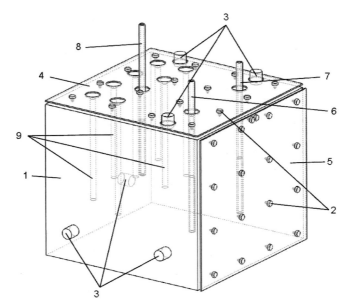

Figure 8.13. Unconfined cement block sensor configuration. (1) Cement block, (2) 34 ea., Ag/AgCl$_2$ electrodes (Biosemi), (3) acoustic emission sensors (Mistras WSa), (4) plastic plate with top array of 16 channels of Ag/AgCl2 electrodes (Biosemi), (5) plastic plate with back array of 16 channels of Ag/AgCl$_2$ electrodes (Biosemi), (6) hole 9 high pressure fluid injection tube, (7) hole 10, high pressure fluid injection tube, (8) hole 6, high pressure fluid injection tube (not used), (9) other holes (no injection).

8.7 Localization of hydromechanical events

In this section, we investigate, through a laboratory experiment, how hydromechanical disturbances associated with fracking can be localized over time by using the remote measurement of the associated electrical field.

8.7.1 Material and method

The porous material we used for the tests is a cement mixture (FastSet Grout Mix). It was cured for about 10 months before the tests proceeded. The porous sample has a cube shape ($x = 30.5$ cm $\times$ $y = 30.5$ cm $\times$ $z = 27.5$ cm; Figure 8.13). After curing, several holes (named #1 to #10 below, 1.5 cm in diameter) were drilled into the block to varying depths such that various tube sealing methods could be tested. Then a stainless steel tube was secured into a chosen position on the block using epoxy or grout as the tube sealing mechanism. The voltage measurement electrodes were attached to the block using a plastic template for each side of the block that was to be monitored. Each plastic template had a 4 × 4 2D electrode position hole matrix drilled into it at the correct relative positions.

The electrical response during the experiment was measured using a very sensitive multichannel voltmeter manufactured by Biosemi, Inc., designed for EEG (http://www.biosemi.com/). In the experiments, the electrical potential measurements were acquired with 32 amplified non-polarizing silver–silver chloride (Ag/AgCl) electrodes. The electrode potentials were measured using the BioSemi ActiveTwo data acquisition system, which is self-contained, battery powered, galvanically isolated and digitally multiplexed with a single high sensitivity analog-to-digital converter per measurement channel. The analog-to-digital converters used in the system were based on a Sigma-Delta architecture with a 24 bit resolution. The system has a typical sampling rate of 2048 Hz with an overall response of d.c. to 400 Hz (system filtered). The digitization part of the system had a sampling process with <10 ps skew among channels and 200 ps sample rate jitter. The digitally multiplexed signals were subsequently serialized into a bit-wide data stream and sent through a fiber optic cable (to achieve galvanic isolation) to a USB based computer interface. This measurement system has a scaled quantization level of 31.25 nV (LSB) with 0.8 μV rms noise at a full bandwidth of 400 Hz with a specified $1/f$ noise of 1 μV pk-pk from 0.1 to 10Hz. The common mode rejection ratio was higher than 100 dB at 50 Hz, and the amplified non-polarizing electrode input impedance was 300 MOhm at 50 Hz (10^{12} Ohm // 11 pF) (see http://www.biosemi.com/, and Crespy *et al.* (2008) and Ikard *et al.* (2012) for further explanations).

The voltage reference for the measurements is contained within the measurement area, and was designed into the measurement system to be a part of the common mode sense and common mode range control electrodes (CMS and DRL electrodes; see Ansari-Asl *et al.* (2007) for further information on the common mode control used in the BioSemi System and Kappenman and Luck (2010) for the effect of electrode impedance on the measured response). All voltages measured by the Biosemi are relative to the CMS electrode used as the reference electrode (zero voltage). The entire system, including the computer, is operated on batteries to minimize conductive coupling with the electrical power system. The flow chart used to analyze the raw electrical data is shown in Figure 8.14.

The experiments were conducted on the porous block in equilibrium with the atmosphere of the laboratory (~30% relative humidity). The experiments used saline water (no proppant) containing 10 g of NaCl dissolved into 1000 ml of deionized water as the injecting liquid (conductivity of 1.76 S m^{-1} at 25 °C). The fluid injection and pressure control system that injects fluid through stainless steel tubes that are secured into holes in the cement block (Figure 8.13). The injection tube is pressurized to 2 MPa with the fracturing fluid and maintained at that pressure for a period of time to be sure that the system was maintaining pressure and to

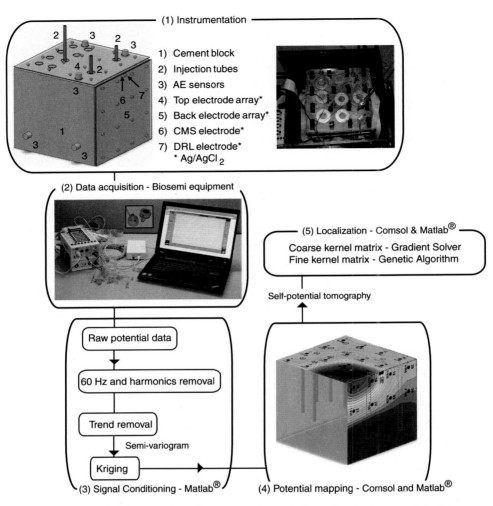

Figure 8.14. Flow chart for the processing of the electrical potential data. (1) Instrumentation of the porous block. (2) Data acquisition showing the BioSemi EEG system and the laptop computer. (3) Signal condition of the raw data. (4) Mapping the voltage response using ordinary kriging. (5) Localization of the causative sources in the block.

measure the leak off rate, then a constant fluid flow rate was imposed on the system. Under constant flow, eventually something would break, either the cement block, or the tubing seal.

The fluid injection and pressure control system was used to control the fluid injection rate and pressure applied to the injection tubes indicated by items labeled 2 in Figure 8.13. The fracturing fluid is injected into 9.5 mm stainless steel tubing secured into the blocks using a computer controlled Teledyne Isco 100DX syringe pump that is able to control flow rate, and pressure. The system has a total fluid capacity of 103 ml, is capable of achieving pressures of up to 68.9 MPa, and maintaining constant flow rates of 0.001–60 ml min^{-1}. Additional components of this part of the test system are control valves, pressure and temperature transducers as well as temperature control.

8.7.2 Experimental procedure

The test procedure began by preparing the cement block for high pressure injection (see Frash and Gutierrez (2012) for details). The injector was filled with the slightly saline solution described above, the hole was filled with the injectant and the injector was coupled to the injection tube. The injection system was purged of air, and then put into a constant pressure control program for about 30 min to be sure that there was no pressure loss. For the fracture experiment associated with hole 10 (prior to the injection in hole 9 that is described below in detail), a constant pressure of 2 MPa, followed by a 5 ml min^{-1} constant flow rate was used. For the fracture experiment associated with hole 9, a 60 s pre-injection measurement period was acquired for the first phase of the measurements (Phase 0, Figure 8.15) to establish individual channel offsets and drift trends for use during post acquisition signal processing. Following Phase 0, a constant pressure fluid injection at 2 MPa (Phase I) was initiated at T0, and terminated at T1 (see Figure 8.15a), followed by Phase II, a 1 ml min^{-1} constant flow rate initiated at T2 (Figure 8.15a). Fluid injection was terminated when further fracturing was no longer occurring through acoustic emission observations or when seal failure was confirmed through the appearance of water on the surface of the block near the injection hole. For this experiment, self-potential data acquisition terminated prior to completion of Phase II injection due to an accidental interruption of the streaming data because of a poor USB 2.0 connection. This did not affect the data saved in the raw data file or any subsequent use of the data.

8.7.3 Observations and interpretation

Figure 8.15 shows the temporal evolution of the electrical potential for all of the electrodes. This figure shows the occurrence of bursts in the electrical potential that are similar in shape (but much larger in amplitude) to the electrical field bursts observed by Haas and Revil (2009) for Haines jumps during the drainage of an initially water-saturated sandbox. There are seven main events of which three are highlighted (events E1 through E3), and two will be used below to test our localization procedure. These events are shown in the time series of Figure 8.15b, c and d. Many other events are also seen in the data, and these events have a variety of magnitudes and temporal evolutions. Each major event is more or less characterized

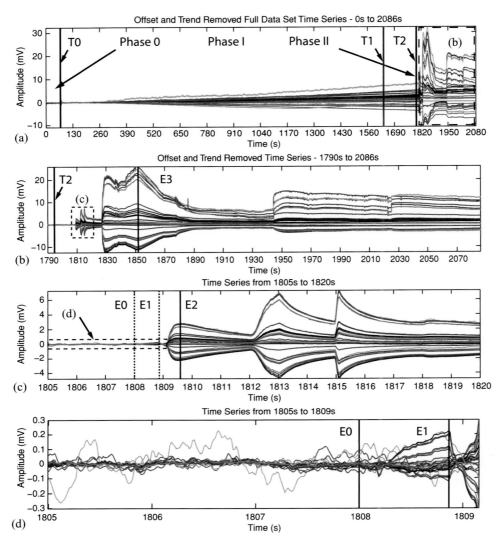

Figure 8.15. Self-potential time series related to hole #9 water injection. (a) Full
time series data set showing the different fluid injection time periods during data
acquisition (T0, T1, and T2). Note the significant change in electrical response
after T2 that is bounded by region (b). (b) Zoom in region (b) highlighting the
background normalized electrical response showing distinct electrical impulses
related to the start of constant flow injection at T2 with selected peak events E5,
E6, and E7. (c) Zoom in region (c) highlighting the first series of impulsive signals
with selected peak events E2, E3, and E4 with temporal reference to E0 and E1.
(d) Zoom in region (d) showing the temporal noise leading up to event E1, with a
voltage background time slice at E0. Note the change in potential after E1.

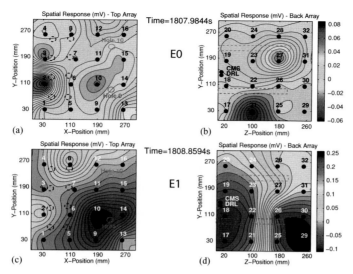

Figure 8.16. Self-potential spatial voltage distributions for the snapshot E0 and the event E1. Each panel is a kriged contoured distribution of the electrical potential on the top and back panels (ordinary kriging). The black dots denote the position of the electrodes. (a) and (b) Spatial electrical potential distribution for snapshot E0 showing the spatial variations associated with the background noise. Note the very small color bar voltage scale (+0.08 mV to −0.06 mV). (c) and (d) Voltage distribution for event E1 showing the burst of the electrical field associated with the first hydraulic pulse taking place during constant flow injection. Note the voltage polarities in the spatial distributions and the much larger color bar voltage scale (+0.25 mV to −0.1 mV). See also color plate section.

by a rapid change in the electrical potential time series followed by an exponential-type relaxation of the potential with time with a characteristic time comprised between several seconds to several tens of seconds. Because the relaxation of the potential distribution is relatively slow after each event, a sequence of overlapping events causes a superposition of the potentials from each event in the sequence to varying degrees (see Figures 8.15b and c). We call the superposition of a past event decay response with a new event a residual potential superposition. It can be clearly seen from Figure 8.15 that the degree of residual potential superposition is dependent on event physics (hydroelectric coupling), event magnitudes, event spatial distribution, time of occurrence and event decay rate. Clearly each of these factors is variable, and to localize and characterize individual impulsive events, the influence of residual potential superposition must be accounted for, and removed to complete a comprehensive analysis of the data.

Figure 8.16 and Figure 8.17 show the spatial evolution of the electrical potential on the monitored faces of the test block, starting with snapshot #E0 taken prior to

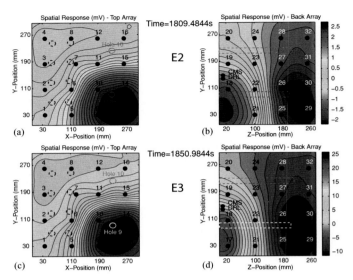

Figure 8.17. Self-potential spatial voltage distributions of events E2 and E3. (a, b) Event E2 voltage distribution showing the peak voltage associated with the second hydraulic pulse during constant flow injection. Note the voltage polarities in the spatial distributions and the color bar voltage scale (+2.5 mV to −2.0 mV). Event E2 represents the first of a series of electrical field bursts. Panels (a) and (b) show a reversal in polarity and increase in peak magnitude relative to panels (c) and (d). (c, d) Event E3 spatial voltage distribution showing the peak voltage associated with the highest magnitude pulse during constant flow injection. See also color plate section.

the occurrence of events E1 and E3. It can be seen in Figure 8.15d and Figures 8.16a and b that the snapshot E0 is dominated by random fluctuations with no particular persistent spatial distribution pattern. In these figures, Channel 13 is noisier with respect to the rest of the channels possibly because of a bad contact between the electrode and the porous block.

Event E1 in Figure 8.16c and d shows an initial voltage distribution with a small negative potential showing up on the top surface of the block while a bipolar signal showing up on the back side of the block. This voltage distribution implies that there is a current source density possibly near hole 9 (see position in Figure 8.13) that is pointing mostly downward into the block. The time series in Figure 8.15d shows the onset of this small peak (event E1), followed by a quick decay and reversal of the polarity of the current source density as indicated by event E2 as shown in Figure 8.15c and Figure 8.17a and b. As we will see later, the direction of the current density corresponding to event E2 is mostly pointing up toward the

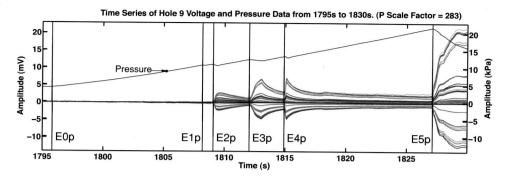

Figure 8.18. The water pressure data shows that the decreases in the fluid pressure agree with the time of occurrence of the onset of the bursts in the electrical field. The local minima in the fluid pressure minimums agree with the maxima of the bursts of the electrical field. This is what is expected from a valve behavior.

top surface and grows in magnitude in an impulsive manner as the fluid injection proceeds further. Event E3, later on, has a higher magnitude than the previous events but maintains the spatial voltage distribution of event E2. It continues until the termination of the data acquisition process. This implies that fluid is moving upward in a persistent manner, somewhere in the vicinity of hole 9 during and after event E2.

The fluid pressure rate was sampled at five samples per second with a fluid transducer located close to the injector. The fluid pressure time series is shown in Figure 8.18. Figure 8.18 shows that the observed burst in the electrical field are directly related to pressure changes that were measured in the injection system. The pressure data indicate that there were some brutal changes in the flow regime inside hole 9 and the leakages were occurring only inside the block (the occurrence of electrical data shows that the fluid that moved was in contact with porous media, no electrokinetic phenomena would occur outside the block and directly in the hole). The pressure data shows only small pressure fluctuations in the early phases of the seal failure around hole 9 while the pressure continues to build. However, the electrical data show more detail of the early development of the seal failure process and we will show that the electrical field can be used to localize these events indicating an imminent seal failure. Only after the pressure reverses for a long period of time can seal failure be determined from the pressure data.

The persistent voltage distribution shown in Figure 8.17 indicates the effects of upward fluid migration somewhere near hole 9. We believe that this set of observations provides a leading indicator of the borehole seal failure. This seal failure was further confirmed through fluid pressure measurements (see previous subsection) and the leakage along the borehole was visually confirmed through the observation of water flow at the top surface of the block in the vicinity of hole 9. The temporal electrical signatures in Figure 8.15 shows numerous impulsive events that get larger as the seal failure progresses. The hypothesis for the explanation of this data considers that the seal failed in a plastic manner beginning with the onset of seal failure with subsequent repeated blockage and breakthrough events and valve behavior until the end of the data acquisition. The seal failure occurs in the annulus of the borehole and as more and more fluid contacts the cement walls of the borehole with higher and higher velocities, the magnitude of the electrical response grew accordingly. The approximate position of the fluid contact with the borehole wall can be determined from the data. The position of the positive anomaly recorded by the top array is not centered on hole 9, but is displaced from the center of the hole.

The data from the back electrical potential array also contain source location and orientation information, indicating that the fluid flow encountered porous media somewhere well above the bottom of the borehole, also a potential indication of borehole seal failure. The observations imply that the fluid flow is occurring along a pathway following the borehole and close to the lower right corner of the top array. The electrical boundary conditions in the borehole is insulating between the borehole wall and the stainless steel tubing, causing the reflection of the electrical potential away from the borehole center. These electrical potential measurements are consistent with the subsequent observations of fluid leakage at the test block surface near hole 9 due to borehole seal failure. These electrical observations occurred several minutes before surface fluid leakage was visually observed on the top surface.

8.7.4 Localization

We use a gradient-based approach to invert the electrical potential distribution at a given time in order to look for the position of the causative source. This gradient-based approach is combined with compactness to minimize the support of the source. The kernel used to perform the inverse problem is shown in Figure 8.19 and the results of the inversion, without and with compactness, are shown in Figure 8.20. The source is recovered in the vicinity of hole 9 as expected. This indicates how the record of trensient electrical fields can be used to locate the dynamics of hydromechanical events in a porous material and by extension in the subsurface of the Earth.

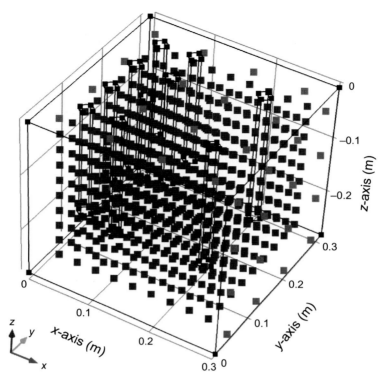

Figure 8.19. Comsol model showing the distribution of points used to compute the coarse kernel matrix (dark points, 729 positions) and the voltage measurement points (light gray points, 32 positions). The model geometry includes all of the holes that were drilled into the block (black cylinders) considered to be perfect insulators. The arrows labeled x, y, and z indicate the positive directions of the corresponding axes in the figure for the Cartesian coordinate system.

8.8 Seismic beamforming and the formation of electrical bursts

We have recently explored with Paul Sava (Center for Wave Phenomena, Colorado School of Mines; see Sava and Revil (2012)), the possibility of beamforming seismic waves at any point of the subsurface and of monitoring the electrical field at a set of electrical stations (dipoles or antennas) located at the ground surface, at the sea floor or in a set of boreholes. The idea is to look at changes in the electrokinetic conversion over time, which, according to the theory described above, is sensitive to the presence of heterogenities or, in the case of repeated beamforming, to changes in water saturation. This opens up the possibility of scanning many points of an oil reservoir for instance and of monitoring the changes in oil saturation with an approach that is higher in resolution than existing methods.

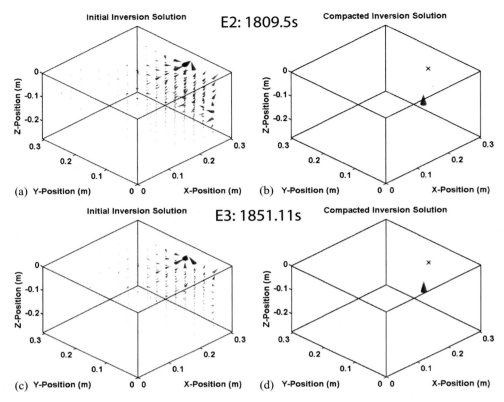

Figure 8.20. Initial and compacted dipole inversion results for events E2 and E3. In these cases, the compact inversion of events E2 and E3 yields one dipole, which dominate the response through a higher magnitude. Note that event 3 is localized at a shallower depth than event 2 showing that the causative source is moving closer to the top surface of the block in the vicinity of well 9. (a, c) Results of the source distribution after gradient-based inversion. (b, d) After 40 iterations in compacting the support of the source. Note that the source moves upward over time.

Figures 8.21 and 8.22 show the beamforming of seismic waves (using the acoustic approximation) and the resulting burst in the electrical field associated with the electrokinetic coupling. A time series of the electrical potential at one electrode located in one of the two wells show the occurrence of the burst in the electrical field when the seismic wave have focused. Such an approach can be used to determine the change of the properties at the point where the seismic energy is focused. This can be performed over a grid of points to look at the change of properties (such as water saturation) over time for an entire reservoir or for vadose zone hydrology. It can be used virtual electrical current injection at any point of a reservoir or to add additional informtaion in cross-well d.c. resistivity and traveltime tomography.

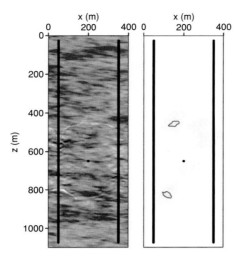

Figure 8.21. Electrical field produced at an early stage of the seismic wave beamforming. The waves are produced by a set of seismic sources located in the two wells. The dot in the reservoir denotes the position where the seismic field is converging. The electrical field is produced by the electrokinetic coupling.

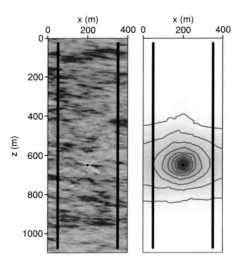

Figure 8.22. Electrical field produced at the final stage of the seismic wave beamforming. We see a strong burst in the electrical field that can be remotely observed in the two wells using a network of electrodes. This infomation can be used to determine the properties of the point where the seismic field has been focused. This operation can be performed for any point of the reservoir and the operation can be repeated over time.

Because of the symmetry in the constitutive equations developed in Chapter 2, there is the possibility, still unexplored, for us to create a seismic source at any point of the Earth by modulating electromagnetic fields from a set of artificial sources. That said, this may prove difficult in practice, as the electrical conductivity of porous rock is more heterogeneous than the sesmic velocity field and at low frequencies, electromagnetic disturbances obey a diffusion equation (not a propagation equation).

8.9 Conclusions

Previous modeling attempts of the electroseismic and seismoelectric coupling effects were restricted to the thin double layer assumption (thickness of the double layer much smaller than the size of the throats) and to fully water-saturated conditions. In addition, these models were not accounting for the induced polarization of the material. We have provided the first electroseismic and seismoelectric model valid for unsaturated clayey soils, sands and sandstones, including the modeling of the induced polarization of this type of clayey materials. This model is valid for porous materials with macropores and micropores with respect to the thickness of the electrical double layer. It couples the Maxwell equations to a very simple version of Biot theory for partially saturated porous material. Finally, we have shown that the seismoelectric method could be used to locate the oil water encroachment front during the water flooding of a contaminated aquifer or the production of an oil reservoir.

Two important observations have come from a set of experimental data performed in the laboratory. First, borehole seal failure during hydraulic fracturing generates electrical currents when fracturing fluid flowed into the annulus between the hole wall and the casing and came in contact with the borehole wall. The associated electrical field can be measured remotely. Second, we have shown that the source current density associated with the seal leak can be located using a self-potential (gradient-based) tomography technique. From these data, the concept of detecting non-intrusively well system leakages can be developed for field applications to monitor, for instance, potential leakages along a borehole during fracking in gas shales.

The last part of this chapter has opened a radically new way to perform geophysics: we can beamform seismic waves at any points of the Earth, vibrate this point, and look at the occurrence of the resulting electrical field remotely. This brings information regarding this point and, if the operation is repeated over time, we can monitor an entire reservoir. This approach may open a new way to monitor saturation changes in the reservoir with a very high resolution.

Exercises

Exercise 8.1. Drained and undrained self-potential response to loading. We have seen above the perfectly undrained response and the drained response of a porous rock. We need now to see the complete form of the dynamic equations involving fluid flow in response to deformation, fluid pressure or confinement pressure. We are going to consider the pore fluid as being uncompressible. The five equations required to analyze the dynamic case are:

$$T_{ij} = \left(K - \frac{2}{3}\mu \right) \varepsilon_{kk}\delta_{ij} + 2\mu\varepsilon_{ij} - \alpha p\delta_{ij}, \qquad (8.155)$$

$$\Delta m = \alpha \rho_f \left(\varepsilon_{kk} + \frac{\alpha p}{K_u - K} \right), \qquad (8.156)$$

$$\dot{\mathbf{w}}_i = -\frac{k}{\eta_f} p_{,i} \quad \left(\text{or} \quad \frac{d\mathbf{w}}{dt} = -\frac{k}{\eta_f}\nabla p \right), \qquad (8.157)$$

$$T_{ij,j} = 0 \ (\nabla \cdot \overline{\overline{T}} = 0), \qquad (8.158)$$

$$(\rho_f \dot{\mathbf{w}}_i)_{,i} = -\dot{m} \quad \left(\text{or} \quad \nabla \cdot (\rho_f \dot{\mathbf{w}}) = -\frac{\partial m}{\partial t} \right), \qquad (8.159)$$

where T_{ij} the components of the stress tensor $\overline{\overline{T}}$, K and μ denote the bulk modulus and the shear modulus, respectively, p denotes the pore fluid pressure, α denotes Biot's coefficient of effective stress, m is the mass of pore fluid per unit volume of porous material, ε_{kk} is the bulk deformation, $\dot{\mathbf{w}}$ is the Darcy velocity, k is the permeability, η_f and ρ_f are the dynamic viscosity and fluid density of the pore fluid, respectively, and K and K_u denote the drained bulk modulus (bulk modulus of the skeleton) and the undrained bulk modulus, respectively. Equation (8.155) corresponds to Hooke's law for poroelasticity, Eq. (8.156) corresponds to the mass of fluid exchanged by a representative element volume of the porous rock in response to a change of deformation and fluid pressure, Eq. (8.157) corresponds to the Darcy law (momentum conservation equation for the pore fluid), Eq. (8.158) corresponds to Newton's law (momentum conservation equation for the porous skeleton), and Eq. (8.159) denotes a mass conservation equation for the pore fluid.

(1) The first part of this problem is to show that the mass of the pore fluid per unit volume of porous material obeys a diffusion equation and to determine the diffusivity.

(a) Taking Eqs. (8.155) and (8.158), demonstrate the following relationship:

$$\left(K + \frac{1}{3}\mu \right) \varepsilon_{kk,i} + \mu u_{i,jj} - \alpha p_{,i} = 0. \qquad (8.160)$$

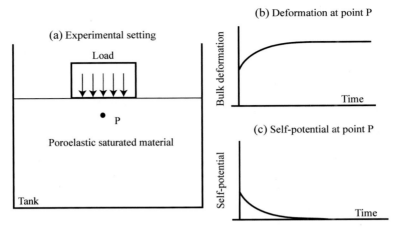

Figure 8.23. Poroelastic defomration when a load is applied to a poroelastic material. (a) Example of tank experiment. (b) Deformation and associated self-potential signmal at point P as a function of time.

(b) Starting with Eqs. (8.157) and (8.159), demonstrate

$$\frac{k\rho_f}{\eta_f}\nabla^2 p = \frac{\partial m}{\partial t}. \tag{8.161}$$

This is a pseudo-diffusion equation with m and p. We need to eliminate p as a function of m.

(c) Taking Eqs. (8.156) and (8.161), demonstrate the following relationship

$$\frac{K_u - K}{\alpha\rho_f}\nabla^2 m = (K_u - K)\nabla^2\varepsilon_{kk} + \alpha\nabla^2 p. \tag{8.162}$$

(d) Taking the derivative of Eq. (8.160) with respect to x_i, show that

$$\left(K + \frac{4}{3}\mu\right)\nabla^2\varepsilon_{kk} = \alpha\nabla^2 p. \tag{8.163}$$

(e) Eliminate $\nabla^2 p$ between Eqs. (8.162) and (8.163) and use Eq. (8.161) to show that m obeys a diffusion equation given by

$$c\nabla^2 m - \frac{\partial m}{\partial t} = 0, \tag{8.164}$$

$$c = \left(\frac{K_u - K}{\alpha^2}\right)\left(\frac{k}{\eta_f}\right)\left(\frac{K + \frac{4}{3}\mu}{K_u + \frac{4}{3}\mu}\right). \tag{8.165}$$

The coefficient c denotes a hydraulic diffusivity.

(2) We consider now a tank as shown in Figure 8.23a. A load is added at $t = 0$ on the top of the sand filling the tank. The sand is consider to be poroelastic. The deformation and self-potential data at a point P located just below the load are shown in Figure 8.23b and c, respectively. Explain these trends. Explain how you would develop a strategy to invert the poroelastic coefficients using time-lapse self-potential measurements.

Appendix A

A simple model of the Stern layer

Here we propose an alternative to the model reported by Revil (2012) and describing the salinity dependence of the partition coefficient f. We first start considering the sorption of sodium in the Stern layer of clays and the dissociation of protons according to

$$>SH^0 \Leftrightarrow >S^- + H^+, \tag{A1}$$

$$>S^- + Na^+ \Leftrightarrow >S^-Na^+, \tag{A2}$$

where $>S$ refers the surface sites attached to the crystalline framework, H^0 are protons (which are assumed to be immobile) while weakly sorbed Na^+ are considered to be mobile in the Stern layer. The equilibrium constants are given by

$$K_{Na} = \frac{\Gamma^0_{SNa}}{\Gamma^0_{S^-}[Na^+]^0}, \tag{A3}$$

$$K_H = \frac{\Gamma^0_{S^-}[H^+]^0}{\Gamma^0_{SH}}, \tag{A4}$$

and a conservation equation for the surface species

$$\Gamma^0_S = \Gamma^0_{SNa} + \Gamma^0_{SH} + \Gamma^0_{S^-} + \Gamma^0_X, \tag{A5}$$

where Γ^0_S denotes the (known) total surface site density (including the charge associated with isomorphic substitutions in the crystalline framework), Γ^0_{SNa}, $\Gamma^0_{S^-}$ and Γ^0_{SH} represent the surface charge density of the sites $> S^-Na^+$, $> S^-$ and $> SH^0$, respectively, and Γ^0_X represents the number of equivalent sites corresponding to isomorphic substitutions (all expressed in sites m^{-2}). To simplify the notation, we write $pH = -\log_{10} [H^+]$ and $[Na^+] = C_f$ denotes the salinity. The resolution of the previous set of equations yields

$$\Gamma_{SH}^0 = \frac{\Gamma_S^0 - \Gamma_X^0}{1 + \frac{K_H}{10^{-pH}}(1 + C_f K_{Na})}, \tag{A6}$$

$$\Gamma_{S^-}^0 = \frac{(\Gamma_S^0 - \Gamma_X^0) K_H / 10^{-pH}}{1 + \frac{K_H}{10^{-pH}}(1 + C_f K_{Na})}, \tag{A7}$$

$$\Gamma_{SNa}^0 = \frac{(\Gamma_S^0 - \Gamma_X^0) C_f K_{Na} K_H / 10^{-pH}}{1 + \frac{K_H}{10^{-pH}}(1 + C_f K_{Na})}. \tag{A8}$$

All the charged sites that are not compensated in the Stern layer need to be compensated in the diffuse layer. Therefore, the fraction of the counterions in the Stern layer is defined by the following sorption isotherm:

$$f = \frac{\Gamma_{SNa}^0}{\Gamma_{SNa}^0 + \Gamma_{S^-}^0 + \Gamma_X^0}. \tag{A9}$$

$$f = f_M \left\{ \frac{C_f K_{Na}}{f_M(1 + C_f K_{Na}) + (1 - f_M)\left[1 + C_f K_{Na} + \frac{10^{-pH}}{K_H}\right]} \right\}, \tag{A10}$$

$$f_M = 1 - \frac{\Gamma_X^0}{\Gamma_S^0}. \tag{A11}$$

The value of f_M for kaolinite, illite and smectite can be found in Revil (2012). An average value for clay minerals is $f_M = 0.90$. At high pH values, the partition coefficient is given by

$$f \approx f_M \left(\frac{C_f K_{Na}}{1 + C_f K_{Na}} \right). \tag{A12}$$

We look now for the pH dependence of the CEC. The CEC is defined by

$$CEC = \frac{e}{f_M} (\Gamma_{SNa}^0 + \Gamma_{S^-}^0 + \Gamma_X^0) S_{sp}, \tag{A13}$$

where S_{sp} corresponds to the specific surface area (in $m^2\ kg^{-1}$) and e the elementary charge. After some algebraic manipulations and simplification, we obtain

$$CEC(pH) = CEC_M(1 - g(C_f, pH)), \tag{A14}$$

where $CEC_M = e\Gamma_S^0 S_{sp}$ corresponds to the maximum CEC at value at high pH values and the function g is defined by

$$g(C_f, \text{pH}) = \frac{f_M\left(\dfrac{10^{-\text{pH}}}{K_1}\right)}{1 + C_f K_{Na} + \dfrac{10^{-\text{pH}}}{K_1}}. \tag{A15}$$

The quadrature conductivity is given by

$$\sigma'' = -\frac{2}{3}\rho_g \beta_{(+)}^S f \, \text{CEC}, \tag{A16}$$

where the product of f by the CEC is given by $f \, \text{CEC} = e\Gamma_{SNa}^0 S_{sp}$ (the quadrature conductivity is controlled by the density of weakly sorbed sodium counterions in the Stern layer). After some algebraic manipulations we get

$$\sigma'' = \sigma_M'' \left(\frac{C_f K_{Na} K_H}{10^{-\text{pH}} + K_H(1 + C_f K_{Na})}\right), \tag{A18}$$

$$\sigma_M'' = -\frac{2}{3}\left(\rho_g \beta_{(+)}^S f_M e\Gamma_S^0\right) S_{sp}. \tag{A19}$$

where σ_M'' denotes the quadrature conductivity reached at high salinity and pH. The pH dependence of the quadrature conductivity is pretty weak, which means that Eq. (A19) can be approximated by

$$\sigma'' \approx \sigma_M'' \left(\frac{C_f K_{Na}}{1 + C_f K_{Na}}\right). \tag{A20}$$

Appendix B

The **u**–p formulation of poroelasticity

The field equation of poroelasticity can be set up either in terms of six unknowns (the three components of the displacement of the solid phase **u** and the three components of the displacement of the filtration displacement **w**) or in terms of four unknowns (the three components of **u** and the fluid pressure p). We investigate this second formulation in this Appendix. We start with the Darcy equation with the electroosmotic coupling term neglected;

$$\frac{\eta_f}{k(\omega)}\dot{\mathbf{w}} = -\nabla p - \rho_f \ddot{\mathbf{u}} + \mathbf{F}_f, \tag{B1}$$

where $\mathbf{F}_f$ is the body force acting on the pore fluid phase. The fact that the electro-osmotic term can be safely neglected has been discussed by a number of authors (see, for instance, Revil *et al.* (1999a, b)). The dynamic permeability is written as

$$\frac{1}{k(\omega)} \equiv \frac{1 - j(\omega/\omega_c)}{k_0}. \tag{B2}$$

The critical frequency is $\omega_c = \eta_f/(k_0\rho_f F) = b/\tilde{\rho}_f$, where $b = \eta_f/k_0$ and $\tilde{\rho}_f = \rho_f F$. Neglecting the body force acting on the fluid phase, Eq. (B1) can be used to express the filtration displacement, **w**, as a function of the pore fluid pressure p and the displacement of the solid phase **u**,

$$\mathbf{w} = k_\omega(\nabla p - \omega^2 \rho_f u), \tag{B3}$$

where k_ω is defined by

$$k_\omega = \frac{1}{\omega^2 \tilde{\rho}_f + j\omega b}. \tag{B4}$$

Equation (B3) can be used in Newton's law to give

$$-\omega^2 \rho_\omega^s \mathbf{u} - \omega^2 \rho_f k_\omega \nabla p = \nabla \cdot \mathbf{T}, \tag{B5}$$

$$\rho_\omega^s = \rho \omega^2 \rho_f^2 k_\omega, \tag{B6}$$

where ρ_ω^s is an apparent mass density for the solid phase. Equation (B5) is a partial differential equation between $\mathbf{u}$ and p, but the stress tensor $\mathbf{T}$ depends also on $\mathbf{w}$. By using the relationships between the stress and the strain, we obtain the following relationship between the divergence of the filtration displacement, $\nabla \cdot \mathbf{w}$, and the divergence of the displacement of the solid phase, $\nabla \cdot \mathbf{u}$,

$$\nabla \cdot \mathbf{w} = -\frac{1}{M}(p + S) - \alpha \nabla \cdot \mathbf{u}, \tag{B7}$$

where α is the classical Biot coefficient of poroelasticity. We use Eq. (B7) in the stress/strain relationships to remove the dependence of the stress tensor $\mathbf{w}$. This yields

$$\mathbf{T} = \lambda(\nabla \cdot \mathbf{u})\mathbf{I} + G[\nabla \mathbf{u} + \nabla \mathbf{u}^T] - \alpha p \mathbf{I}, \tag{B8}$$

$$\lambda = K - \frac{2}{3}G, \tag{B9}$$

where λ is the Lamé modulus of the skeleton. The effective stress tensor is written as

$$\hat{\mathbf{T}} = \lambda(\nabla \cdot \mathbf{u})\mathbf{I} + G[\nabla \mathbf{u} + \nabla \mathbf{u}^T], \tag{B10}$$

$$\mathbf{T} = \hat{\mathbf{T}} - \alpha p \mathbf{I}. \tag{B11}$$

The effective stress tensor is the equivalent stress tensor of the skeleton without fluid (in vacuo). Using Eqs. (B6)–(B9) and (B3), we obtain an equation connecting the solid displacement and the fluid pressure assuming that the Biot coefficient is constant,

$$-\omega^2 \rho_\omega^s \mathbf{u} + \theta_\omega \nabla p = \nabla \cdot \hat{\mathbf{T}}, \tag{B12}$$

$$\theta_\omega = \alpha \rho_f \omega^2 k_\omega, \tag{B13}$$

where θ_ω is a volumetric hydromechanical coupling coefficient. Equation (B12) corresponds to Newton's law for a poroelastic body. This equation is similar to the classical Newton's law for an elastic solid except for the coupling term $\theta_\omega \nabla p$, which accounts for the dynamic coupling between the pore fluid and the solid phase. Regarding the description of the filtration displacement, we obtain

$$\frac{1}{M}(p + S) + \nabla \cdot \{k_\omega[\nabla p - \omega^2(\rho_f + \alpha)\mathbf{u}]\} = 0. \tag{B14}$$

Equation (B14) is the classical diffusion equation for the pore fluid pressure with a source term related to the harmonic change of displacement of the solid phase. This yields

$$-\omega^2 \rho_\omega^s \mathbf{u} + \theta_\omega \nabla p = \nabla \cdot \hat{\mathbf{T}}, \tag{B15}$$

$$\frac{1}{M}(p + S) + \nabla \cdot [k_\omega(\nabla p - \omega^2(\rho_f + \alpha)u)] = 0. \tag{B16}$$

References

Ahmad, M. U. (1964). A laboratory study of streaming potentials. *Geophysical Prospecting*, **12**(1), 49–64.

Ahmed, S. & Carpenter P. J. (2003). Geophysical response of filled sinkholes, soil pipes and associated bedrock fractures in thin mantled karst, east central Illinois. *Environmental Geology*, **44**, 705–716.

Aizawa, K., Ogawa, Y. & Ishido, T. (2009). Groundwater flow and hydrothermal systems within volcanic edifices: delineation by electric self-potential and magnetotellurics. *Journal of Geophysical Research*, **114**, B01208, doi:10.1029/2008JB005910.

Al-Saigh, N. H., Mohammed, Z. S. & Dahham, M. S. (1994). Detection of water leakage from dams by self-potential method. *Engineering Geology*, **37**(2), 115–121.

Ansari-Asl, K., Chanel, G. & Pun, T. (2007). Channel selection method for EEG classification in emotion assessment based on synchronisation likelihood. In *15th European Signal Processing Conference (EUSIPCO 2007)*, Poznan, Poland, September 3–7, 2007, copyright by EURASIP, pp. 1241–1245.

Antelman, M. S. (1989). *The Encyclopedia of Chemical Electrode Potentials*. New York, Plenum Press.

Araji, A., Revil, A., Jardani, A., Minsley, B. J. & Karaoulis, M. (2012). Imaging with cross-hole seismoelectric tomography. *Geophysical Journal International*, **188**, 1285–1302.

Archie, G. E. (1942). The electrical resistivity log as an aid in determining some reservoir characteristics. *Transactions of the American Institute of Mining, Metallurgical and Petroleum Engineers*, **146**, 54–62.

Arora, T., Linde, N., Revil, A. & Castermant, J. (2007). Non-intrusive determination of the redox potential of contaminant plumes by using the self-potential method. *Journal of Contaminant Hydrology*, **92**, 274–292.

Asfahani, J., Radwan, Y. & Layyous, I. (2010). Integrated geophysical and morphotectonic survey of the impact of Ghab extensional tectonics on the Qastoon Dam, northwestern Syria. *Pure Applied Geophysic*, **167**, 323–338.

Atalla, N., Panneton, R. & Debergue, P. (1998). A mixed displacement-pressure formulation for poroelastic materials. *Journal of the Acoustical Society of America*, **104**, 1444–1452.

Aubert, M. & Atangana, Q. Y. (1996). Self-potential method in hydrogeological exploration of volcanic areas. *Ground Water*, **34**, 1010–1016.

Aubert, M., Dana, I. I. N. & Gourgaud, A. (2000). Internal structure of the Merapi summit from self-potential measurements. *Journal of Volcanology and Geothermal Research*, **100**, 337–343.

348

Avena, M. J. & De Pauli, C. P. (1998). Proton adsorption and electrokinetics of an Argentinean Montmorillonite. *Journal of Colloid Interface Science*, **202**, 195–204.

Barrash, W. & Clemo, T. (2002). Hierarchical geostatistics and multifacies systems: Boise Hydrogeophysical Research Site, Boise, Idaho. *Water Resources Research*, **38**(10), 1196, doi:10.1029/2002 WR001436, 2002.

Barrash, W. & Reboulet, E. C. (2004). Significance of porosity for stratigraphy and textual composition is subsurface coarse fluvial deposits, Boise Hydrogeophysical Research Site. *Geological Society of America Bulletin*, **116**(9/10), 1059–1073, doi:10.1130/B25370.1.

Barrash, W., Clemo, T. & Knoll, M. D. (1999). Boise Hydrogeophysical Research Site (BHRS): objectives, design, initial geostatistical results. In: *Proceedings of the Symposium on the Application of Geophysics to Engineering and Environmental Problems*, pp. 389–398. Oakland, California: Environmental & Engineering Geophysical Society.

Barus, C. (1882). *On the Electrical Activity of Ore Bodies*. In G. F. Becker (ed.), Geology of the Comstock Lode and the Washoe District, U. S. Geol. Surv. Monog., 309–367, 400–404.

Batchelor, G. K. (1972). *An Introduction to Fluid Dynamics*. Cambridge University Press.

Bear, J. & Verruijt, A. (1987). *Modeling Groundwater Flow and Pollution*. D. Reidel Publ. Co., Dordrecht, The Netherlands.

Berenger, J. P. (1994). A perfectly matched layer for the absorption of electromagnetic waves. *Journal of Computational Physics*, **114**, 185–200.

Bernabé, Y. (1998). Streaming potential in heterogeneous networks. *Journal of Geophysical Research – Solid Earth*, **103**(B9), 20 827–20 841, doi:10.1029/98JB02126.

Bernabé, Y. & Revil, A. (1995). Pore-scale heterogeneity, energy dissipation and the transport properties of rocks. *Geophysical Research Letters*, **22**, 12, 1529–1552.

Bigalke, J. & Grabner, E. W. (1997). The geobattery model: a contribution to large scale electrochemistry. *Electrochimica Acta*, **42**, 3443–3452.

Bigalke, J., Junge, A. & Zulauf, G. (2004). Electronically conducting brittle-ductile shear zones in the crystalline basement of Rittsteig (Bohemian Massif, Germany): evidence from self-potential and hole-to-surface electrical measurements. *International Journal of Earth Sciences*, **93**, 44–51.

Biot, M. (1941). General theory of three-dimensional consolidation. *Journal of Applied Physics*, **12**(2), 155–164.

Biot, M. (1956a). Theory of propagation of elastic waves in a fluid-saturated porous solid, I. Low-frequency range. *Journal of the Acoustical Society of America*, **28**, 168–178.

Biot, M. A. (1956b). Theory of propagation of elastic waves in a fluid-saturated porous solid, II. Higher-frequency range. *Journal of the Acoustical Society of America*, **28**, 178–191.

Biot, M. A. and Willis, D. G. (1957). The elastic coefficients of the theory of consolidation. *Journal of Applied Mechanics*, **24**, 594–601.

Bockris, J. O'M. & Reddy, A. K. N. (1970). *Modern Electrochemistry*, Vol. **II**. New York: Plenum Press.

Bogoslovsky, V. A. & Ogilvy, A. A. (1973). Deformation of natural electric fields near drainage structures. *Geophysical Prospecting*, **21**, 716–723.

Bogoslovsky, V. A. & Ogilvy, A. A. (1977). Geophysical methods in the investigations of landslides. *Geophysics*, **42**, 562–571.

Bolève, A., Crespy, A., Revil, A., Janod, F. & Mattiuzzo J. L. (2007a). Streaming potentials of granular media: influence of the Dukhin and Reynolds numbers. *Journal of Geophysical Research*, **112**, B08204, doi:10.1029/2006JB004673.

Bolève, A., Revil, A., Janod F., Mattiuzzo, J. L. & Jardani, A. (2007b). Forward modeling and validation of a new formulation to compute self-potential signals associated with ground water flow. *Hydrology and Earth System Sciences*, **11**, 1–11.

Bolève, A., Revil, A., Janod, F., Mattiuzzo, J. L. & Fry J.-J. (2009). Preferential fluid flow pathways in embankment dams imaged by self-potential tomography. *Near Surface Geophysics*, **7**(5), 447–462.

Bolève, A., Janod, F., Revil, A., Lafon, A. & Fry, J.-J. (2011). Localization and quantification of leakages in dams using time-lapse self-potential measurements associated with salt tracer injection. *Journal of Hydrology*, **403**(3–4), 242–252.

Börner, F. D. (1992). Complex conductivity measurements of reservoir properties. In *Proceedings of the Third European Core Analysis Symposium*, pp. 359–386, Advances in Core Evaluation v. 3: Reservoir Management – Reviewed Proceedings of the Society for Core Analysis Third European Core Analysis Symposium – Paris, eds. P. F. Worthington and Catherine Chardaire-Riviere. Harwood Academic, Neward, New Jersey.

Bou Matar, O., Prebrazhensky, V. & Pernod, P. (2005). Two-dimensional axisymmetric numerical simulation of supercritical phase conjugation of ultrasound in active solid media. *Journal of the Acoustical Society of America*, **118**, 2880–2890.

Brooks, R. H. & Corey, A. T. (1964). Hydraulic properties of porous media. *Hydrology Papers*, **3**. Colorado State University, Ft. Collins, Colorado.

Bussian, A. E. (1983). Electrical conductance in a porous medium. *Geophysics*, **48**(9), 1258–1268, doi:10.1190/1.1441549:1983.

Casagrande, L. (1983). Stabilization of soils by means of electro-osmosis: state of art. *Journal of Boston Society of Civil Engineers*, **69**(2), 255–302.

Castermant, J., Mendonça, C. A., Revil, A. *et al.* (2008). Redox potential distribution inferred from self-potential measurements during the corrosion of a burden metallic body. *Geophysical Prospecting*, **56**, 269–282.

Chen, J., Hubbard, S. & Rubin, Y. (2001). Estimating the hydraulic conductivity at the South Oyster Site from geophysical tomographic data using Bayesian techniques based on the normal linear regression model. *Water Resources Research*, **37**(6), 1603–1613.

Chew, W. C. & Weedon W. H. (1994). A 3-D perfectly matched medium from modified Maxwell's equations with stretched coordinates. *Microwave and Optical Technology Letters*, **7**, 599–604.

Childers, S. E., Ciufo, S. & Lovley, D. R. (2002). *Geobacter metallireducens* accesses insoluble Fe(III) oxide by chemotaxis. *Nature*, **416**, 767–769.

Christensen, T. H., Bjerg, P. L., Banwart S. A. *et al.* (2000). Characterization of redox conditions in groundwater contaminant plumes. *Journal of Contaminant Hydrology*, **45**, 165–241.

Clavier, C., Coates, G. & Dumanoir, J. (1984). The theoretical and experimental basis for the dual-water model for the interpretation of shaly sands. *Society of Petroleum Engineers Journal*, **24**(2), 153–169.

Clayton, R. & Engquist, B. (1977). Absorbing boundary conditions for acoustic and elastic wave equations. *Bulletin of the Seismological Society of America*, **67**, 1529–1540.

Closson, D., Karaki, N. A., Hussein, M. J. *et al.* (2003). Subsidence et effondrements le long du littoral jordanien de la mer Morte: apport de la gravimétrie et de l'interférométrie radar différentielle. *Comptes Rendus Geosciences*, **335**, 869–879.

Corwin, R. F. & Hoover, D. B. (1979). The self-potential method in geothermal exploration. *Geophysics*, **44**(2), 226–245.

Coussy, O. (1995). *Mechanics of Porous Continua*. Wiley, Chichester.

Crespy, A., Revil, A., Linde, N. *et al.* (2008). Detection and localization of hydromechanical disturbances in a sandbox using the self-potential method. *Journal of Geophysical Research*, **113**, B01205, doi:10.1029/2007JB005042.

Cruden, D. M. & Varnes, D. J. (1996). Landslide types and processes and mitigation. In Turner AK and Schuster R. L. (eds.) *Landslides – Investigation and Mitigation. Transportation Research Board, Spec. Rep. 247.* National Academy of Sciences, Washington, DC, pp. 36–75.

Curie, P. (1894). Sur la symétrie dans les phénomènes physiques, symétrie d'un champ électrique et d'un champ magnétique. *Journal de Physique (Paris)*, **3**, 393–415.

Degond P., Génieys, S. & Ansgar Jüngel, A. (1998). A steady-state system in non-equilibrium thermodynamics including thermal and electrical effects. *Mathematical Methods in the Applied Sciences*, **21**, 1399–1413.

de Groot, D. V. (1961). Non-equilibrium thermodynamics of systems in an electromagnetic field. *J. Nucl. Energy. Part C: Plasma Physics*, **2**, 188–193.

de Groot, D. V. & Mazur, P. (1984). *Non-Equilibrium Thermodynamics*. Dover, New York.

de la Peña, L. A. & Puente, C. I. (1979). *The geothermal field of Cerro Prieto*. In: Elders, W. A. (ed.) *Geology and geothermics of the Salton Trough*: Riverside, University of California, Inst. of Geophysics and Planetary Phys., Report UCR/IGPP-79/23, 20–35.

Descostes, M., Blin, V., Bazer-Bachi, F. *et al.* (2008). Diffusion of anionic species in Callovo-Oxfordian argillites and Oxfordian limestones (Meuse/Haute–Marne, France). *Applied Geochemistry*, **23**, 655–677.

Detournay, E. & Cheng, A. H.-D. (1993). Fundamentals of poroelasticity. Chapter 5 in *Comprehensive Rock Engineering: Principles, Practice and Projects*, Vol. **II**, Analysis and Design Method, ed. C. Fairhurst, Pergamon Press, pp. 113–171.

Deutsch, C. V. & Journel, A. G. (1992). *GSLIB: Geostatistical Software Library and User's Guide*. Oxford University Press, New York.

Deutsch, C. V. & Journel, A. G. (1998). *GSLIB: A Geostatistical Software Library and User's Guide*, 2nd edn. Oxford University Press, London.

Dimick, N. J. (2007). The ability to predict ground water flow in a structurally faulted river valley with naturally occurring hot springs using multivariate geochemical analyses. MS Thesis, Colorado School of Mines.

Draper, D. (1995). Assessment and propagation of model uncertainty. *Journal of the Royal Statistical Society, Series B*, **57**, 45–97.

Dullien, F. A. L. (1992). *Porous Media: Fluid Transport and Pore Structure*, 2nd edn, 574 pp. Academic, San Diego.

Ellingsrud, S., Eidesmo, T. & Strack, K.-M. (2008). *CSEM: a fast growing technology*, SEG Annual Meeting, Las Vegas.

Erchul, R. A. & Slifer, D. W. (1987). The use of spontaneous potential in the detection of groundwater flow patterns and flow rate in karst areas. *Karst Hydrogeology*, Proc. 2nd Conference, Orlando, pp. 217–226.

Ermakova L. E., Sidorova, M., Jura, N. & Savina, I. (1997). Adsorption and electrokientic characteristics of micro- and macroporous glasses in 1:1 electrolytes. *Journal of Membrane Science*, **131**, 125–141.

Fitterman, D. V. (1976). Theoretical resistivity variations along stressed stike-slip faults, *Journal of Geophysical Research*, **81**, 4909–4915.

Fitterman, D. V. & Corwin, R. F. (1982). Inversion of self-potential data from the Cerro Prieto geothermal field, Mexico. *Geophysics*, **47**, 938–945.

Fourie, F. D. (2003). Application of electroseismic techniques to geohydrological investigations in Karoo Rocks. PhD Thesis, University of the Free State, Bloemfontein, South Africa.

Fournier, C. (1989). Spontaneous potentials and resistivity surveys applied to hydrogeology in a volcanic area: case history of the Chaîne des Puys (Puy-de-Dôme, France). *Geophysical Prospecting*, **37**, 647–668.

Fox, R. W. (1830). On the electromagnetic properties of metalliferous veins in the mines of Cornwall. *Philosophical Transactions of the Royal Society*, **120**, 399–414.

Frash, L. P. & Gutierrez, M. (2012). Development of a new temperature controlled true-triaxial apparatus for simulating enhanced geothermal systems (EGS) at the laboratory scale. In *Proceedings 37th Workshop on Geothermal Reservoir Engineering*, Stanford University, Stanford, CA, January 30th–February 1st, 2012, SGP-TR-194.

Frenkel, J. (1944). On the theory of seismic and seismoelectric phenomena in a moist soil. *Journal Physics* (Soviet), **8**(4), 230–241.

Friborg, J. (1996). Experimental and theoretical investigations into the streaming potential phenomenon with special reference to applications in glaciated terrain. PhD thesis, Lulea University of Technology, Sweden.

Furini, S., Zerbetto, F. & Cavalcanti, S. (2006). Application of the Poisson–Nernst–Planck theory with space-dependent diffusion coefficients to KcsA. *Biophysical Journal*, **91**(9), 3162–3169.

Gallipoli, M., Lapenna, V., Lorenzo, P. *et al.* (2000). Comparison of geological and geophysical prospecting techniques in the study of a landslide in Southern Italy. *European Journal of Environmental and Engineering Geophysics*, **4**, 117–128.

Gaucher, E., Robelin, C., Matray, J. M. *et al.* (2004). ANDRA underground research laboratory: interpretation of the mineralogical and geochemical data acquired in the Callovo-Oxfordian Formation by investigative drilling. *Physics and Chemistry of the Earth*, **29**, 55–77, doi:10.1016/j.pce.2003.11.006.

Gelman, A. G., Roberts, G. O. & Gilks, W. R. (1996). Efficient Metropolis jumping rules, In *Bayesian Statistics V*, pp. 599–608, J. M. Bernardo *et al.* (eds.). Oxford University Press, New York.

Gelman, A. G., Carlin, J. B., Stern, H. S. and Rubin, D. B. (2004). *Bayesian Data Analysis*, 2nd edn. London CRC Press, London.

Gex, P. (1980). Electrofiltration phenomena associated with several dam sites. *Bulletin of the Society Vaud Science and Nature*, **357**, 39–50.

Gex, P. (1993). Electrofiltration measurements on the Frasse landslide, the Pre-Alps of western Switzerland. *Hydrogeologie*, **3**, 239–246.

Goldie, M. (2002). Self-potentials associated with the Yanacocha high sulfidation gold deposit in Peru. *Geophysics* **67**, 684–689.

Gómez-Ortiz, D. & Agarwal, B. N. P. (2005). 3DINVER.M: a Matlab® program to invert the gravity anomaly over a 3D horizontal density interface by Parker-Oldenburg's algorithm. *Computers & Geosciences*, **31**(4), 513–520.

Gonçalvès, J., Rousseau-Gueutin, P. & Revil, A. (2007). Introducing interacting diffuse layers in TLM calculations: a reappraisal of the influence of the pore size on the swelling pressure and the osmotic efficiency of compacted bentonites. *Journal of Colloid and Interface Science*, **316**(1), 92–99.

Grech, R. T., Cassar, J., Muscat, K. P. *et al.* (2008). Review on solving the inverse problem in EEG source analysis. *Journal of NeuroEngineering and Rehabilitation*, **5**, 25, doi:10.1186/1743-0003-5-25.

Green, D. J. (2000). Discussion: geophysical studies at Kartchner Caverns State Park, Arizona. *Journal of Cave and Karst Studies*, **62**(1), 27.

Gouveia, W. P. & Scales, A. J. (1998). Bayesian seismic waveform inversion: parameter estimation and uncertainty analysis. *Journal of Geophysical Research*, **103**(B2), 2759–2779, doi:10.1029/97JB02933.

Haario, H., Saksman, E. & Tamminen, J. (2001). An adaptive Metropolis algorithm. *Bernoulli*, **7**, 223–242.

Haario, H., Laine, M., Lehtinen, M., Saksman, E. & Tamminen, J. (2004). McMC methods for high dimensional inversion in remote sensing, *Journal of the Royal Stastistical Society, Series B*, **66**, 591–607.

Haas, A. & Revil, A. (2009). Electrical signature of pore scale displacements. *Water Resources Research*, **45**, W10202, doi:10.1029/2009WR008160.

Hack, R. (2000). Geophysics for slope stability. *Surveys in Geophysics*, **21**, 423–448.

Hallenburg, J. K. (1998). *Standard Methods of Geophysical Formation Evaluation*, PennWell Books, Tulsa, Okla.

Hansen, P. C. (1998). *Rank-Deficient and Discrete Ill-Posed Problems: Numerical Aspects of Linear Inversion*. SIAM, Philadelphia.

Helmholtz, H. (1879). Study concerning electrical boundary boundary layers. *Weidemann Annal Physik Chemie*, **7**, 337–382, 3rd Ser.

Hildebrand, F. B. (1965). *Methods of Applied Mathematics*, 2nd edn. Englewood Cliffs, Prentice-Hall.

Hostetler, J. D. (1984). Electrode electrons, aqueous electrons, and redox potentials in natural waters. *American Journal of Science*, **284**, 734–759.

Hutchinson, D. J., Harrap, R., Diederichs, M., Villeneuve, M. & Kjelland, N. (2003). Geotechnical rule development for ground instability assessment using intelligent GIS and networked monitoring sensors. In *3rd Canadian Conference on Geotechnique and Natural Hazards*. Edmonton, Alberta, Canada. June 9 and 10, 2003.

Huang, Q. & Liu, T. (2006). Earthquakes and tide response of geoelectric potential field at the Niijima station. *Chinese Journal of Geophysics*, **49**(6), 1745–1754.

Huygue, J. M. & Janssen, J. D. (1997). Quadriphasic mechanics of swelling incompressible porous media. *International Journal of Engineering Science*, 35(8), 793–802.

Ikard, S. J., Revil, A., Jardani, A. *et al.* (2012). Saline pulse test monitoring with the self-potential method to nonintrusively determine the velocity of the pore water in leaking areas of earth dams and embankments. *Water Resources Research*, **48**, W04201, doi:10.1029/2010WR010247.

Ishido, T. (1989). Self-potential generation by subsurface water flow through electrokinetic coupling. *Lecture Notes in Earth Science*, **27**, 121–133.

Ishido, T. (2004). Electrokinetic mechanism for the "W"-shaped self-potential profile on volcanoes. *Geophysical Research Letters*, **31**, L15616, doi:10.1029/2004GL020409.

Ishido, T. & Pritchett, J. W. (1999). Numerical simulation of electrokinetic potentials associated with subsurface fluid flow. *Journal of Geophysical Research*, **104**(B7), 15 247–15 259.

Iuliano, T., Mauriello, P. & Patella, D. (2002). Looking inside Mount Vesuvius by potential fields integrated probability tomographies. *Journal of Volcanology and Geothermal Research*, **113**, 363–378.

Ivanov, A. G. (1939). Effect of electrization of earth layers by elastic waves passing through them. *Doklady Akademii nauk SSSR*, **24**, 42–45.

Jackson, D. B. & Kauahikaua, J. (1987). Regional self-potential anomalies at Kilauea Volcano. *USGS Professional Paper*, **1350**, 947–959.

Jackson, M. D. (2008). Characterization of multiphase electrokinetic coupling using a bundle of capillary tubes model. *Journal of Geophysical Research*, **113**, B04201, doi:10.1029/2007JB005490.

Jackson, M. D. (2010). Multiphase electrokinetic coupling: insights into the impact of fluid and charge distribution at the pore scale from a bundle of capillary tubes model. *Journal of Geophysical Research*, **115**, B07206, doi:10.1029/2009JB007092.

Jardani, A. & Revil, A. (2009). Stochastic joint inversion of temperature and self-potential data. *Geophysical Journal International*, **179**(1), 640–654, doi:10.1111/j.1365–246X.2009.04295.x, 2009.

Jardani, A., Dupont, J. P. & Revil, A. (2006a). Self-potential signals associated with preferential groundwater flow pathways in sinkholes. *Journal of Geophysical Research*, **111**, B09204, doi:10.1029/2005JB004231.

Jardani. A., Revil, A. & Dupont, J. P. (2006b). Self-potential Tomography applied to the determination of cavities. *Geophysical Research Letters*, **33**, L13401, doi:10.1029/2006GL026028.

Jardani, A., Revil, A., Santos, F., Fauchard, C. & Dupont, J. P. (2007a). Detection of preferential infiltration pathways in sinkholes using joint inversion of self-potential and EM-34 conductivity data. *Geophysical Prospecting*, **55**, 1–11, doi:10.1111/j.1365–2478.2007.00638.x.

Jardani, A., Revil, A., Bolève, A. *et al.* (2007b). Tomography of groundwater flow from self-potential (SP) data. *Geophysical Research Letters*, **34**, L24403, doi:10.1029/2007GL031907.

Jardani, A., Revil, A., Bolève, A. & Dupont, J. P. (2008). 3D inversion of self-potential data used to constrain the pattern of ground water flow in geothermal fields. *Journal of Geophysical Research*, **113**, B09204, doi:10.1029/2007JB005302.

Jardani, A., Revil, A., Barrash, W. *et al.* (2009). Reconstruction of the water table from self-potential data: a bayesian approach. *Ground Water*, **47**(2), 213–227.

Jardani, A., Revil, A., Slob, E. & Sollner, W. (2010). Stochastic joint inversion of 2D seismic and seismoelectric signals in linear poroelastic materials. *Geophysics*, **75**(1), N19–N31, doi:10.1190/1.3279833.

Jardani, A., Revil, A. & Dupont, J. P. (2012). Stochastic joint inversion of hydrogeophysical data for salt tracer test monitoring and hydraulic conductivity imaging. *Advances in Water Resources*, **52**, 62–77.

Jardine, P. M., Wilson, G. V & Luxmoore, R. J. (1988). Modeling the transport of inorganic ions through undisturbed soil columns from two contrasting watersheds. *Soil Science Society of America Journal*, **52**, 1252–1259.

Jardine, P. M., Jacobs, G. K. & Wilson, G. V. (1993a). Unsaturated transport processes in undisturbed heterogeneous porous media. I. Inorganic contaminants. *Soil Science Society of America Journal*, **57**, 945–953.

Jardine, P. M., Jacobs, G. K. & O'Dell, J. D. (1993b). Unsaturated transport processes in undisturbed heterogeneous porous media II. Co-contaminants. *Soil Science Society of America Journal*, **57**, 954–962.

Johnson, D. L. & Sen, P. N. (1988). Dependence of the conductivity of a porous medium on electrolyte conductivity. *Physical Review* B, **37**, 3502–3510.

Jougnot, D., Revil, A. & Leroy, P. (2009). Diffusion of ionic tracers in the Callovo-Oxfordian clay-rock using the Donnan equilibrium model and the electrical formation factor. *Geochimica et Cosmochimica Acta*, **73**, 2712–2726.

Jougnot, D., Ghorbani, A., Revil, A., Leroy, P. & Cosenza, P. (2010b). Spectral induced polarization of partially saturated clay-rocks: a mechanistic approach. *Geophysical Journal International*, **180**(1), 210–224, doi:10.1111/j.1365–246X.2009.04426.x.

Jougnot, D., Linde, N., Revil, A. & Doussan, C. (2012). Derivation of soil-specific streaming potential electrical parameters from hydrodynamic characteristics of partially saturated soils. *Vadose Zone Journal*, **11**(1), doi:10.2136/vzj2011.0086.

Kappenman, E. S. & Luck, S. J. (2010). The effects of electrode impedance on data quality and statistical significance. *ERP Recordings*, **47**(5), 888–904, doi:10.1111/j.1469–8986.2010.01009.x.

Karaoulis M., Revil, A., Werkema, D. D. *et al.* (2011). Time-lapse 3D inversion of complex conductivity data using an active time constrained (ATC) approach. *Geophysical Journal International*, **187**, 237–251, doi:10.1111/j.1365–246X.2011.05156.x.

Karaoulis, M., Revil, A., Zhang, A. J. & Werkema, D. D. (2012). Time-lapse cross-gradient joint inversion of cross-well DC 1 resistivity and seismic data: a numerical investigation. *Geophysics*, **77**, D141–D157, doi:10.1190/GBO2012–0011.1.

Kawakami, N. & Takasugi, S. (1994). SP Monitoring during hydraulic fracturing using the TG-2 well. *European Assoc. of Exploration Geophysicists*; 56th meeting and tech. exhibition, Vienna, Austria.

Kemper, W. D. & Rollins, J. B. (1966). Osmotic efficiency coefficients across compacted clays. *Soil Science Society of America Proceedings*, **30**(5), 529–534.

Kim Y. J., Moon, J. W., Roh, Y. & Brooks, S. C. (2009). Mineralogical characterization of saprolite at the FRC background site in Oak Ridge, Tennessee. *Environmental Geology*, **58**, 1301–1307.

Koch, K., Kemna, A., Irving, J. & Holliger, K. (2011). Impact of changes in grain size and pore space on the hydraulic conductivity and spectral induced polarization response of sand. *Hydrology and Earth System Sciences*, **15**, 1785–1794, doi:10.5194/hess-15-1785-2011.

Kooner, Z. S., Jardine, P. M. & Feldmen, S. (1995). Competitive surface complexation reactions of SO_4^{2-} and natural organic carbon on soil. *Journal of Environmental Quality*, **24**, 656–662.

Kowalsky, M. B., Finsterle, S., Peterson, J. *et al.* (2005). Estimation of field-scale soil hydraulic and dielectric parameters through joint inversion of GPR and hydrological data. *Water Resource Research*, **41**(11), W11425.

Kulessa, B., Murray, T. & Rippin, D. (2006). Active seismoelectric exploration of glaciers. *Geophysical Research Letters*, **33**, L07503, doi:10.1029/2006GL025758.

Kuwano, O., Nakatani, M. & Yoshida, S. (2006). Effect of the flow state on streaming current. *Geophysical Research Letters*, **33**, L21309, doi:10.1029/2006GL027712.

Laignel, B., Dupuis, E., Rodet, J., Lacroix, M. & Massei, N. (2004). An example of sedimentary filling in the chalk karst of the Western Paris Basin: characterization, origins, and hydrosedimentary behavior. *Z. F. Géomorphologie*, **48**(2), 219–243.

Lake, C. B. & Rowe, R. K. (2000). Diffusion of sodium and chloride through geosynthetic clay liners. *Geotextiles and Geomembranes*, **18**, 2–4, 103–131.

Lange, A. L. (1999). Geophysical studies at Kartchner caverns State Park, Arizona. *Journal of Cave and Karst Studies*, **61**(2), 68–72.

Lange, A. L. (2000). Reply: Geophysical studies at Kartchner caverns State Park, Arizona. *Journal of Cave and Karst Studies*, **62**(1), 28–29.

Lapenna, V., Lorenzo, P., Perrone, A. *et al.* (2003). High-resolution geoelectrical tomographies in the study of the Giarrossa landslide (Potenza, Basilicata). *Bulletin of Engineering Geology and the Environment*, **62**, 259–268.

Lapenna, V., Lorenzo, P., Perrone, A. *et al.* (2005). 2D electrical resistivity imaging of some complex landslides in Lucanian Apennine chain, southern Italy. *Geophysics*, **70**(3), B11–B18.

Lappin-Scott, H. M. & Costerton, J. W. (1995). *Microbial Biofilms*. Cambridge University Press.

Leroy, P. & Revil, A. (2004). A triple layer model of the surface electrochemical properties of clay minerals. *Journal of Colloid and Interface Science*, **270**(2), 371–380.

Leroy, P. & Revil, A. (2009). Spectral induced polarization of clays and clay-rocks. *Journal of Geophysical Research*, **114**, B10202. doi:10.1029/2008JB006114.

Leroy, P., Revil, A., Altmann, S. & Tournassat, C. (2007). Modeling the composition of the pore water in a clay-rock geological formation (Callovo-Oxfordian, France). *Geochimica et Cosmochimica Acta*, **71**(5), 1087–1097, doi:10.1016/j.gca.2006.11.009.

Leroy, P., Revil, A., Kemna, A., Cosenza, P. & Gorbani, A. (2008). Spectral induced polarization of water-saturated packs of glass beads. *Journal of Colloid and Interface Science*, **321**(1), 103–117.

Levenston, M. E., Frank, E. H. & Grodzinsky, A. J. (1999). Electrokinetic and poroelastic coupling during finite deformations of charged porous media. *Journal of Applied Mechanics*, **66**, 323–333.

Limbach, F. W. (1975). The Geology of the Buena Vista area, Chaffee County Colorado. MS thesis, Colorado School of Mines, T-1692.

Linde, N. & Revil, A. (2007). Inverting residual self-potential data for redox potentials of contaminant plumes. *Geophysical Research Letters*, **34**, L14302, doi:10.1029/2007GL030084.

Linde, N., Jougnot, D., Revil, A. *et al.* (2007a). Streaming current generation in two-phase flow conditions. *Geophysical Research Letters*, **34**(3), L03306, doi:10.1029/2006GL028878.

Linde, N., Revil, A., Bolève, A. *et al.* (2007b). Estimation of the water table throughout a catchment using self-potential and piezometric data in a Bayesian framework. *Journal of Hydrology*, **334**, 88–98.

Lipsicas, M. (1984). Molecular and surface interactions in clay intercalates, in *Physics and Chemistry of Porous Media*, edited by D. L. Johnson and P. N. Sen, pp. 191–202, American Institute of Physics, College Park, Md.

Lippmann, M. J. & Bodvarsson, G. S. (1982). *Modeling Studies on Cerro Prieto*, Presented at the 4th Symposium on the Cerro Prieto Geothermal Field, pp. CP–25, Lawrence Berkeley Lab., Guadalajara, Mexico, August 10–12, 1982, Report LBL-14897.

Lippmann, M. J. & Bodvarsson, G. S. (1983). Numerical studies of the heat and mass transport in the Cerro Prieto geothermal field, Mexico. *Water Resource Research*, **19**, 753–767.

Lo, W.-C., Sposito, G. & Majer, E. (2002). Immiscible two-phase fluid flows in deformable porous media. *Advances in Water Resources*, **25**(8–12), 1105–1117.

Lo, W.-C., Sposito, G. & Majer, E. (2005). Wave propagation through elastic porous media containing two immiscible fluids. *Water Resources Research*, **41**, W02025, doi:10.1029/2004WR003162.

Lockhart, N. C. (1980). Electrical properties and the surface characteristics and structure of clays, II. Kaolinite: a nonswelling clay. *Journal of Colloid Interface Science*, **74**, 520–529.

Loke, M. H. & Barker, R. D. (1996). Rapid least-squares inversion of apparent resistivity pseudosections by a quasi-Newton method. *Geophysical Prospecting*, **44**, 131–152.

Lu, N., Godt, J. W. & Wu, D. T. (2010). A closed-form equation for effective stress in unsaturated soil. *Water Resources Research*, **46**, W05515, doi:10.1029/2009WR008646.

Lyons, D. J. & van de Kamp, P. C. (1980). *Subsurface geological and geophysical study of the Ceerro Prieto geothermal field, Berkeley*. Lawrence Berkeley Laboratory, LBL-10540.

Ma, C. & Eggleton, R. A. (1999). Cation exchange capacity of kaolinite. *Clays and Clay Minerals*, **47**, 174–180.

Macaskill, J. B. & Bates, R. G. (1978). Standard potential of the silver–silver chloride electrode. *Pure and Applied Chemistry*, **50**, 1701–1706.

Mahardika, H., Revil, A. & Jardani, A. (2012). Waveform joint inversion of seismograms and electrograms for moment tensor characterization of tracking events. *Geophysics*, **77**(5), 1D23–1D39, doi:10.1190/GEO2012-0019.1,2012.

Maineult, A., Bernabé, Y. & Ackerer, P. (2006). Detection of advected, reacting redox fronts from self-potential measurements. *Journal of Contaminant Hydrology*, **86**, 32–52.

Maineult, A., Strobach, E. & Renner, J. (2008). Self-potential signals induced by periodic pumping tests. *Journal of Geophysical Research*, **113**, B1: B01203.

Malusis, M. A. & Shackelford, C. D. (2004). Predicting solute flux through a clay membrane barrier. *Journal of Geotechnical and Geological Engineering*, **130**(5), 477–487, doi:10.1061/(ASCE)1090–0241.

Malusis, M. A., Shackelford, C. D. & Olsen, H. W. (2003). Flow and transport through clay membrane barriers. In: 3rd British Geotechnical Society Geoenvironmental Engineering Conference, Edinburgh, Scotland. *Engineering Geology*, **70**(3–4), 235–248, doi:10.1016/50013-7952(03)00092-9.

Manon, A., Mazor, E., Jimenez, M. *et al.* (1977). *Extensive geochemical studies in the geothermal field of Cerro Prieto, Mexico.* LBB report 7019.

Mariner, R. H., Swanson, J. R., Orris, G. J., Presser, T. S. & Evans, W. C. (1980). *Chemical and isotopic data for water of thermal springs of Oregon.* USGS open file report, 80–737.

Martínez-Pagán, P., Jardani, A., Revil, A. & Haas. (2010). A. self-potential monitoring of a salt plume: a sandbox experiment. *Geophysics*, **75**(4), WA17–WA25.

Matheron, G. (1965). *Les variables régionolisées et lewr estimation: une application de la théorie des fonctions aléatoires auk science de la nature.* Masson, Paris.

Matteucci, M. C. (1865). Sur les courants electriques de la terre [On the electrical currents of the earth]. *Annales de chimie et de physique*, **4**(4), 177–192.

Mauritsch, H. J., Seiberl, W., Arndt, R. *et al.* (2000). Geophysical investigations of large landslides in the Carnic region of southern Austria. *Engineering Geology*, **56**, 373–388.

McCann, D. M. & Forster, A. (1990). Reconnaissance geophysical methods in landslide investigations. *Engineering Geology*, **29**, 59–78.

McKay, L. D., Driese, S. G., Smith, K. H. & Vepraskas, M. J. (2005). Hydrogeology and pedology of saprolite formed from sedimentary rock, eastern Tennessee, USA. *Geoderma*, **126**, 27–45.

Melkior, T., Yahiaoui, S., Thoby, D., Motellier, S. & Barthes, V. (2007). Diffusion coefficients of alkaline cations in Bure mudrock. *Physics and Chemistry of the Earth*, **32**, 453–462.

Mendonça, C. A. (2008). Forward and inverse self-potential modeling in mineral exploration. *Geophysics*, **73**(1), F33–F43.

Merkler, G. P, Militzer, H., Hötzl, H., Armbruster, H. & Brauns, J. (1989). Detection of subsurface flow phenomena. *Lecture Notes in Earth Sciences*, **27**. Springer, Berlin/Heidelberg. ISBN 9783540518754.

Meyer, K. H. & Sievers, J. F. (1936). La perméabilité des membranes. I. Théorie de la perméabilité ionique. *Helvetica Chimica Acta*, **19**, 649.

Michalak, A. M. (2008). A Gibbs sampler for inequality-constrained geostatistical interpolation and inverse modeling. *Water Resources Research*, **44**, W09437, doi:10.1029/2007WR006645.

Migunov, N. & Kokorev A. (1977). Dynamic properties of the seismoelectric effect of water saturated rocks. *Izvestiya, Earth Physics*, **13**, 443–446.

Mikhailov, O. V., Haartsen, M. W. & Toksoz, M. N. (1997). Electroseismic investigation of the shallow subsurface: field measurements and numerical modelling. *Geophysics*, **62**, 97–105.

Mikhailov, O., Queen V. J. & Toksöz, M. N. (2000). Using borehole electroseismic measurements to detect and characterize fractured (permeable) zones. *Geophysics*, **65**, 1098–1112.

Miller, M. G. (1999). Active breaching of a geometric segment boundary in the Sawatch Range normal fault, Colorado, USA. *Journal of Structural Geology*, **21**, 769–776.

Minsley, B. J. (2007). *Modeling and inversion of self-potential data.* PhD Thesis, Massachusetts Institute of Technology, Cambridge, Massachusetts.

Minsley, B. J., Sogade, J. & Morgan, F. D. (2007a). Three-dimensional source inversion of self-potential data. *Journal of Geophysical Research*, **112**, B02202, doi : 1029/2006JB004262.

Minsley, B. J., Sogade, J. & Morgan, F. D. (2007b). Three-dimensional self-potential inversion for subsurface DNAPL contaminant detection at the Savannah River Site, South Carolina. *Water Resources Research*, **43**, W04429, doi:10.1029/2005WR003996.

Minsley, B. J., Coles D. A., Vichabian, Y. & Morgan, F. D. (2008). Minimization of self-potential survey mis-ties acquired with multiple reference locations. *Geophysics*, **73**(2), F71–F81, doi:10.1190/1.2829390.

Minsley, B. J., Burton, L. B., Ikard, S. & Powers, H. M. (2011). Hydrogeophysical Investigations at Hidden Dam, Raymond, California. *Journal of Environmental & Engineering Geophysics*, **16**(4), 145–164.

Mitchell, J. K. (1993). *Fundamentals of Soil Behavior*. New York, John Wiley & Sons.

Moore, J. R., Boleve, A., Sanders, J. W. & Glaser, S. D. (2011). Self-potential investigation of moraine dam seepage. *Journal of Applied Geophysics*, **74**, 277–286.

Morency, C. & Tromp, J. (2008). Spectral-element simulations of wave propagation in porous media. *Geophysical Journal International*, **175**, 301–345.

Mualem, Y. (1986). Hydraulic conductivity of unsaturated soils: prediction and formulas. Methods of soil analysis. Part 1. Physical and mineralogical methods, 2nd edn, *Agronomy*, A. Klute (ed.), Am. Soc. of Agronomy, Inc. and Soil Sci. SOC. of Am. Inc. Madison. Wis., 799–823.

Naudet, V. & Revil, A. (2005). A sandbox experiment to investigate bacteria-mediated redox processes on self-potential signals. *Geophysical Research Letters*, **32**, L11405, doi:10.1029/2005GL022735.

Naudet, V., Revil, A., Bottero, J. –Y. & Bégassat, P. (2003). Relationship between self-potential (SP) signals and redox conditions in contaminated groundwater. *Geophysical Research Letters*, **30**(21), 2091, doi:10.1029/2003GL018096.

Naudet, V., Revil, A., Rizzo, E., Bottero, J. Y. & Bégassat, P. (2004). Groundwater redox conditions and conductivity in a contaminant plume from geoelectrical investigations. *Hydrology and Earth System Sciences*, **8**(1), 8–22.

Newman, J. S. (1991). *Electrochemical Systems*, 2nd edn, Prentice Hall, Englewood Cliffs.

Nitao, J. J. & Bear, J. (1996). Potentials and their role in porous media. *Water Resources Research*, **32**(2), 225–250, doi:10.1029/95WR02715.

Nourbehecht, B. (1963). *Irreversible thermodynamic effects in inhomogeneous media and their application in certain geoelectric problems*. PhD Thesis, MIT Cambridge.

Ntarlagiannis, D., Atekwana, E. A., Hill, E. A. & Gorby, Y. (2007). Microbial nanowires: is the subsurface "hardwired"? *Geophysical Research Letters*, **34**, L17305, doi:10.1029/2007GL030426.

Nyquist, J. E. & Corry, C. E. (2002). Self-potential: the ugly duckling of environmental geophysics. *Leading Edge*, **21**(5), 446–451.

Ogilvy, A. A., Ayed, M. A. & Bogoslovsky, V. A. (1969). Geophysical studies of water leakage from reservoirs. *Geophysical Prospecting*, **22**, 36–62.

Oltean, C. & Buès, M. A. (2002). Infiltration of salt solute in homogeneous and saturated porous media – an analytical solution evaluated by numerical simulations. *Transport in Porous Media*, **48**, 61–78.

Pain, C. C., Saunders, J. H., Worthington, M. H. *et al.* (2005). A mixed finite-element method for solving the poroelastic Biot equations with electrokineticcoupling. *Geophysical Journal International*, **160**, 592–608.

Panthulu T. V., Krishnaiah, C. & Shirke, J. M. (2001). Detection of seepage paths in earth dams using self-potential and electrical resistivity methods. *Engineering Geology*, **59**, 281–295.

Pascal-Marquis, R. D., Esslen, M., Kochi, K. & Lehmann, D. (2002). Functionnal imaging with low resolution brain electromagnetic tomography (LORETA): review, new comparaisons, and new validation. *Japanese Journal of Clinical Neurophysiology*, **30**, 81–94.

Parker, R. L. (1973). The rapid calculation of potential anomalies. *Geophysical Journal Royal Astronomical Society*, **31**, 447–455.

Patchett, J. G. (1975). An investigation of shale conductivity, Society of Professional Well Logging Analysis. *16th Logging Symposium*, Paper U.

Patella, D. (1997). Self-potential global tomography including topographic effects, *Geophysical Prospecting*, **45**, 843–863.

Paul, K. (1965). Direct interpretation of self-potential anomalies caused by inclined sheets of infinite horizontal extensions. *Geophysics*, **30**, 418–423.

Pengra, D. B., Li, S. X. & Wong, P.-Z. (1999). Determination of rock properties by low-frequency AC electrokinetics. *Journal of Geophysical Research*, **104**(B12), 29 485–29 508.

Perrier, F., Petiau, G., Clerc, G. *et al.* (1997). A one-year systematic study of electrodes for long period measurements of the electric field in geophysical environments. *Journal of Geomagnetism and Geoelectricity*, **49**, 1677–1696.

Perrier, F., Trique, M., Lorne, B. *et al.* (1998). Electrical potential variations associated with yearly lake level variations. *Geophysical Research Letters*, **25**, 1955–1959.

Perrone, A., Iannuzzi, A., Lapenna, V. *et al.* (2004). High-resolution electrical imaging of the Varco d'Izzo earthflow (Southern Italy). *Journal of Applied Geophysics*, **56/1**, 17–29.

Petiau, G. (2000). Second generation of lead-lead chloride electrodes for geophysical applications. *Pure and Applied Geophysics*, **157**, 357–382.

Plona, T. J. (1980). Observation of a second bulk compressional wave in a porous medium at ultrasonic frequencies. *Applied Physics Letters*, **36**, 259–261.

Poldini, E. (1938). Geophysical exploration by spontaneous polarization methods. *Mining Magazine London*, **59**, 278–282.

Polemio, M. & Sdao, F. (1998). Heavy rainfalls and extensive landslides occurred in Basilicata, southern Italy, in 1976. In *Proc. 8th Int. Cong. EEGS, Vancouver, Canada,* pp. 1849–1855.

Pride, S. (1994). Governing equations for the coupled electromagnetics and acoustics of Porous Media. *Phys. Rev. B*, **50**(21), 15 678–15 696.

Prigogine, I. (1947). *Etude Thermodynamique des Phénomènes Irréversibles*. Desoer, Liège.

Quarto, R. & Schiavone, D. (1996). Detection of cavities by the SP method. *First Break*, **48**(1), 76–86.

Quincke, G. (1859). Concerning a new type of electrical current. *Annalen der Physics und Chemie (Poggendorff's Annal., Ser. 2)* **107**, 1–47.

Raboute, A., Revil, A. & Brosse, E. (2003). In situ mineralogy and permeability logs from downhole measurement: application to a case study of chlorite-coated sandstone. *Journal of Geophysical Research*, **108**(B9), doi:10.1029/2002JB002178.

Rañada Shaw, A., Denneman, A. I. M. & Wapenaar, C. P. A. (2000). Porosity and permeability effects on seismo-electric reflection. In *Proceedings of the EAGE conference*. Paris, France.

Rao, A. D. & Babu, R. H. V. (1984). Quantitative interpretation of self-potential anomalies due to two-dimensional sheet like bodies. *Geophysics*, **48**, 1659–1664.

Reguera, G., McCarthy, K. D., Metha, T. *et al.* (2005). Extracellular electron transfer via microbial nanowires. *Nature*, **435**, 1098–1101.

Revil, A. (1999). Ionic diffusivity, electrical conductivity, membrane and thermoelectrical potentials in colloids and granular porous media: a unified model. *Journal of Colloid and Interface Science*, **212**, 503–522.

Revil, A. (2007a). Thermodynamics of transport of ions and water in charged and deformable porous media. *Journal of Colloid and Interface Science*, **307**(1), 254–264.

Revil, A. (2007b). Comment on "Permeability prediction from MICP and NMR data using an electrokinetic approach" (P. W. J. Glover, I. I. Zadjali, and K. A. Frew, 2006, Geophysics, 71, F49–F60), *Geophysics*, **72**(4), X3–X4, doi:10.1190/1.2743006.

Revil, A. (2012). Spectral induced polarization of shaly sands: influence of the electrical double layer. *Water Resources Research*, **48**, W02517, doi:10.1029/2011WR011260.

Revil, A. & Cathles, L. M. (1999). Permeability of shaly sands. *Water Resources Research*, **35**(3), 651–662.

Revil, A. & Florsch, N. (2010). Determination of permeability from spectral induced polarization data in granular media. *Geophysical Journal International*, **181**, 1480–1498, doi:10.1111/j.1365–246X.2010.04573.x.

Revil, A. & Jardani, A. (2010a). Stochastic inversion of permeability and dispersivities from time lapse self-potential measurements: a controlled sandbox study. *Geophysical Research Letters*, **37**, L11404, doi:10.1029/2010GL043257.

Revil, A. & Jardani, A. (2010b). Seismoelectric response of heavy oil reservoirs. Theory and numerical modelling. *Geophysical Journal International*, **180**, 781–797, doi:10.1111/j.1365–246X.2009.04439.x.

Revil, A. & Leroy, P. (2001). Hydroelectric coupling in a clayey material. *Geophysical Research Letters*, **28**, 8, 1643–1646.

Revil, A. & Leroy, P. (2004). Constitutive equations for ionic transport in porous shales. *Journal of Geophysical Research*, **109**, B03208, doi:10.1029/2003JB002755.

Revil, A. & Linde, N. (2006). Chemico-electromechanical coupling in microporous media. *Journal of Colloid and Interface Science*, **302**, 682–694.

Revil, A. & Mahardika, H. (2013). Coupled hydromechanical and electromagnetic disturbances in unsaturated clayey materials. *Water Resources Research*, **49**, doi:10.1002/wrcr.20092.

Revil, A. & Pezard, P. A. (1998). Streaming potential anomaly along faults in geothermal areas. *Geophysical Research Letters*, **25**(16), 3197–3200.

Revil, A., Darot, M. & Pezard, P. A. (1996). Influence of the electrical diffuse layer and microgeometry on the ionic diffusion coefficient in porous media. *Geophysical Research Letters*, **23**(15), 1989–1992.

Revil, A., Cathles, L. M., Losh, S. & Nunn, J. A. (1998). Electrical conductivity in shaly sands with geophysical applications. *Journal of Geophysical Research*, **103**(B10), 23 925–23 936.

Revil, A., Pezard, P. A. & Glover, P. W. J. (1999a). Streaming potential in porous media. 1. Theory of the zeta-potential. *Journal of Geophysical Research*, **104**(B9), 20 021–20 031.

Revil, A., Schwaeger, H., Cathles, L. M. & Manhardt, P. (1999b). Streaming potential in porous media. 2. Theory and application to geothermal systems. *Journal of Geophysical Research*, **104**(B9), 20 033–20 048.

Revil, A., Ehouarne, L. & Thyreault, E. (2001). Tomography of self-potential anomalies of electrochemical nature. *Geophysical Research Letters*, **28**(23), 4363–4366.

Revil, A., Naudet, V., Nouzaret, J. & Pessel, M. (2003). Principles of electrography applied to self-potential electrokineticsources and hydrogeological applications. *Water Resources Research*, **39**(5), 1114, doi:10.1029/2001WR000916.

Revil, A., Leroy, P. & Titov, K. (2005). Characterization of transport properties of argillaceous sediments. Application to the Callovo-Oxfordian Argillite. *Journal of Geophysical Research*, **110**, B06202, doi:10.1029/2004JB003442.

Revil, A., Linde, N., Cerepi, A. *et al.* (2007). Electrokinetic coupling in unsaturated porous media. *Journal of Colloid and Interface Science*, **313**(1), doi. 315–327, 10.1016/j.jcis.2007.03.037.

Revil, A., Mendonça, C. A., Atekwana, E. *et al.* (2010). Understanding biogeobatteries: where geophysics meets microbiology. *Journal of Geophysical Research*, **115**, G00G02, doi:10.1029/2009JG001065.

Revil, A., Woodruff, W. F. and Lu, N. (2011a). Constitutive equations for coupled flows in clay materials. *Water Resources Research*, **47**, W05548, doi:10.1029/2010WR010002.

Revil, A., Jardani, A., Hoopes, J. *et al.* (2011b). Non intrusive estimate of the flow rate of thermal water along tectonic faults in geothermal fields using the self potential method. *FastTIMES*, **16**, 4.

Revil, A., Koch, K. & Holliger, K. (2012a). Relating grain size distribution to permeability and spectral induced polarization relaxation times in sands. *Water Resources Research*, **48**, W05602, doi:10.1029/2011WR011561.

Revil, A., Karaoulis, M., Johnson, T. & Kemna, A. (2012b). Review: some low-frequency electrical methods for subsurface characterization and monitoring in hydrogeology. *Hydrogeology Journal*, **20**(4), 617–658, doi:10.1007/s10040–011-0819-x.

Richards, L. A. (1931). Capillary conduction of liquids through porous media. *Physics*, **1**, 318–333.

Richards, K., Revil, A., Jardani, A. *et al.* (2010). Pattern of shallow ground water flow at Mount Princeton Hot Springs, Colorado, using geoelectrical methods. *Journal of Volcanology and Geothermal Research*, **198**, 217–232.

Risgaard-Petersen N., Revil, A., Meister, P. & Nielsen. L. P. (2012). Sulfur, iron-, and calcium cycling associated with natural electric currents running through marine sediment. *Geochimica et Cosmochimica Acta*, **92**, 1–13.

Rizzo, E., Suski, B., Revil, A., Straface, S. & Troisi, S. (2004). Self-potential signals associated with pumping-tests experiments. *Journal of Geophysical Research*, **109**, B10203, doi:10.1029/2004JB003049.

Roden, J. A. & Gedney, S. D. (2000). Convolution PML, (CPML): An efficient FDTD implement of CFS-PML for arbitrary media. *Microwave & Optic, Technological Letters*, **27**, 334–339.

Rosanne, M., Mammar, N., Koudina, N. *et al.* (2003). Transport properties of compact clays. II. Diffusion. *Journal of Colloid and Interface Science*, **260**, 195–203.

Rousseau-Gueutin, P., Gonçalvès, J., Cruchaudet, M., de Marsily, G. and Violette, S. (2010). Hydraulic and chemical pulse tests in a shut-in chamber imbedded in an argillaceous formation: numerical and experimental approaches. *Water Resources Research*, **46**, W08516.

Rozycki, A. (2009). Evaluation of the streaming potential effect of piping phenomena using a finite cylinder model. *Engineering Geology*, **104**, 98–108.

Rozycki, A., Fonticiella, J. M. R. & Cuadra, A. (2006). Detection and evaluation of horizontal fractures in Earth dams using self-potential method. *Engineering Geology*, **82**(3), 145–153.

Rubino, J. G. & Holliger, K. (2012). Seismic attenuation and velocity dispersion in heterogeneous partially saturated porous rocks. *Geophysical Journal International*, **188**(3), 1088–1102, doi:10.1111/j.1365–246X.2011.05291.x.

Rust, W. M. (1938). A historical review of electrical prospecting methods. *Geophysics*, **3**(1), 1–6.

Salvatia, R. & Sasowskyb, I. (2002). Development of collapse sinkholes in areas of groundwater discharge. *Journal of Hydrology*, **264**, 1–11.

Samson, E., Marchand, J. & Snyder, K. A. (2003). Calculation of ionic diffusion coefficients on the basis of migration test results. *Material and Structures*, **36**, 156–165.

Sato, M. & Mooney, H. M. (1960). The electrochemical mechanism of sulfide self-potentials. *Geophysics*, **25**, 226–249.

Sava, P. & Revil, A. (2012). Virtual electrode current injection using seismic focusing and seismoelectric conversion. *Geophysical Journal International*, **191**(3), 1205–1209, doi:10.1111/j.1365.246X.2012.05700.x,2012.

Schenk, O., Bollhoefer, M. & Roemer, R. (2008). On large-scale diagonalization techniques for the Anderson model of localization. *SIAM Review*, **50**, 91–112.

Schlumberger, C. (1920). Etude sur la prospection électrique du sous-sol [Study on underground electrical prospecting]. *Gauthier-Villars et Cie, Paris*.

Schlumberger, C., Schlumberger, M. & Leonardon, E. G. (1932). Electrical coring: a method of determining bottom-hole data by electrical measurements. *American Institute of Mining and Metallurgical Engineers, Technical Publication*, **462**.

Schlumberger C., Schlumberger M. & Leonardon, E. G. (1933). A new contribution to subsurface studies by means of electrical measurements in drill holes. *American Institute of Mining and Metallurgical Engineers, Technical Publication*, **503** (also in 1934, *Trans.*, 110, 159–182), AIME, Englewood, CO.

Schmutz, M., Guerin, R., Andrieux, P. *et al.* (2009). Determination of the 3D structure of an earthflow by geophysical methods The case of Super Sauze, in the French southern Alps, *Journal of Applied Geophysics*, **68**(4), 500–507, doi:10.1016/j.jappgeo.2008.12.004.

Schwartz, L. M., Sen, P. N. & Johnson, D. L. (1989). Novel geomatrical effects in electrolytic conduction in porous media. *Physics A*, **157**(1), 493–496, doi:10.1016/0378-4371(89)90348-8.

Semenov, A. S. (1980). *Elektrorazvedka metodom estestvennogo elektricheskogopolia* (Electrical prospecting with the natural electric field method, 2nd edn, in Russian). Nedra, Leningrad, 341–364.

Shainberg, I., Alperovitch, N. & Keren, R. (1988). Effect of magnesium on the hydraulic conductivity of Na-smectite-sand mixtures. *Clays and Clay Minerals*, **36**, 432–438.

Sheffer M. (2007). Forward modeling and inversion of streaming potential for the interpretation of hydraulic conditions from self-potential data. PhD thesis, The University of British Columbia.

Sheffer M. R. & Howie, J. A. (2001). Imaging subsurface seepage conditions through the modeling of streaming potential. In: *Proceedings of 54th Canadian Geotechnical Conference, Calgary*, pp. 1094–1101.

Sheffer, M. R. & Howie, J. A. (2003). A numerical modelling procedure for the study of the streaming potential phenomenon in embankment dams. In *Symposium on the Application of Geophysics to Engineering and Environmental Problems, San Antonio, Texas, USA*, pp. 475–487.

Sheffer, M. R. & Oldenburg, D. W. (2007). Three-dimensional modelling of streaming potential. *Geophysical Journal International*, **109**(3), 839–848.

Sharma, P. S. (1997). *Environmental and Engineering Geophysics*. Cambridge University Press.

Sill, W. & Killpack, T. (1982). *SPXCPL: Two-dimensional modeling program of self-potential effects from cross-coupled fluid and heat flow, user's guide and documentation for version 1.0*, NASA STI/Recon Tech. Rep., **83**, 13,400.

Sill, W. R. (1983). Self-potential modeling from primary flows. *Geophysics*, **48**. 76–86.

Sinitsyn, V. A., Aja, S. U., Kulik, D. A. & Wood, S. A. (2000). Acid-base surface chemistry and sorption of some lanthanides on K+-saturated Marblehead illite. I. Results of an experimental investigation. *Geochimica and Cosmochimica Acta*, **64**, 185–194.

Skold, M., Revil, A. & Vaudelet, P. (2012). The pH dependence of spectral induced polarization of silica sands: experiment and modeling. *Geophysical Research Letters*, **38**, L12304, doi:10.1029/2011GL047748.

Slater, L., Ntarlagiannis, D., Yee, N. *et al.* (2008). Electrodic voltages in the presence of dissolved sulfide: implications for monitoring natural microbial activity. *Geophysics*, **73**(2), F65–F70.

Slob, E., Snieder, R. & Revil, A. (2010). Retrieving electrical resistivity data from self-potential measurements by cross-correlation. *Geophysical Research Letters*, **37**, L04308, doi:10.1029/2009GL042247.

Spies, B. R. (1996). Electrical and electromagnetic borehole measurements: a review. *Surveys in Geophysics*, **17**(4), 517–556, doi:1007/BF01901643.

Spinelli, L. (1999). *Analyse Spatiale de l'Activité Electrique Cérébrale: Nouveaux Développements*, PhD Thesis (in French), Université Joseph Fourier-Grenoble I.

Stern, O. (1924). Zur Theorie der elektrolytischen Doppelschicht (The theory of the electrolytic double shift). *Zeitschrift Fur Elektrochemie Und Angewandte Physikalische Chemie*, **30**, 508–516.

Stoll, J., Bigalke, J. & Grabner, E. W. (1995). Electrochemical modeling of self-potential anomalies. *Surveys in Geophysics*, **16**(1), 107–120.

Straface, S., Falico, C., Troisi, S., Rizzo, E. & Revil, A. (2007). Estimating of the transmissivities of a real aquifer using self-potential signals associated with a pumping test. *Ground Water*, **45**(4), 420–428.

Su, Q., Feng, Q. & Shang, Z. (2000). Electrical impedance variation with water saturation in rock. *Geophysics*, **65**, 68–75.

Sumner, M. E. & Miller, W. P. (1996). Cation exchange capacity and exchange coefficients. In: Page, D. L. (ed.) *Methods of Soil Analysis Part 3: Chemical Methods, Soil Science Society of America*, Madison, WI.

Suski, B., Revil, A., Titov, K. *et al.* (2006). Monitoring of an infiltration experiment using the self-potential method. *Water Resources Research*, **42**, W08418, doi:10.1029/2005WR004840.

Tarantola, A. (2005). *Inverse Problem Theory and Methods for Model Parameter Estimation*. SIAM, Philadelphia.

Tarantola, A. & Valette, B. (1982). Inverse problem = quest for information. *Journal of Geophysics- Zeitschrift Fur Geophysik*, **50**, 3, 159–170.

Teja, A. S. & Rice, P. (1981). Generalized corresponding states method for viscosities of liquid mixtures. *Industrial & Engineering Chemistry Fundamentals*, **20**, 77–81.

Teorell, T. (1935). An attempt to formulate a quantitative theory of membrane permeability. *Proc. Soc. Exp. Biol. Med.*, **33**, 282.

Thompson, A. H. & Gist, G. A. (1993). Geophysical applications of electrokineticconversion. *The Leading Edge*, **12**, 1169–1173.

Thompson, A. H., Hombostel, S., Burns J. *et al.* (2007). Field tests of electroseismic hydrocarbon detection. *Geophysics*, **72**, 1, N1–N9.

Thorstenson, D. C. (1984). *The concept of electron activity and its relation to redox potentials in aqueous geochemical systems*, U.S. Geological Survey Open-File Report, 84–072.

Tikhonov, A. N. & Arsenin, V. Y. (1977). *Solutions of Ill-Posed Problems*. John Wiley & Sons, Washington.

Timm, F. & Möller, P. (2001). The relation between electrical and redox potential: evidence from laboratory and field measurements. *Journal of Geochemical Exploration*, **72**(2), 115–128.

Titov, K., Revil, A., Konasovsky, P., Straface, S. & Troisi, S. (2005). Numerical modeling of self-potential signals associated with a pumping test experiment. *Geophysical Journal International*, **162**, 641–650.

Tosha, T., Matsushima, N. & Ishido, T. (2003). Zeta potential measured for an intact granite sample at temperatures to 200 °C. *Geophysical Research Letters*, **30**(6), doi:10.1029/2002GL016608.

Trique, M., Perrier, F., Froidefond, T., Avouac, J. P. & Hautot, S. (2002). Fluid flow near reservoir lakes inferred from the spatial and temporal analysis of the electric potential. *Journal of Geophysical Research*, **107**(B10), 2239, doi:10.1029/2001JB000482.

Trolard, F., Bourrie, G., Abdelmoula, M., Refait, P. & Feder, F. (2007). Fougerite, a new mineral of the pyroaurite-Iowaite group: description and crystal structure. *Clays and Clay Minerals*, **55**, 324–335.

Truesdell, C. (1969). *Rational Thermodynamics*. McGraw-Hill, New York, XII. 208 S 1969.

Trujillo-Barreto, N. J., Aubert-Vásquez, E. & Valdès-Sosa, P. A. (2004). Bayesian model averaging in EEG-MEG imaging. *NeuroImage*, **21**, 1300–1319.

van Genuchten, M. T. (1980). A closed-form equation for predicting the hydraulic conductivity of unsaturated soils. *Soil Science Society of America Journal.*, **44**, 892–898.

van Schoor, M. (2002). Detection of sinkholes using 2D electrical resistivity imaging. *Journal of Applied Geophysics*, **50**, 393–399.

Vaudelet, P., Revil, A., Schmutz, M., Franceschi, M. & Bégassat, P. (2011a). Induced polarization signature of the presence of copper in saturated sands. *Water Resources Research*, **47**, W02526, doi:10.1029/2010WR009310.

Vaudelet, P, Revil, A., Schmutz, M., Franceschi, M. & Bégassat, P. (2011b). Changes in induced polarization associated with the sorption of sodium, lead, and zinc on silica sands. *Journal of Colloid and Interface Science.* **360**, 739–752.

Vinegar, H. J. & Waxman, M. H. (1984). Induced polarization of shaly sands. *Geophysics*, **49**, 1267–1287.

von Smoluchowski, M. (1903). Contribution à la théorie de l'endosmose électrique et de quelques phenomènes corrélatifs. *Bull. Int. Acad. Sci., Cracovie*, **8**, 182–200.

Wahba, G. & Wang, Y. (1995). Behavior near zero of the distribution of the GCV smoothing parameter estimates. *Statistics and Probability Letters*, **25**, 105–111.

Wan, C. F. & Fell, R. (2008). Assessing the potential of internal instability and suffusion in embankmentdams and their foundations. *Journal of Geotechnical and Geoenvironmental Engineering*, **134**(3), 401–408.

Wanfang, Z., Beck, B. F. & Stephenson, J. B. (1999). Investigation of groundwater flow in karst areas using component separation of natural potential measurements. *Environmental Geology*, **37**(1–2), 19–25.

Wang, M. & Revil, A. (2010). Electrochemical charge of silica surface at high ionic strength in narrow channels. *Journal of Colloid and Interface Science*, **343**, 381–386.

Watanabe, T. & Katagishi, Y. (2006). Deviation of linear relation between streaming potential and pore fluid pressure difference in granular material at relatively high Reynolds numbers. *Earth Planets Space*, **58**(8), 1045–1051.

Waxman, M. H. & Smits L. J. M. (1968). Electrical conductivities in oil bearing shaly sands. *Society of Petroleum Engineers Journal*, **8**, 107–122.

Wedekind, J. E., Osten, M. A., Kitt, E. & Herridge, B. (2005). Combining surface and downhole geophysical methods to identify karst conditions in North-central Iowa. *Geotechnical Special Publication*, **144**, 616–625.

Weller, A., Breede, K., Slater, L. & Nordsiek, S. (2011). Effect of changing water salinity on complex conductivity spectra of sandstones. *Geophysics*, **76**(5), 315–327.

Westermann-Clark, G. B. & Christoforou, C. C. (1986). The exclusion-diffusion potential in charged porous membranes. *Journal of Electroanalitical Chemistry*, **198**(2), 213–231.

Wilt, M. J. & Butler, D. K. (1990). *Geotechnical applications of the self-potential (SP) method; Report 4: Numerical modelling of SP anomalies: documentation of program SPPC and applications.* Technical Report REMR-GT-6. US Army Corps of Engineers, Waterways Experiment Station.

Wilt, M. J. & Corwin, R. F. (1989). Numerical modeling of self-potential anomalies due to leaky dams: Model and field examples. In *Detection of Subsurface Flow Phenomena. Lecture Notes in Earth Sciences*, G. P. Merkler (ed.), 27, 73–89. Springer-Verlag.

Wilt, M. J. & Goldstein, N. E. (1981). Results of two years of resistivity monitoring at Cerro Prieto. In *Third Symposium on the Cerro Prieto Geothermal field, Baja California, Mexico*, March 24–26, 1981, Proceedings/Actas CONF-810399–27, pp. 372–376.

Wood, A. W. (1955). *A Textbook of Sound.* MacMillan Publishing Company.

Woodruff, W. F. & Revil, A. (2011). CEC-normalized clay-water sorption isotherm. *Water Resources Research*, **47**, W11502, doi:10.1029/2011WR010919.

Woodruff, W. F., Revil, A., Jardani, A., Nummedal, D. & Cumella, S. (2010). Stochastic inverse modeling of self-potential data in boreholes. *Geophysical Journal International*, **183**, 748–764, doi:10.1111/j.1365-246X.2010.04770.x.

Wurmstich, B., Morgan, F. D., Merkler, G.-P. & Lytton, R. (1991). Finite element modelling of streaming potential due to see page: study of a dam. *Soc. Explor. Geophysicists Technical Program Expanded Abstracts*, **10**, 542–544.

Zablocki, C. J. (1976). Mapping thermal anomalies on an active volcano by the self-potential method, Kilauea, Hawaii. In *Proc. 2nd U. N. Symp. on the Development and Use of Geothermal Resources*, **2**, 1299–1309.

Zhang, G.-B. & Aubert, M. (2003). Quantitative intepretation of self-potential anomalies in hydrogeological exploration of volcanic areas: a new approach. *Near Surface Geophysics*, **1**, 69–75.

Zhou, W., Beck, B.F. & Adams, A. L. (2002). Effective electrode array in mapping karst hazards in electrical resistivity tomography. *Environmental Geology*, **42**(8), 922–928.

Zhu Z. & Toksöz, N. (2012). Experimental measurements of streaming potential and seismoelectric conversion in Berea sandstone. *Geophysical Prospecting* (in press).

Zimmermann, E., Kemna, A., Berwix, J. *et al.* (2008). A high-accuracy impedance spectrometer for measuring sediments with low polarizability. *Measurement Science and Technology*, **19**, doi:10.1088/0957–0233/19/10/105603.

Zukoski, C. F. & Saville, D. A. (1986a). The interpretation of electrokinetic measurements using a dynamic model of the Stern layer. I. The dynamic model. *Journal of Colloid and Interface Science*, **114**(1), 32–44.

Zukoski, C. F. & Saville, D. A. (1986b). The interpretation of electrokinetic measurements using a dynamic model of the Stern Layer. II. Comparisons between theory and experiments. *Journal of Colloid and Interface Science*, **114**(1), 45–53.

Zundel, J. P. & Siffert, B. (1985). Mécanisme de rétention de l'octylbenzene sulfonate de sodium sur les minéraux argileux. In: *Solid-Liquid Interactions in Porous Media*, pp. 447–462, Technip, Paris.

Index

algorithm
  adaptive Metropolis, 190, 257
  Parker, 215, 219
Ampère's law, 315
Archie's law, 49, 90, 173, 182, 187, 288, 302, 306

bacteria, 14, 70, 71, 75, 233, 234
Bayesian inference, 256
beamforming, 284, 335, 336
biofilm, 14, 75
biogeobattery, 23, 243
bulk modulus, 43, 288, 298, 311, 314, 315, 339

capillary, 44, 45, 46, 48, 49, 51, 53, 54, 55, 57, 78, 80, 81, 107, 129, 130, 180, 182, 194, 198, 221, 300, 308, 310, 311, 312
catchment, 192, 197, 223, 225, 226, 230, 232
cation exchange capacity, 22, 66, 82, 84, 90, 95, 106, 224, 254, 304, 317
cave, 167
cavity, 168, 169
cementation exponent, 49, 62, 90, 106, 134, 173, 288
clay, 13, 46, 51, 58, 62, 65, 66, 67, 68, 83, 84, 93, 94, 134, 160, 161, 162, 164, 166, 167, 173, 203, 224, 264, 266, 285, 317, 343
compactness, 122, 123, 124, 334
compressibility, 40, 309
conductivity
  complex, 82, 85, 87, 88, 89, 90, 305, 306, 316
  electrical, 13, 16, 17, 51, 53, 56, 57, 71, 74, 84, 85, 90, 104, 105, 106, 110, 113, 114, 133, 134, 135, 136, 137, 152, 173, 179, 182, 192, 193, 195, 212, 220, 224, 229, 231, 232, 233, 235, 242, 244, 246, 247, 248, 254, 255, 258, 261, 268, 277, 285, 290, 298, 306, 316, 338
  quadrature, 88, 91, 92, 93, 106, 306, 307, 344
  surface, 13, 53, 56, 90, 106, 306, 317, 318
  thermal, 247, 248, 254, 255, 258, 264
conservation
  charge, 78

  mass, 24, 25, 79, 133, 339
  momentum, 77, 313, 339
contaminant plume, 14, 70, 71, 75, 111, 233, 234, 236, 238, 242, 243
convoluted perfect matched layer, 322
corrosion, 14, 17, 70, 72, 73, 99, 102, 103, 104
co-seismic, 285, 286, 294, 298, 323, 325
cross-correlation, 111, 113, 169, 190
current density, 15, 16, 17, 23, 25, 26, 52, 57, 58, 59, 70, 74, 75, 77, 82, 88, 94, 95, 96, 106, 110, 111, 114, 115, 120, 123, 124, 127, 128, 132, 135, 151, 152, 168, 172, 173, 177, 184, 185, 190, 192, 194, 195, 212, 213, 214, 233, 235, 249, 268, 269, 270, 275, 280, 286, 290, 297, 305, 315, 325, 332, 338

dam, 171, 186, 188, 189, 244
Darcy
  law, 50, 86, 87, 96, 106, 112, 132, 186, 212, 246, 251, 301, 305
  velocity, 8, 36, 50, 52, 55, 57, 58, 60, 61, 87, 95, 112, 132, 133, 135, 137, 173, 174, 187, 194, 246, 249, 259, 268, 274, 279, 282, 301, 305, 339
detection
  cavities, 154, 167, 171, 191
  contaminants, 234
  geothermal fluids, 15, 245, 251, 255, 261, 262, 265, 266, 271, 274, 277, 280, 282
  sinkholes, 154, 160, 161, 162, 163, 164, 166, 167, 191
diffuse layer, 13, 21, 22, 46, 51, 55, 57, 62, 64, 66, 68, 78, 81, 87, 90, 92, 94, 95, 97, 106, 135, 163, 172, 194, 268, 285, 290, 300, 304, 306, 317, 343
diffusion
  current, 15, 75, 135, 172, 190
  potential, 9, 179, 184, 185
dispersion, 57, 58, 132, 133, 136, 137, 142, 172, 179, 182, 184, 190
dispersivities, 188
dispersivity, 57

dissipation, 30, 31, 33, 34, 35, 37, 38, 77, 300
Dukhin number, 90, 173

electrical double layer, 8, 11, 13, 19, 43, 44, 62, 157,
    172, 246, 268, 285, 298, 304, 317, 338
electrode
    drifting, 3, 8, 9, 10, 11, 79, 177, 205, 231, 232, 329
    non-polarizing, 1, 8, 14, 18, 99, 102, 107, 163, 168,
        174, 185, 205, 218, 268, 271, 290, 294, 327
    Petiau, 1, 8, 19, 185, 205, 232, 271
    reference, 1, 3, 4, 5
electroencephalography, 8, 114
electrokinetic, 15, 122, 151, 163, 180, 184, 192, 219,
    231, 233, 234, 236, 246, 254, 284, 286, 290, 319,
    333, 335, 336, 353, 359, 360, 361, 364, 366
electromigration, 51, 72, 95, 311, 316
electron, 14, 20, 23, 71, 74, 75, 77, 104, 134, 173, 234
embankment, 171, 185, 243

finite element method, 115, 286, 291
flow
    inertial laminar, 194, 196, 200, 302, 304, 308
    viscous laminar, 27, 194, 195, 199, 302, 304, 308
formation factor, 37, 49, 52, 55, 64, 68, 90, 104, 106,
    133, 134, 173, 181, 194, 285, 288, 290, 301, 306,
    317
fracking, 284, 326, 338

Gauss–Newton method, 268, 271, 272, 276
geobattery, 14, 23, 82, 99, 102, 104, 107, 111, 233
geothermal field, 245, 261, 262, 265, 266, 271, 277,
    282
Green function, 169, 269

heat flow equation, 247
hydraulic conductivity, 85, 87, 113, 132, 135, 136,
    137, 138, 139, 141, 143, 144, 148, 174, 180, 181,
    187, 194, 195, 210, 212, 244, 246, 247, 311, 312
hydrothermal, 245, 262, 266, 274

impedance, 1, 3, 9, 87, 88, 102, 107, 135, 163, 168,
    214, 218, 254, 272, 327
induced polarization, 13, 51, 82, 88, 191, 338, 355,
    356, 360, 361, 364
inversion
    coupled, 143, 146, 148, 256, 258, 261, 267, 294
    Gauss–Newton, 268, 271, 272, 276, 280
    gradient-based, 110, 111, 245, 282, 334, 338
    stochastic, 110, 114, 138, 188, 245, 282

kernel, 115, 118, 122, 123, 195, 196, 209, 268, 270,
    334
kriging, 177, 221, 222, 226, 238, 242, 280

landslide, 154, 155, 157

magnetic field, 27, 290, 315
magnetic induction vector, 315

magnetic permeability, 290, 315
Maxwell equations, 4, 16, 47, 74, 286, 289, 290, 307,
    315, 338
misfit, 118, 119, 120, 270, 271
modeling
    forward, 105, 111, 115, 123, 131, 141, 177, 195,
        196, 209, 215, 247, 249, 256, 258, 265, 269,
        283
    inverse, *see* inversion
moisture capacity, 180, 310, 311, 312
monitoring, 3, 8, 9, 136, 154, 167, 171, 185, 191, 195,
    197, 232, 243

Newton's law, 27, 287, 288, 313, 339, 346
noise, 6, 9, 17, 19, 120, 123, 128, 138, 141, 153, 163,
    175, 188, 197, 200, 205, 232, 271, 276, 281, 294,
    327
non-equilibrium thermodynamics, 15, 19, 23, 32, 33

ore body, 14, 71, 72, 73, 111

permeability, 37, 50, 55, 62, 64, 84, 86, 87, 98, 104,
    106, 132, 135, 137, 157, 162, 167, 174, 180, 181,
    183, 185, 187, 188, 190, 210, 214, 219, 236, 238,
    246, 247, 248, 251, 253, 254, 255, 259, 260, 261,
    266, 267, 268, 274, 285, 286, 288, 289, 290, 293,
    301, 302, 303, 304, 310, 312, 315, 316, 319, 321,
    339, 345
permittivity, 290, 315
phase, 20, 23, 34, 43, 48, 49, 52, 71, 83, 87, 88, 89,
    91, 92, 106, 107, 172, 177, 209, 284, 285, 287,
    288, 289, 293, 296, 300, 301, 306, 309, 310, 313,
    314, 319, 321, 329, 345, 346, 347
Poisson equation, 110, 111, 115, 129, 173, 193, 246,
    268, 293, 320
Poisson–Boltzmann equation, 20, 46
poroelasticity, 287, 314, 339, 346
porosity, 34, 40, 43, 49, 52, 54, 58, 59, 62, 85, 90, 95,
    99, 104, 107, 132, 133, 137, 144, 173, 180, 187,
    190, 286, 288, 293, 298, 301, 302, 306, 319, 321
potential
    electrochemical, 19, 20, 47, 51, 80, 88
    electrostatic, 16, 19, 20, 47, 78, 290, 315
    redox, 14, 70, 71, 72, 73, 74, 75, 82, 99, 102, 103,
        104, 110, 111, 233, 234, 236, 237, 238, 242
    zeta, 64, 96, 172, 285
pressure, 19, 29, 31, 34, 36, 40, 42, 46, 47, 48, 50, 54,
    55, 61, 75, 76, 77, 78, 79, 87, 104, 180, 194, 210,
    213, 246, 265, 283, 287, 289, 293, 300, 301, 305,
    307, 308, 309, 310, 311, 312, 313, 314, 321, 327,
    328, 329, 333, 334, 339, 345, 346, 347, 348, 352,
    365
pumping test, 2, 9, 15, 192, 197, 198, 243

resistivity, 13, 17, 19, 74, 82, 87, 104, 111, 113, 114,
    115, 123, 124, 128, 132, 133, 134, 135, 137, 139,
    140, 141, 144, 148, 160, 161, 166, 167, 168, 169,
    180, 182, 183, 184, 185, 190, 191, 193, 195, 198,

201, 205, 206, 209, 210, 214, 217, 220, 224, 225, 229, 236, 245, 254, 260, 262, 264, 268, 269, 271, 272, 274, 275, 276, 277, 279, 280, 282, 306, 325, 336

Richards equation, 180, 182, 311

salt tracer test, 136, 141, 154, 185
seismoelectric conversion, 123, 285, 286, 294, 298, 319, 323, 325
self-potential method
  equipement, 3
  history, 14
  noise, 17
sinkhole, 163, 164, 166
Stern layer, 13, 22, 44, 48, 51, 53, 57, 62, 63, 64, 66, 67, 68, 80, 88, 90, 93, 95, 106, 172, 284, 300, 306, 317, 318, 342, 343, 344
stochastic inversion, 245
streaming
  current, 15, 75, 76, 81, 94, 95, 106, 113, 127, 131, 135, 137, 151, 172, 192, 193, 194, 195, 198, 209
  potential, 15, 60, 62, 63, 64, 70, 82, 84, 85, 87, 96, 104, 110, 111, 151, 163, 167, 177, 179, 182, 184, 193, 194, 206, 211, 213, 214, 220, 221, 225, 229, 233, 234, 235, 237, 238, 241, 246, 268, 274, 283, 294, 298, 307, 308, 316

potential coupling coefficient, 60, 62, 63, 64, 84, 85, 87, 96, 104, 151, 165, 177, 179, 192, 193, 194, 195, 206, 208, 211, 212, 213, 214, 220, 221, 225, 229, 231, 233, 246, 268, 274, 275, 298, 307, 308, 316, 346
stress tensor, 24, 26, 27, 30, 33, 37, 38, 42, 43, 287, 288, 312, 313, 314, 339, 346

temperature profile, 253, 264
triple layer model, 13

vadose zone, 73, 110, 164, 167, 180, 183, 185, 192, 209, 210, 211, 212, 213, 214, 215, 219, 220, 221, 224, 225, 228, 229, 230, 234, 236, 243, 336
voltmeter, 1, 3, 8, 9, 88, 102, 107, 129, 135, 163, 168, 205, 218, 272, 327
volumetric charge density, 44, 46, 54, 62, 64, 66, 95, 97, 246, 251, 268, 282, 298, 304, 310, 316, 317, 319

water table, 72, 180, 182, 192, 198, 199, 205, 211, 212, 216, 219, 220, 221, 222, 224, 226, 228, 229, 230, 231, 232, 233, 234, 235, 238, 249, 274, 312
wave
  compressional, 286
  shear, 286, 297

Printed in the United States
By Bookmasters